Turbulence in Economics

An Evolutionary Appraisal of Cycles and Complexity in Historical Processes

Francisco Louçã

Edward Elgar
Cheltenham, UK • Lyme, US

Published by
Edward Elgar Publishing Limited
8 Lansdown Place
Cheltenham
Glos GL50 2HU
UK

Edward Elgar Publishing, Inc.
1 Pinnacle Hill Road
Lyme
NH 03768
US

A catalogue record for this book
is available from the British Library

Library of Congress Cataloguing in Publication Data
Louçã, Francisco.
Turbulence in economics: an evolutionary appraisal of cycles and complexity in historical processes / Francisco Louçã.
Includes bibliographical references and index.
1. Economics—Methodology. 2. Business cycles. 3. Economic history. 4. Turbulence. 5. Complexity (Philosophy) 6. Paradigms (Social sciences) I. Title
HB131.L68 1997
330—dc21 97–13455
CIP

ISBN 1 85898 563 3

Printed and bound in Great Britain by
Biddles Ltd, Guildford and King's Lynn

Contents

PART THREE: BOUNDED HERESIES

PART FOUR: DR. PANGLOSS HUNTED BY THE SNARKS

CONCLUSION: COMPLEXITY, THE CONDITION OF THE WORLD

Figures

Tables

Introduction

How is it that apparently simple phenomena create so evidently complex relations? And how is it that those complex relations can still generate simple, coordinated and relatively stable patterns of evolution? The history of modern science is being written by the contradictory explanations of these paradoxes. Meteorologists study the irregularity of weather and the continuity of the climate; sociologists study the irreducible individuality of human action and the forms of gregarious behaviour; biologists study the ultimate components of the genetic code and the consequent extraordinary variety in nature; a large number of economists study the diversity of firms, agents and institutions in order to argue that there is some sort of visible or invisible hand preventing the system from falling apart.

But for some centuries the contrast and connection between simplicity and complexity was not even recognized as a legitimate enigma for science: reductionism and mechanicism, the twin souls of positivism, claimed that the whole universe was explainable as a clock with its precise functions and dedicated parts. A rather esoteric debate continued for some time about the good faith and the will of the watchmaker, but then a broad agreement about the nature of the essential mechanism emerged from the impressive progresses in Newtonian science. The grammar of science was established by the very concept of deterministic laws and by the widespread trust in the power of mathematics to represent all phenomena and to establish a common language for all research, strengthened by the certainties of methodological individualism. Nevertheless, some disturbing questions resisted those disciplinary methods and drastic epistemological conditions.

Turbulence has been one of those enigmas: why and how does the behaviour of a fluid change under acceleration? The problem was ignored for a long time, since it could not even be conceived of in the universe of mechanical forces leading to equilibrium, until it came under scrutiny by one of the most brilliant mathematicians of the late nineteenth-century, Henri Poincaré. Fascinated by the unsolved problems of Newtonian physics, such as the three-bodies problem (the determination of the trajectories of three gravitationally interacting celestial bodies), Poincaré was the forerunner of the theories of complexity and, in 1892, he published a *Théorie des Tourbillons*, although he was only able to indicate the nature of the question and not to solve it.

Some fifty years later, Landau, in 1944, and Hopf, in 1948, suggested an explanation for the phenomenon: after a disturbance, a previously stationary fluid was supposed to acquire a periodical movement if going through a bifurcation, which establishes one mode of oscillation at one precise frequency. If the fluid goes through successive bifurcations, other frequencies are added and turbulence results from the cumulation of all those modes of oscillation. The periodical movements are therefore conceived of as originated by independent oscillators, which do not interact. Complexity is defined as the juxtaposition of several distinct modes of behaviour, and the reductionist legend is safe: these cycles can be causally identified and are regular.

Yet, some years later, in 1971, Ruelle and Takens challenged that definition, which had not been experimentally confirmed. Their argument was, simply put, that after the third bifurcation the interaction among the oscillators would lead to chaos: a strange attractor may exist in the neighbourhood of the torus formed at that dimension or at a superior one by the motion of the system, and there is sensitive dependence to the initial conditions. That was indeed the insight of Poincaré, that small events can provoke large effects. So, if the oscillators are not independent, the paradigm of equilibrium collapses: there are distinct oscillators, but they interact and are no longer fully identifiable back from their effects to their origins; they generate frequencies, but superimposed in a continuous spectrum and not as isolated phenomena — the whole is more than the sum of its parts. In other words, complexity is an emergent property of turbulent fluids. There is chaos, but the attractor describes a certain regime — although the sensitivity to initial conditions and to changes in the trajectories moves the system away from equilibrium, there is order, too.

David Ruelle was aware of the implications of his model for the development of hydrodynamics as well as for other scientific disciplines. In a more recent book, he argues that the evolution of economic and social relations may be represented by the interaction of several oscillators: turbulence is a consequence of the development of technological innovations and new industries or simply of the coupling of economic systems, for instance through trade (Ruelle, 1991: 111).

The argument of this book is that turbulence is an adequate and obvious metaphor for economics, helping to tune our conjectures about entrained cycles in real social evolutionary processes. Furthermore, another striking feature of the metaphor is that there is a close resemblance between this story and the evolution of the science itself, namely in the case of the major changes in its interpretations of cycles and fluctuations.

In fact, economics was born as the expression of a genuine confidence in progress, such as that encapsulated by the Industrial Revolution: growth, division of labour, specialization; all movements were articulated and combined in the general trend of development — the oscillations were ignored or despised as produced by factors alien to the scope of science. But the next generation of the founding fathers of economics rejected such a certainty: Malthus or Ricardo were rather pessimistic and Marx rather critical of that vision, while simultaneously Charles Dickens vividly described those times of social upheaval; the clock no longer controlled the rhythm of time. Nevertheless, shortly after that paradigmatic crisis, the neo-classical revolution redefined the discipline and provided new concepts and new metaphors to economics: equilibrium emerged thereafter as the central reference for all theories and models. The rocking-horse metaphor, examined in this book, has been the most powerful tool for the creation of new methods of inquiry in this domain, for the mathematization of the economic relations and for the description of cycles and evolution — and here is where the first version of the turbulent story takes shape.

The rocking horse supposes that there are just two possible modes of behaviour, those of the stabilizing mechanism and of the impulse system, so that non-correlated shocks create serially correlated cycles, and imposes severe restriction on those modes. Therefore, the basic frequencies — the growth trend, the cycles such as the Kitchin, Juglar, Kuznets or others — are considered as independent and separable oscillatory systems, and this reductionist and mechanistic approach is obviously comparable to Landau's and Hopf's theories about turbulence. But if otherwise the emergent phenomena of cycles are interpreted as attractors of related social and economic movements, the comparison falls apart and the world of complexity becomes recognizable. And this is the story turbulence tells in economics.

There are, of course, close analogies which will be explored later on, but also decisive differences: turbulence is not a persistent phenomenon with long memory and low frequency dependence, unlike economic growth — turbulence is just a metaphor for social complexity. Social history is the basic characteristic of economic evolution and its *differentia specifica* with other sciences: the theories of complexity may suggest some common grounds and methods of inquiry, but no physical or biological phenomenon can exhibit the same type of human intentional and institutionally built determination that constitutes the object of social sciences — social phenomena are more complex than complex organic phenomena. A plea for this second generation of complexity models and theories is the theme of the present book.

*

In the recent decades, the scientific discontent with traditional positivism has mounted to unprecedented levels, opening a major paradigmatic shift. Empiricism and Rationalism, the co-founders of modernity, generally claimed that positivism was the decisive condition for the existence of science — but this has been challenged by the intromission of the concepts of uncertainty, bounded rational behaviours, different forms of rationality, evolution in society, bifurcation in history and others. New research programs have been developed in the direction of the incorporation of change, choice, contradiction and dialectics; causality could no longer be restricted to deterministic processes, and the very pretension of natural sciences to be the purest form of cognition has been weakened in a broader concept of plurality of knowledge. If the purpose of science is to understand and explain real and complex processes, or life, then a change must be imposed on its traditional methods and philosophy. Indeed, this change is under way.

Simultaneously, and most crucially, the discovery of new phenomena or the inquiry into familiar events under new lights led to the analysis of processes out of equilibrium in thermodynamics, of turbulence in physics and meteorology, of morphogenesis in biology and chemistry: at the very core of the hard sciences, history, unpredictability, evolution and change are nowadays considered to be relevant topics for inquiry. Drawing from Maxwell or Poincaré, the complexity approach has been broadened by Lorenz, Jantsch, Smale, Prigogine and Stengers, Ruelle, Kauffman, Bak and so many others. The science of chaos is based on the paradoxical evolution of deterministic processes creating intrinsic indetermination; still more generally, the science of emergence, complexity, articulates deterministic and indeterministic processes as well as intrinsic and extrinsic randomness.

In economics, this crisis of positivism is particularly dramatic. Economics has been the most formalized of the social sciences, and consequently the closest to the positivist paradigm: mechanical equilibrium has been considered as the epistemological condition for modelling and for explanation; prediction has been defined as the very purpose of theory, and the normative character of the discipline has been generally rejected in favour of the exclusive positivistic definition of the assumptions and conclusions. Rejecting Adam Smith's original definition of economics as a 'moral science' — or political economy — the current mainstream grounded economics solidly in positivism. Axiomatization took formalism to unprecedented levels and, for many generations, economics' most considerable resources were dedicated to the development of the general equilibrium paradigm, originally inspired by the physics of the 1840s but rapidly superseded in that science itself. This paradigm incorporated later on the probabilistic revolution, maturing by the 1930s and the 1940s, at the root of econometrics, extended its core hypotheses in several new directions and tried to cope with the concepts of uncertainty and change. In its most sophisticated forms it is indeed a monument to scientific ingenuity and skill. Yet, it has been challenged ever since: by Sismondi and Marx, by Veblen and Keynes, by Schumpeter and Georgescu-Roegen, as well as, at some point, by many of its most distinguished practitioners such as Frisch, Leontief, Solow, Friedman, Hahn or Arrow.

This evolution follows from the fact that economics lived for a long time in a very paradoxical situation. In spite of its inspiration in classical positivism and permanent search for metaphors and analogies from physics, which made possible the access to a high level of formal elegance and asserted the pretension of the discipline to have an acceptable experimental capacity, economics could not and would not overcome its exhaustive Cartesianism, which protected the most unacceptable hypotheses with the doubtful argument that they still lead to plausible deductions. The history of the failure of successive attempts to impose infirmationist methodologies is an example of the difficulty of matching the research practices with epistemological claims: hypotheses and theories developed with no counterparts in reality, a forest of *ceteris paribus* clauses, heroic assumptions protected by a castle of metaphysical claims about rational expectations and perfect information, everything conspired to prevent reasonable tests. For at least a dozen years, the crisis of positivism has been firmly installed in the province of economics as a consequence of its own success.

This book is an inquiry into the crisis of orthodoxy in economics; it is aimed at defining the scientific conditions for the research on the characteristics of evolution and the systemic mutations that organize the history of capitalism. It tells the legend of the rocking-horse, that powerful metaphorization which led the early econometric program, and argues that its premises and methods are inadequate to capture the essential features of social evolution. In fact, the existence and differentiation of long periods of structural evolution is intuitive and indeed is accepted by most of the economic historians, although defined in several different ways: after the Industrial Revolution, the 'hungry forties', the Victorian boom, the Great Depression, the Belle Epoque and the revolution of steel and electricity, the crises of the wars, the Fordist thirty golden years, and now the years of crises — the rhythm of economic time is marked by the succession of those long expansions and depressions separated by deep structural changes. Those long periods represent distinct forms of organization of social relations, of science and technology, different cultural trends and political and social institutions. The paradox is that this description is accepted in history but rejected in orthodox economic science, since it is not compatible with one mode of explanation which denies by definition the very notion of structural mutation. The argument of this book is that such a divorce between economics and history, as well as the separation of the context of growth from that of fluctuations, are dramatic limits on the ability of economics to explain reality. The study of the distinct periods of capitalist development is therefore useful and necessary, since it explains concretely the evolution of the economies and their environment and since it defies economics to accept realism and therefore new methods of inquiry.

The scientific conditions for that study are defined on three levels.

First, the general epistemological conditions are dealt with in Part One, where the crisis of positivism is discussed. The role of metaphor in the definition of new scientific hypotheses, in the general conceptualization of the theories and in persuasion is highlighted following Aristotle, Peirce, Black, Ricoeur, Hesse, Kuhn. It will be easily verified that metaphors have a pervasive and important role in every major paradigm in economics: such diverse authors as Marx, Marshall, Walras, Edgeworth, Fisher, Veblen, Van Gelderen, Kondratiev, Schumpeter, Keynes, Robinson, Shackle, Leontief, Samuelson, Goodwin, Solow and many others dealt extensively with the adequacy and influence of metaphors in economic thought. It is difficult to find any single main author who did not use important constitutive metaphors, and even who did not explicitly refer to the essential role of metaphor in his own work.

But these metaphors are unequal in rigour and quality. They may direct research into interesting and valid hypotheses, just as they may mislead generations of scientists. They may feed productive conjectures or insinuate superficial rhetoric, and some discrimination must be introduced. The two main families of metaphors in economics, those derived from physics and those derived from evolutionary biology, are discussed in Part One of the book: the error of the oblique transfer, the deduction of a causal relation on the basis of evidence of similitude between two systems, is presented as a very general feature of economic theories and statistical inference. As a conclusion, it is suggested that economics must draw inspiration from the study of dissipative and self-organizing structures which are so common in nature and in society, and that this implies an alternative epistemology based on realism and evolutionary concepts rather than on instrumentalism or relativism and mechanical concepts.

Parts Two and Three deal with the theoretical conditions for such a reconstruction. The first of these consists of a brief review of the literature about growth and cycles, a thematic inquiry into some of the methods and theories dealing with economic mutation. The centre of the argument is the critique of the mechanical metaphor, the dominant feature of the mainstream paradigm for the interpretation and analysis of cycles, as it was defined by Frisch in 1933. In fact, Wicksell, Ackerman and Slutsky provided preliminary versions of the rocking-horse metaphor for cycles and the former even coined its name, but it was up to Frisch and also to Tinbergen to develop afterwards the statistical methods and the theoretical interpretation that defined and refined the consistency of the program, which progressed later on under the influence of Haavelmo, Koopmans, the Cowles Commission and the Econometric Society. The intellectual environment and some of the major protagonists of the introduction of the probabilistic approach in economics, which still lacks its history, as well as the early and surprisingly powerful controversies about that program are discussed in this book.

The theoretical conditions for the study of economic development and fluctuations are defined as the rejection of such a powerful metaphor, which established simple, aesthetically perfect, powerful and yet wrong rules for the inquiry into social evolution. One of its contradictory rules is the antinomic definition of all quantifiable relations either as endogenous or as exogenous variables, for the sake of the mathematical language driving the demonstration of the theory. But it amounts to impossible and absurd results which simultaneously deny the heuristic power of the model; it led and goes on leading to the reduction of the science to the status of a logical game, it undermines the stability of the paradigm and it reduces economics to instrumentalism and the theory to inconsequential accounting. The notion of 'semi-autonomous variables' is presented as a solution for the problem of this dichotomic definition, as it is imposed by the rocking horse model of models.

The introduction of evolutionary concepts is far from trivial and that is why it took so long in spite of the initial affair between economics and Darwinism: it supposes the abolition of the structural rigidity of the equilibrium concepts, and as a consequence it requires the framework of coevolutionary history. Rejecting the conceptual acupuncture which prevailed in the orthodox paradigm for the interpretation of the economic processes, evolutionary economics must be developed as a new program. The argument of this book is about some of its building-blocks: the concept of semi-autonomous variables in order to incorporate the inter-disciplinary fertilization in historical models, which is also the condition for not abdicating from the rigour which is required in economics; the concepts of morphogenesis, structural mutation or the emergence of order out of equilibrium; the twin concepts of complexity and coordination.

Part Three examines some of these theoretical conditions, in the light of the critiques of Schumpeter and Keynes of the equilibrium paradigm, which are part of the strategic heritage which is decisive for this new program. The scope and limits of both assessments are discussed: these authors contributed, from sometimes opposed points of view, to the critique of the static, equilibrium, stability, certainty and rationality postulates of the orthodox theory; although both authors were famous the core of their propositions is frequently ignored or misrepresented. The balance sheet of their contributions is indicated, and it is part of the theoretical conditions for the reincorporation of history into economics: these conditions include the postulates of disequilibrium, bounded rationality, structural instability, uncertainty and dynamic processes of evolutionary economics.

Part Four examines the analytic conditions for the study of cycles and history in economics. In particular, it extends the critique of the rocking horse metaphor to those econometric and statistical methods which are based on unsustainable assumptions about the sampling character of the available series and their relation to hypothetical universes, about linearity and namely about the additivity of trend and cycle as independent modes of oscillation. It is indicated that there are strong theoretical reasons to believe that nonlinearities are pervasive in economics, that complex phenomena must naturally be more common than simple and linear events, and that the frequent and brutal switches of regimes in economic series indicate the presence of factors responsible for nonlinearity. It is also argued that there are simultaneously strong statistical reasons for believing that this is really the case, namely if alternative tools are used and the a priori assumptions of traditional econometrics are replaced by a more realist approach. The analytic condition for the study of historical capitalism and its constitutive feature of bursts of structural change is therefore the use of statistical methods appropriate to the object of the inquiry, namely of methods derived from the emerging paradigm of complexity, so that formal precision can be maintained in the models while not compromising the evolutionary perspective of the theory. In particular, it is suggested that specific forms of coordination, or modes of social cohesion, explain the recurrent but irregular oscillations in economic history.

This implies the return to the evolutionary metaphor. Back to turbulence, where complexity, unpredictability and attractors coexist and determine each other, this book is therefore part of an effort to rebuild economics as an evolutionary and realist science, to reintroduce history as one of its best controlled methodologies with higher heuristic power, and to develop new methods of nonlinear and complex dynamics for the research on concrete economies.

*

The origin of this book was a doctoral dissertation presented to the University of Lisbon under the supervision of Christopher Freeman, Emeritus Professor of the University of Sussex, whose careful advice and sharp critique were far beyond the traditional role of an academic supervisor. I must also extend my thanks to A. Ramos dos Santos and to João Caraça (ISEG, Lisboa) as well as to Carlota Pérez (Caracas and SPRU, Sussex University), who thoroughly followed the preparation of the text, suggesting many improvements. In particular, I must stress my intellectual debt to Richard Goodwin (Siena University) and Ernest Mandel (Brussels University), who died recently.

As acknowledged in the text, it largely benefited from the insights, writings or discussions with Philip Mirowski (Notre Dame University). Other researchers also read and discussed some chapters, namely João Ferreira do Amaral (ISEG, Lisbon), Daniel Bensaid (Philosophy, University of Paris VIII), Norman Clark (SPRU, Sussex), Giovanni Dosi (University of Rome and SPRU), Taborda Duarte (Mathematics, IST, Lisbon), Augusto Mateus (ISEG), Robert May (Zoology, University of Oxford), John Maynard Smith (Biology, University of Sussex), and Charles-Andrè Udry. Correspondence or discussion with W. Brian Arthur (Santa

Fé Institute), the late David Gordon (New School of Social Research, NY), Oscar Lopes (Faculdade de Letras, Porto), Richard Nelson (Columbia University) and Ulrich Witt (Max Planck Institute) was also helpful. The usual disclaimer applies to all them: although implicated by this reference, they are innocent of the mistakes of the following pages.

Finally, I must thank Penelope Duggan, who corrected parts of the English version, as well as António Louçã and Isabel Pargana, who translated some German papers. I must also acknowledge the Portuguese Junta Nacional de Investigação Científica, which partially funded the research, and Blackwell and Macmillan editors for the authorization to quote Schumpeter's and Keynes's works.

PART ONE

The Evolutionary Metaphors in the Reconstruction of Economics

The indiscriminate application of the term 'evolution' however, has led to some unfortunate formulations, if not absurdities. Non biologists who favour the evolutionary conceptualization are often unaware of the Darwinian and Neo-Darwinian theory and may, for instance, promote orthogenetic schemes, such as the theory that human culture automatically passes through a series of stages from that of the hunter-gatherer to that of the urban megapolis. Teleological principles have been very popular among those who have used evolutionary language outside of biology, but when these teleological schemes were refuted, it was thought that this refuted the whole concept of evolution. A study of such literature demonstrates rather painfully that no one should make sweeping claims concerning evolution in fields outside the biological world without first becoming acquainted with the reasoned concepts of organic evolution and, furthermore, without a most rigorous analysis of the concepts he plans to apply. Evolutionary thinking is indispensable in any subject in which a change in the time dimension occurs. However, there are many 'kinds' of evolution, depending on the nature of the causes that are responsible for the change, on the nature of the constraints, and on the nature of the success of the changes. The appropriate analysis of the different kinds of so-called evolution in different areas has not yet been undertaken. Nevertheless, there is no doubt that applying evolutionary principles has greatly enriched many areas of human thought.

Mayr, 1982: 627

1. Introduction to Part One: Simplicity or Complexity

Evolutionary economics is based on a constitutive metaphor inspired by Darwin's biology. Despite its looseness, the metaphor is frequently presented as a powerful antidote against Neoclassical mechanicism and as an irreplaceable tool for a necessary and alternative organic, holistic and systemic epistemology. The first part of the book considers and elaborates those claims, along some familiar but necessary themes.

The argument is organized following the three basic questions: (i) is it really necessary to abandon positivism and, in economics, the derived neoclassical paradigm?; (ii) is metaphor a possible and useful tool for such a paradigmatic rupture?; (iii) is the biological metaphor adequate for such a purpose? The preliminary answers suggest the next steps of the argument. Since that is not the purpose of the current book, no general survey of the epistemological debates and no detailed discussion of the philosophical issues is presented, and the following chapters are limited to the case for the reincorporation of the constructive role of time and history in the methodology of social sciences. This crucial change is the epistemological condition for the research on the stages of development of capitalism.

The next chapter criticizes the positivist research program, which constitutes a central structure of modernity. The program combines two major intellectual influences, Baconian and Newtonian empiricism on the one hand, and Cartesian rationalism on the other hand, both claiming to provide a clear distinction between science and metaphysics and pre-scientific thought and to furnish powerful tools in order to progress in scientific inquiries. In particular, the positivist program has been developed around two non-soluble enigmas: the problem of cognition and the related problem of demarcation. Alternative methodologies, like Darwin's eclectic combination of induction, deduction and analogy, are briefly presented and constitute major points of reference for the epistemological reconstruction for which this part argues. On that basis, a strategy for reorganizing the landscape of the debate is suggested, so that a new and extended criterion of signification addresses both problems.

Neoclassical economics is the brilliant incarnation of positivism in economics. Its main methodological contradictions are discussed in the following chapter, namely the effects of the paradigmatic change which occurred in economics when the marginalist revolution transcribed the concepts and procedures of nineteenth-century energetics to the social realm. As a consequence, economics includes both the extremist version of inductivism and the extremist version of deductivism: no wonder the appearance of its epistemology is clumsy. Then, the fourth chapter deals more generally with metaphoric innovation. Some basic theories of the metaphoric constitution of models are surveyed and discussed, and a general classification and methodology is summarized, emphasizing the metaphoric reconstruction of the deductive corpus of theories as a semantic impertinence at the level of the *explanandum*, which creates new degrees of freedom for radical hypothesizing and is therefore central for the solution of the problem of signification. Peirce's abduction is presented as a particular case of metaphoric innovation.

Metaphors are widespread in every major economic theory, not only as means of rhetoric persuasion but also and mainly as part of the very working of new hypotheses: as Borges wrote, maybe our intellectual universe is just a succession of metaphors. At the frontier of our science there is now a clear conscience of the importance and pervasiveness of new organic metaphors for the reconstruction of methods and knowledge, and that search is explicitly stated in the studies on complexity (for instance Arthur, in Cowan et al., eds, 1994: 680).

A final chapter briefly summarizes the evolution of evolutionary thought — from Lamarck and Darwin to Spencer and Weissmann, from Social-Darwinism to sociobiology — and proceeds to the discussion of its applications for economic theory. It is argued that such a metaphor has been decisive for the assault upon the positivist paradigm of economics, but that it has been in most cases wrongly used and that no model was presented until now arguing for a convincing general and precise analogy between evolutionary biology and economics, namely for the theory of the firm or of innovation. Moreover, that model probably does not exist.

Darwinian evolution represents essentially an allegory for economics: it provides a new vision, escaping from the mechanistic prison, but the attempts to generate precise biological analogies orienting the research in economics are doomed to fail. No economic analogue exists for the replication unit in biology and the discrimination between genotype and phenotype is not relevant in society, neither is social evolution identifiable by natural selection processes. Indeed, an excessive expectation attributed to the metaphor the power of selection of specific hypotheses and of defining models for analysis, but the results were scarce.

Yet, the evolutionary metaphor is not to be dropped. The argument of this book is that it is even more useful than ever, if its merits are considered: coevolution, complexity, order out of equilibrium and coordination are the core concepts of the new evolutionary metaphors for the reconstruction of economics, which are explored in this part.

2. The Impasse of Positivism

The modernization of the world is a long civilizational process beginning in the sixteenth century. One of its constitutive elements was the emergence of a new world-vision, a culture of change and transformation defining science as its most authoritative element — in fact, the central role of science was the creation and development of legitimate and general patterns of communication and social validation, simultaneously breaking with the medieval mysticism and imposing a new general reference for knowledge.

This modern science has been organized around three dogmas:

a. Naturalism, the epistemological association of the whole universe of knowable elements, namely society, to Nature. Such a claim implies: (a) that observability entails reproducible experiments; (b) that observation is the genesis of complete knowledge, since allowing for the description of all phenomena under general laws; (c) that the goal of science is the control of Nature; (d) that there is an univocal relation and submission of the object of inquiry to the inquirer. A more familiar and pervasive form of naturalism, weakening or transforming the second and third assumptions, is scientism. It is, indeed, an ideology.
b. Universalism, that is, the generalization of the laws defined from observation and experiment, in order to include the universe of knowable elements under the same criterion of inquiry. Such a dogma includes the quantification as a necessary and sufficient form of definition and description.
c. Objectivity, that is, the affirmation of the ethical and prescriptive neutrality of the scientific inquiry, neither concerned nor influenced by any kind of social values.

The general consequence of these dogmas is the reductionist or atomist approach still prevailing today in science, including in social sciences. An early example is Hume's recommendation to his readers in the conclusion of his *Enquiry Concerning Human Understanding:*

> When we run over libraries, persuaded of these principles, what havoc must we make? If we take in our hand any volume — of divinity or scholar metaphysics, for instance

> — let us ask, *Does it contain any abstract reasoning concerning quantity or number? No. Does it contain any experimental reasoning concerning matter of fact and existence? No.* Commit it then to the flames for it contains nothing but sophistry and illusion. (Hume, 1748: 165, his emphasis)

In the same sense, Bacon, or later on Kelvin, strongly stressed that all knowledge is to be expressed by numbers — quantification is the goal of science and allows it to interpret the world. In the same vein, Galileo stated that what cannot be measured cannot be counted and does not count. Finally, the most celebrated example of the third claim, objectivity, was Weber's claim about the 'axiologic neutrality' of science.

This modern science proudly formulated Laplace's dream: for such a complete knowledge of all forces acting in Nature, 'nothing will be uncertain, the future and the past will be opened to its regard' (Laplace, 1812: vi-vii). The scientific laws should capture the universal essence and transcend time. Positivism was indeed born as a dogma — or, as Comte put it, as a 'catechism' — abolishing time and historicity, reducing the events and relations to atomistic and quantifiable facts, and submitting facts to a single and indisputable pattern of knowledge. As a consequence, science came to be conceived of as the quest for universal and timeless statements completely describing the world and was therefore supposed to appropriate the ontological attributes of Divinity.

Positivism as such was formulated only in the nineteenth century as the articulation of the main trends in modern science but it was really the heir of this metaphysic claim for complete and exhaustive knowledge. In fact, it established a bold synthesis between distinct methods of thought and methodological prescriptions, from Galilaic, Baconian and Newtonian empiricism to Cartesian rationalism. For some decades, this paradigm was immensely productive and in fact it was the organizer of the transition for modernity. These two main trends will be discussed now.

2.1. The modern clock: birth, rise and crisis of classical positivism

By the beginning of the seventeenth century, modern science was developed enough to challenge some of the religious dogmas of the pre-Enlightenment period, and the first principle of the new positivist approach was the separation of the realm of science and of religion. In a private letter to a friend, Kepler explained the radical new attitude as the development of the metaphor of the clock:

> I am now much engaged in investigating physical causes; my goal is to show that the celestial machine is not the likeness of a divine being, but is the likeness of a clock (he who believes that the clock is animate ascribes the glory of the maker of the thing made). In this machine nearly all the variety of movement flows from one very simple force just as in a clock, all the motions flow from a simple weight. (Kepler, quoted in Olson, 1971: 60)

Science was part of the modernization hurricane which was going on and intended to dominate and to develop it. Indeed, more or less a quarter of century after this letter was written, Galileo was still condemned for adopting the Copernican cosmology — but he knew he was right, and perhaps as important is the fact that the reason for the condemnation was his unwillingness to accept presenting his inquiry as a simple mathematical tool, 'as if' it was just a model with no claims about reality and therefore not questioning the Church's conception of the universe. In that case, as Cardinal Bellarmino suggested, the court would have accepted his apologies. Galileo's condemnation was in fact the first modern skirmish between instrumentalism and realism and it marked the turning point: although Galileo was defeated for choosing the dangerous strategy of defending the truth, it announced the coming period of victory of science over the theological dogmas.

The mechanistic approach was thereby strongly established and constituted the operational version of that extreme naturalist dogma. For Bacon, a contemporary of Galileo, the collection of data — the exclusive and defining preoccupation for the scientist — should avoid four dangers: the seduction by the idols of the tribe (the influence of the innate features of the mind), by the idols of the market place (language), by the idols of the cave (individual education), and by those of the theatre (misleading philosophies). The objectivity and completeness of knowledge was sure when and if these dangers were avoided (Bacon, 1620: 22 f.). In particular, induction was the method to vanquish those demons and to establish true knowledge: 'The formation of ideas and axioms by true induction is no doubt the proper remedy to be applied for the keeping off and clearing away of idols' (ibid.: 48). Laplace, two hundred years later on, fully supported the same method, enthroning Baconian empiricism as the very definition of science (Laplace, 1812: cxxxix-cxl).

In the generation following Bacon, Newton was the most important scientist to apply and develop these ideas.[1] In his *Mathematical Principles* he presented the new method:

> In experimental philosophy we are to look upon propositions inferred by general induction from phenomena as accurately or very nearly true, notwithstanding any contrary hypotheses that may be imagined, until such time as other phenomena occur, by which they may either be made more accurate, or liable to exceptions. This rule we must follow, that the argument of induction may not be evaded by hypotheses. (Newton, 1687: 166)

Or, even more conclusively, his *hypotheses non fingo*:

> Whatever is not deduced from the phenomena is to be called an hypothesis; and hypotheses, whether metaphysical or physical, whether mechanical or of occult qualities, have no place in experimental philosophy. (Newton, ibid.: 204-5).

Newton developed extensively the first dogma of positivism, under its atomistic and mechanistic form: the basic elements of his conceptual system are atoms, their basic properties being mass, extension and shape, and their basic relations being motion and interaction. For studying the forms of those atoms, Newton employed two distinct methods of generalization: (a) by enlargement (generalization of particular instances); and (b) by division (the known primary qualities of bodies being assigned to others too small to be observed; Harré, 1964: 57, 104, 108-9).

But Newton frequently departed from Bacon's discipline. In fact, Newton was aware of the limitations of this general method: it could not logically sustain a general conclusion,[2] but should still be considered as the 'strongest' available method; this was widely shared by other scientists, since their confidence was reinforced by the new scientific developments which established Newton's aura.[3] Anyway, his inductive method was complemented by an axiomatic and systemic approach in most cases: some of his most celebrated findings, such as the suggestion of the existence of Neptune, follow from a principle of universal gravitation which was not coherent with the proclaimed extension of Kepler's laws (Duhem, 1906: 190 f.; Ruse, 1982, 42; Giere, 1988: 73-4); a supplementary hypothesis had been necessary in order to direct the investigation. A largely deductive system defined a new axiomatic (the theory of gravitation, the three laws of motion in Newton's mechanics) and defined a procedure for relating theorems of the axiom system with observations, in order to corroborate the hypothesis and to lead to new findings (Losee, 1980: 86). In this sense, Newton's concrete work was substantially opposed to the pure empiricist epistemology of Bacon, in which only observation and experience were accepted and there was no role for concepts. Furthermore, Newton presented his method in *Principia* as following an exact mathematical model in analogy to an imaginary world, while in other works he considered non-mechanic causality, such as forces acting at a distance. Berkeley, an empiricist whose positivist traits were unambiguous, criticized these contradictions in Newton's work, and most of all his introduction of non-observable invisible quantities, which challenged the inductive dogma. In fact, Newton's unique position in the seventeenth and eighteenth centuries science was due to his methodological eclecticism and bold hypothesizing.[4]

It is nevertheless unquestionable that the strict inductivist account of Newton's work dominated the interpretation of his major achievements. Voltaire, after a travel to Britain, praised Newton for destroying Cartesianism (Voltaire, 1733: 91). The same Voltaire published in 1738 an important piece of popularization along the same lines, *Elements of Newton's Philosophy*. This new path for rigorous inductive science was also cultivated and praised by Saint-Simon or Comte, and became the hallmark of positivism.

Classical positivism developed this general approach: induction was defined as the only criterion for cognition, as it established a general demarcation principle against all kinds of metaphysical pseudo-knowledge, or the four idols of Bacon. The concrete procedures deriving from these methodological assumptions were in consequence: (a) the definition of tests as the valid task to check the laws derived from observation; (b) the existence of a Baconian *experimentum crucis*, a crucial experiment allowing for the final validation of laws derived from observations, and (c) the verification principle establishing the possibility and necessity of testing the law in the whole domain of the function. The continuity of the function was stressed by the general idea according to which *Natura non facit saltum*, common to Leibniz (*Natura abhoret saltum*, nature hates jumps), Linnaeus, Lamarck, Darwin, or later on to Marshall, who printed the motto in the first page of his *Principles*.

Even if this method obtained a large support and was even considered canonical in sciences — including in social sciences, since the birth of sociology under the spell of Comte's dogma (Comte, 1852: 15) — the weakness pointed out by Newton himself suggested a reconsideration. The main argument for such a debate has been the unsolved Hume's Paradox: even if a verification is extensively developed in an inductive process, it is never possible to assure that the first after *n* observations will not refute the general law; there is indeed no logical way of stating that 'the sun will rise tomorrow', from the single fact that it did yesterday and today. The Paradox suggests a general suspicion against inductivism.[5]

Hume was in fact a supporter of inductivism but also a sceptic about the method's capacities to demonstrate scientific propositions. The 'dilemma of objective validity' was the conclusion of his criticism of both deduction and induction: if stating a cause-effect relation is the central task for science, deduction is not enough to define scientific rationality.[6] On the other hand, induction is not logically demonstrative but still it is the only path for scientific progress: 'causes and effects are discoverable not by reason but by experience' (Hume, 1748: 28; also 1740: 8).

This paradox puzzled the scientists and suggested a reconsideration of verificationism and induction, and a requalification of the positivist program.

*

The classical positivist program — based on the articulation of the dogmas of naturalism, objectivity and universalism, and their consequences, the reductionist and atomistic approaches — established a strong principle of demarcation against non-scientific statements, including a cognitive method, induction, and its correlate verificationist procedure. But the method was not logically consistent and was challenged by Hume's Paradox, and furthermore was not complete for practical purposes, as proved by Newton.

During the nineteenth century, three main challenges were defined against this program by Marx, Darwin and Carnot and Clausius, and the alternative which was immediately most successful, Darwinism, created a new pluralistic methodology for biology, which was potentially extensible to other sciences. Nevertheless, by the end of the nineteenth and in the early twentieth centuries, the positivist program was reorganized. This rationalistic reconstruction, originally inspired by Cartesianism, is discussed in the next section.

2.2. The rationalistic reconstruction of positivism

Descartes, contemporary of Bacon, Kepler and Galileo, had already established an alternative to empiricism, which constituted a second building-block of modern science. As he told the legend,[7] he was inspired by his dreams of the night of the 10 November 1619, while he was practising some military exercises, to elaborate a new method, the 'basis for an admirable science', unifying all knowledge and creating a universal mathematics (Descartes, 1637: ix). *Cogito, ergo sum*, the 'first principle' of the new philosophy, did not establish truth, since God would be the only guarantee of truth (Descartes, ibid.: 40, 47; in fact the purpose of that whole section of the text was to prove the existence of divinity). Yet, he claimed that the book presented a true method: since I can think, I do exist, rationality logically precedes existence, and therefore what the human beings can understand very clearly must be true (ibid.: 41). This method consists in the following four recommendations: (a) one should consider true the statements recognized as such by reason; (b) every difficulty should be divided in as many elements as possible; (c) one should order the thoughts, rising slowly from simple to complex statements; and finally, (d) one should omit no detail (ibid.: 22-3).

The three last items are already familiar: they are part of the classical positivist approach. But a striking difference was introduced by the first and central deductive procedure suggested by Descartes. That was the very essence of his method: the cognitive criterion was established by deductive reasoning and not by observation or experiment. Descartes himself did not consider the experimental demonstration. One obvious example of that was Descartes' praise for Harvey's book on blood circulation and the heart, which included convincing evidence for the new theory, but did not change his own theory about the mechanical functioning of the body (Descartes, 1637: 57, 62). The development of Cartesianism organized this methodology in the following terms:

a. A mechanical science was built on absolute knowledge, since deduction (as a discrete and controllable set of logical statements) should be logically complete and irrefutable, unlike induction,[8] which could only be accepted as non-demonstrative. Furthermore: (a1) such a knowledge was fully cumulative, (a2) the logical meaning of every statement was established in relation to the axioms; and accordingly (a3) the statements should be organized in a deductive hierarchy and referred to quantifiable bits of reality. The cosmic and mechanistic analogies for the organization of matter corresponded to the certainty of thought based on the deduction of that organization (Descartes, ibid.: 53, 56): intuition was the primary source, and the pure ideas were supposed to be certain.

b. The context of discovery was separated from that of justification: Descartes' own dream did not justify his assertions on method, and the nature of the discovery process was to be delt with by non-scientific speculations.
c. Finally, the method vindicated Cartesian dualism (body/soul, matter/spirit, *res extensa/res cogitans*, Descartes, ibid.: 40-1), which was also an extension of the dual Aristotelian logic (*tertium non datur*). Since matter could be explained by the same principles of extension and movement, the different sciences, such as biology,[9] physics or mechanics should follow the same principles and methods (ibid.: 56).

These principles had a considerable and durable impact on epistemology and nourished a decisive paradigmatic shift: since all essences can be reduced to magnitudes, mathematical formalization can account for all natural and social phenomena and eventually create a common language for sciences. Furthermore, the second assertion plays a major role in the Popperian theory, and dualism has a pervasive influence in all scientific thought. A whole new departure for positivism was established, since the annoying Hume's Paradox was avoided.

This proud deductive science shared with empiricism the requirement for a perfectly deterministic body of knowledge, and that of a mechanistic functioning of models under general laws ruling reality: the clock interpretation synthesized both trends, in spite of their differences about the cognitive process. But since both cognitive criteria were vulnerable, because of Humean scepticism or because of the tautological character of deduction, a truce was favoured by both armies.

This synthetic model coming out of this development, called the Hypothetico-Deductive Model (HDM), was conceptualized and suggested later on by the authors of the Vienna Circle, which developed an ambitious program searching for a scientific and objective language. The Vienna Circle (Carnap, Godel) developed logical positivism since 1925, suggesting a reassessment of the old criterion of cognitive meaning based on induction and of the procedure of verification. By 1935 the logical positivists adopted a different strategy, what came to be called 'logical empiricism' (Carnap, Ayer, Hempel, Nagel), since including a logical analysis in their construction, in a major departure from pure induction.

A refined HDM was developed, including a correspondence theory of truth (according to Tarski's definition, a true statement was that corresponding to facts), and defined by the quest for some universal scientific language. As a theory was considered to integrate an *explanans*, a general law and the definition of initial and final conditions, and an *explanandum*, the proposition about the events to be explained, the demonstration should be the deduction of the explanandum from the explanans. And since the law was inductively based, a symmetry could be established between this explanation and a prediction also deducted from the explanans; such a prediction could make possible a comparison of theories. The procedure associated with this model was to be called confirmationism, based on the test of the hypothesis against experimental evidence, strengthening the confidence in it if confirmed, or correcting it if refuted. It is still dominant in sciences even if its theoretical foundations are outdated by the conceptual predominance of infirmationism.

It is evident that this elegant construction reintroduced Hume's Paradox, since it did not restrict itself to the Cartesian deductive process and recreated an inductive logic. Inspired by young Wittgenstein, the effort of the logical empiricist was therefore to define science as a coherent linguistic system and to guarantee the logic of the demonstration. Positivism was redefined by postulating the necessary formal symbolic logic and axiomatic system, a language for expressing scientific propositions. Such a language should establish a non-refutable inductive logic, 'a canon of thought whereby conclusions could be established with a specific probability from premises that included only basic logic and mathematics and reports of observations and experiments' (Hausman, 1984: 18).

The evident failure of this mystical search for a completely objective language for science left several old and new unsolved questions. In fact, it proved neither the efficiency of a syntactic reconstruction of epistemology, nor the possibility of overcoming the problem of induction by a *corpus* of undisputed deductive rules for the transformation of observational statements into theoretical statements. For those reasons, the attempt to reconsider a criterion of cognition autonomously from the criterion of demarcation — that is, induction or deduction as procedures independent from the distinction between scientific and non-scientific theories — was abandoned. Different solutions were suggested to this new impasse, and they will be very briefly presented.

2.3. The return of the demarcation principle

So far, the evolution of positivism was described and discussed around two main periods and concepts. The first form of positivism, classical positivism, was empiricist and established a strong demarcation principle — previously defined as the criterion to separate scientific from metaphysical statements and theories. This includes the cognitive criterion — defined as the legitimate form of acquisition of scientific knowledge — and a consequent procedure of inductive inference and thus of verification of the general laws in all the domain of the facts involved. Description, a complete description of phenomena, was therefore equivalent to explanation.

The second positivist current, organized around the HDM and taking the form of logical empiricism, combined the Cartesian approach to the deduction of the explanandum from the explanans with the cognitive inferences from induction. The symmetry explanation-prediction was established by the logical account, and a confirmation principle was introduced as the basis for the theoretical statements and the general laws defined from observational statements. In other words, the cognitive criterion, the inductive inference of laws from observation, was combined with a signification criterion, defined hereby as the semantic and cognitive organization of the theoretical statements. The inductive logic corresponded to the cognition criterion, but the deduction from the axiomatic system corresponded to the signification criterion.

Modern positivism tried to establish science as an ideal linguistic system, making possible a new inductive demonstrative logic, supposed to avoid Hume's central objection and to establish an irrefutable cognitive criterion. But this effort was previously at the centre of the impasse of the logical empiricist current, and the Humean scepticism resuscitated. Consequently, the impasse favoured Popper's alternative: since the time of the transformation of logical positivism into logical empiricism, Popper developed instead a comprehensive and influential new synthesis which challenged some traditional definitions of classical positivism and changed its whole approach. Popper reduced the inductive cognition criterion to a simple non-demonstrative possible inference from facts, and argued for a return to a strongest demarcation criterion comprising the signification criterion as well. The logical primacy of the demarcation principle is the specific positivist characteristic of Popper's philosophy of science, since it reaffirms the reducibility of the object and the irreducibility of the subject.

The demarcation principle was to be established on the basis of the falsifiability of universal theoretical statements. In this approach, the scientificity of a theory was defined by its willingness to submit concrete and bold predictions to refutation by future evidence and testing, and by its resistance to falsification: 'One theory which is really not refuted after the test of new daring and not probable predictions may be considered as confirmed by tests' (Popper, 1982: 245). Bacon's *experimentum crucis* was therefore reintroduced as the concrete form of the simultaneously demarcation and signification criterion. A sophisticated form of relativism was supported and it stressed the provisional and conjectural nature of knowledge but also its radical separation from metaphysics, including in this category all historical or non-experimental sciences unable to formulate predictions (History, Palaeontology; or theories, like Darwinism, Marxism or Psychoanalysis).

This program rebuilt positivism. In fact, it denied its naturalistic foundation

— and Popper strongly argued against it — but strengthened the universalist and objectivist dogmas. As a whole, it provided the scientific community with a new powerful heuristic, the 'corroboration' process being the condition for the search for new hypotheses and 'falsifiability' being the control of the process. Its success guaranteed a new form of self-identification of science as a separate and demarcated structure of knowledge inside society.

The Popperian program constituted therefore the single most important and most successful effort to reorganize positivism in recent times. It wrongly claimed to be a solution of Hume's Paradox and to define a logic of justification allowing for the development of science based on an excluding and exhaustive demarcation principle, but the result was the prolongation of the impasse (Louçã, 1993: 22-7). The main criticisms to this program can be shortly summarized:

a. It excludes scientific fields of inquiry, and not only the single theories supposed to reject falsifiability. According to this conception, all historical sciences are rejected, such as these social or even natural sciences where experiment cannot take place following the physic's canons.[10]
b. The program is not logically viable nor consistent. Some universal existential statements cannot be refuted: for instance, the statement 'There are abominable snowmen' is always valid since no refutation is ever possible (Caldwell, 1982: 21). Of course, in spite of it, this statement is rejected by the Popperian criterion, since it does not even allow for a refutation test: the only acceptable statement would be the negative form, 'there are no abominable snowmen', which is considered a true scientific statement since it can be refuted (if one snowman is found). But this stratagem is not always possible or desirable, since a large part — if not the majority — of scientific inquiries is concerned with hypotheses formulated as universal existential statements: the assertion 'there are quarks' directed a research even if it is irrefutable, but the assertion 'there are no quarks' is either irrelevant or a mere disguise of the previous statement for some ceremonial reasons.

   Speaking for Popper, Samuelson argued for the applicability and triviality of the infirmationist principle with the following example: 'Drilling to find where oil isn't differs not all that much from drilling to find where it is' (Samuelson, 1974: 12) — happily for his career, the author is a Nobel Prize winner and a distinguished professor of economics, and not just a manager of some oil corporation.
c. The method is never conclusive. The argument was stated as early as 1906 by Duhem: rejecting Baconian inductivism and naive belief in the experimentum crucis (Duhem, 1906: xi, 183), Duhem stated that an experiment never condemns an isolated hypothesis but a whole set of hypotheses. This statement, later on known as the Duhem-Quine Thesis,

argued against naive infirmationism:

> In sum, the physicist can never subject an isolated hypothesis to experimental test, but only a whole group of hypotheses; when the experiment is in disagreement with his predictions, what he learns is that at least one of the hypotheses constituting this group is unacceptable and ought to be modified, but the experiment does not designate which one should be changed. (ibid.: 187)

There is no possible refutation indicating the single hypothesis to be falsified.[11] For Duhem, the conclusion was that a new form of confirmationism was needed, so that the confirmation of a prediction may strengthen the confidence in the hypothesis (ibid.: 28). Popper answered to the Duhem-Quine Thesis with a partial retreat: 'In reality, a decisive refutation of a theory can never be presented, since it is always possible to state that it is impossible to believe in experimental results or that the distances between the experimental results and the theory are only apparent and will disappear with the progress of our understanding' (Popper, 1962a: 50, my translation). Blaug presents this position as an acceptance of the D-Q thesis (Blaug, 1982: 16-7). But, as it is obvious from the quotation, Popper only accepted some psychological limitations (the declarations or beliefs of scientists) and reduced the whole case to a question of rhetoric.

d. Finally, the method is not practically applicable, since the hypotheses are not strictly separable, some initial conditions are not verifiable (for example, in economics, the state of information of the agents), not all exogenous variables are known and controllable, and no general law can be submitted to test without complementary hypotheses. It is not applicable and it is not applied: the majority of the scientists looks for confirmation, not for refutation — Mendel was trying to prove the laws of heredity and to support his hypothesis, not to refute it, just as Newton looked for his planets and not for the cosmic emptiness, just like physicists look for their quarks.

Faced with the failure of positivism, Popper inverted the inductivist equation: instead of grounding the authority of science in an empiricist demarcation criterion, or in a cognitive principle supporting the formulation of general laws or the elaboration of the axiomatic system, he suggested a strictly deductive procedure for the formulation of hypotheses, to be submitted to the falsification test of comparison of predictions against inductive evidence. As a consequence, in spite of claiming to have solved the Paradox, Popper reintroduced it in the final stage of his epistemology. In the real world, facts are indeed more stubborn than theories.

The failure of the verificationist methodology (which does not constitute a foundation for universal laws inferred from observational statements) and of the falsificationist methodology (which does not accept universal existential statements) maintained the most modest confirmationist methodology as the only practical procedure to be generally employed in science, and opened the way to radical challenges or even to nihilistic attitudes, such as Feyerabend's 'anything goes' or McCloskey's reduction of science to the status of rhetoric conversations.

2.4. An early critique of positivism: Darwin

The positivist synthesis is only a product of the nineteenth century, but its main trends — empiricism and Cartesian rationalism — dominated the horizon of modern science for a long time. It is therefore not surprising that the main alternatives to positivism were exiled to the boundaries of the paradigm. One of them emerged in a new science, biology.

Darwin considered himself a good pupil of Baconian methodology or of positivist prescriptions; in spite of it, he introduced new concepts of causality and new methods of inquiry with lasting consequences for science in general. In fact, Darwin claimed to follow Bacon and Newton (Ruse, 1982: 42 f.; Brooks and Wiley, 1986: x), but that was certainly contradictory with his own experience. In the *Autobiography* prepared for his grand-sons, Darwin candidly remarked that in July 1837, when beginning his first notebook, 'I worked on true Baconian principles, and without any theory collected facts on a wholesale scale' (Darwin, 1876: 119). But, according to the same *Autobiography*, only fifteen months were needed to give to those facts a meaning and to formulate a theory since when, rereading Malthus in 1838 'for amusement', Darwin was able to incorporate the concept of natural selection. 'Here, then, I had at last got a theory with which to work' (ibid.: 120). A constitutive metaphor and a theory had been necessary in order to interpret the facts and to define a scientific program. Newton's influence was more consistent, since it justified the shift of the theory from internal to external forces, from 'transformational drives to environmental forces' (Depew and Weber, 1996: 115), in opposition to Lamarck's finalism (ibid.: 115, 127).

Some authors, like Blaug,[12] present Darwin as a sophisticated positivist and a supporter of the later Hypothetico-Deductive Model (Blaug, 1982: 8-9). Mayr presents a similar argument: the HD model followed by Darwin included three steps: (a) the generation of hypotheses from pure 'speculation'; (b) the experiments to test the hypotheses; and consequently (c) the formulation of a theory (Mayr, 1982: 28, 850). But, of course, the first step is not a purely deductive one: it is 'abductive' and thus non-positivist. The HDM interpretation of Darwin is too narrow; in fact, Darwin's method combined four different and innovative approaches:

a. A new concept of local causation. Darwin rejected both essentialism and Lamarckian teleological evolution by intrinsic drive to perfection, and supported a separate concept of causation for each natural process: variation through a process of genetic or adaptive change (a 'black box' which would be opened only when Mendel's genetic laws were rediscovered by the end of the century), or differential survival and reproduction in the selection process through the interaction of the genetic structure and the environment. The causal assertions were established by a unique combination of observation and deduction. The contradictory influences of Malthus and of Adam Smith's concept of unintended consequences of individual behaviour,[13] were important to establish a general framework for these local processes.

b. The use of heuristic devices, the theory of common descent being the main one: the scientists should look for the 'missing links' (Darwin, 1859: 341), and in fact that activity would become a source of confirmation and development of the evolutionary theory.
c. An organic approach. In the context of modern science, this was one of the most important contributions by Darwin, since the first Aristotelian organicist approach had been buried with medieval metaphysics: Darwin presented a new comprehensive system which included the indeterminacy of biological phenomena, an understanding of complex and hierarchical processes and the affirmation of the impossibility of absolute predictions ('I believe in no law of necessary development', Darwin, 1859: 348).
d. And, finally, a large, original and crucial use of metaphorical and speculative reasoning. The most evident example is the sequence of the first and the second chapters of *The Origin*: Darwin dealt with variation under domestication and then, by analogy, discussed variation in Nature.[14] As Darwin himself stated, he used metaphor and analogy as a complementary heuristic to define hypotheses about the common descent of all species (Darwin, 1859: 454-5), and argued for the legitimacy of speculation based solely on some facts, even if it leads to a false conclusion (which should be corrected, and then one way for error would have been closed, Darwin, 1871: 926 f.), and on the recourse to imagination for the interpretation of facts (Darwin, 1859: 263).

In a letter of December 1859 to Huxley, just after the publication of *The Origin of the Species*, Darwin explained why did he insist on the explicit presentation of analogies in his demonstration:

> Not only sir H. Hollan, but several others, have attacked me about analogy leading to belief in one primordial *created* form. (By which I mean only that we know nothing as yet of how life originates) .... I answered that though perhaps it would have been more prudent not to have it put in, I would not strike it out, as it seemed to me probable, and I give it no other grounds. You will see in your mind the kind of arguments which made me think it probable, and no more fact had so great effect on me as your most curious remarks on the apparent homologies of the head of Vertebrate and Articulate. (Darwin, quoted in Carabelli, 1988: 265; this letter was also noted and quoted by Keynes)

This rather important break with positivist traditions was quite obvious in the discussions of Darwin with some of the most distinguished epistemologists of his time: Herschel and Whewell. Herschel argued in 1830 and 1831 for the use of analogy as a criterion of good science: 'If the analogy of two phenomena be very close and striking, while, at the same time, the cause of one is very obvious, it becomes scarcely possible to refuse to admit the action of an analogous case in the other, though it is not obvious in itself' (Herschel, quoted in Ruse, 1982: 45-6). But Herschel, an astronomer, was a firm believer in Newton's methods: science was vindicated as the statement of simple, constant and uniform laws. As a consequence, he rejected *The Origin*, since natural selection was neither a reversible process nor a deterministic mechanism (Depew and Weber, 1996: 66, 147, 153).

Whewell was also a supporter of the use of analogy and defined theories as statements of unifying analogies between two different domains of phenomena (Hesse, 1974: 221-2). But analogy should be built on extensive and complete induction, the ultimate source of every reasoning, and Darwin was suspected of having abandoned the rigorous Newtonian method, as Sedgwick emphatically accused him: 'You have deserted ... the true method of induction, and started in machinery as wild ... as the locomotive that was to sail us to the moon' (quoted in Depew and Weber, 1996: 141).

It is quite obvious that Darwin's method surpassed these traditional Newtonian approaches. The map of Darwin's concept of natural selection includes therefore two directions of analogical reasoning (the Malthusian theory of population in a theoretical level and the evidence of artificial selection at an empirical level), induction from so diverse fields as the study of the Instincts, Palaeontology, Embryology or Biogeography, and the deduction of a general law. Darwin's use of deduction, induction and analogy illustrates quite well his approach to causality and to science: since there are infinite possibilities of mutation, no prediction is accurate and no univocal determinism in the relation genotype-environment can be stated. The traditional antinomy contingency-necessity was therefore challenged: mechanicism and reductionism were consequently annihilated.

2.5. Causality and determination

The positivist program established strict conditions for causal claims to be accepted as legitimate: cause and effect must be identifiable and separable, and the nature of determination must be unequivocally univocal. Furthermore, these conditions could be arguably established in the inductivist version, since reproducibility of the experiment and control of its exact conditions define the determination, as well as in the deductivistic account, since the singular instances can be proved to integrate a general law.

Both these versions provoked intense arguments since their formulation, and an alternative organicist thought challenged such a perspective since very early with Kant, who established in the *Critique or Judgement* the difference between mechanic drive and intentionality: 'An organised being is then not a mere machine, for that has merely *moving* power, but it possesses in itself *formative* power of a self-propagating kind which it communicates to its material though they have it not of themselves; it organises them, in fact, and this cannot be explained by the mere mechanical faculty of motion' (Kant, 1790: 278, his italic). In the same mood, Hegel criticized this mechanical analogy implied both in the Newtonian paradigm and in the definition of positivist laws, and argued that mechanisms should be subordinated to the organic conception, functioning as a regulatory system. Hegel formulated a clear critique of the causal chains in the mechanic interpretations of reality, particularly of its extensions to the realm of organic wholes or other contexts where qualitative phenomena should instead be considered:

> Thus decidedly must we reject the mechanical mode of inquiry when it comes forward and arrogates itself the place of rational cognition in general, and seeks to get mechanism accepted as an absolute category. But we must not on that account forget expressly to vindicate for mechanism the right and import of a general logic category. It would be, therefore, a mistake to restrict it to the special physical department from which it derives its name. There is no harm done, for example, in directing attention to mechanical actions, such as that of gravity, the lever, etc., even in departments, notably in physics and in physiology, beyond the range of mechanics proper. It must however be remembered that within these spheres the laws of mechanism cease to be final or decisive, and sink, as it were, to a subservient position. To which it may be added that in Nature, when the higher or organic functions are in any way checked or disturbed in their normal efficiency, the otherwise subordinate category of mechanism is immediately seen to take the upper hand. (Hegel, 1830: 262)

In economics in particular, those debates appeared since very early. Sismondi, before Marx, sharply criticized Ricardo's deductive method, arguing for realist assumptions and historical methods:

> [My method] is not founded on dry calculations, nor on a mathematical chain of theorems, deduced from some obscure axioms, given as incontestable truth ... Political economy is founded on the study of man and men; human nature must be known, and also the condition and life of societies in different times and in different places. One must consult the historians, and the travellers; one must look into one's self; not only study the laws, but also know how they are executed; not only examine the tables of exportation and importation, but also know the aspect of the country .... (Sismondi, 1819: xv)

Sismondi added that sequential and reciprocal causality — when 'each effect becomes cause' (Sismondi, 1819: 125) — cannot be either isolated or ignored. The advice was not followed by late nineteenth-century economics, neither has it been followed by mainstream social sciences, which mimicked the dominant scientism: according to Saint-Simon, history should be as certain as astronomy (Grossman, 1943: 389).

These severe assaults upon the positivist program of mechanical causality — Kant's or Hegel's consideration of quality and change, and Sismondi's argument of complex causality — converged to Darwin's and Marx's new organic epistemologies, which established the connection between *contingency* (mutations, choice) in the domain of external causation and *necessity* (internal development of contradictions and therefore lawfulness, crises) in the domain of internal mediation.

Mainstream science resisted these challenges: positivism is the encapsulation of naturalism, universalism and objectivism, and defines causality as the action of impulses in a mechanical world ruled by rigid laws postulated by the certainty of the repetition of causal relations, allowing for prediction. Of course, classical positivism had already pointed to some of the difficulties of this approach, the most important of all being Hume's paradox: since induction does not support a demonstrative argument, there is no proof of the 'ultimate connection' between cause, defined as a spatio-temporal homogeneity under the conditions of constant conjunction and contiguity, and its effect (Hume, 1740: 119).

The impasse led Mach and the Vienna Circle to reject any investigation about fundamental causes, since they were supposed to be unobservable, and to consider instead a stricter version of cause, redefined as the 'change in the independent variable', and of the cause-effect as a 'functional relation' (Olson, 1971: 244). This change was considerable and proved very powerful, since it made possible a general formalization of causality as exogenous impulses in a mathematically treatable model of reality. It is indeed the dominant form of production of causal claims in the positivist paradigm to this day.

But this concept of *causa efficiens*, instrumentally defined as exogeneity of the explanatory variable, is logically incoherent and ontologically deficient. As Feigl argued, the concept is arbitrary, since the inverse of the function cannot be defined as an inverse causal relation (Feigl, 1953: 417); moreover, if events are unique or historical by nature, causality cannot be formalized as a functional relation of dependence (Bunge, 1959: 115; Feigl, 1953: 410). On the other hand, the positivist concept of causation is inseparable from the prediction corollary, since 'causes' are replaced by general causal 'laws' which allow for prediction — and prediction is considered to be the salvation of positivist epistemology under the light of Popperian reconstruction. Still, prediction is inconclusive and not demonstrative:

> successful prediction with the help of a set of laws is not a test of their causal nature. ... Predictability and artificial reproducibility are empirical criteria for testing the truth of law statements; the attempt to derive all the meaning of law statements from their use and from the procedures of their verification amounts to confusing truth with one of its criteria, semantics with pragmatics. (Bunge, 1959: 331-2)

Bunge suggested that causality — in this sense of cause-effect functional relation — be considered as a specific part of a general category of determinism, since it is not universally valid and may be considered only as a convenient approximation whenever there is an extreme asymmetry between cause and effect in the framework of negligible feedbacks (ibid.: 170), and added that other forms of determinism should be considered, such as dialectics and holism.[15] As indicated, Marx and Darwin fall into these categories and argued for causal determinations which cannot be formulated as univocal relations in timeless environments: every causal explanation, if it is to be realistic, must be based on a constellation of determinations and not on a single abstract representation as a functional relation in a model.

The introduction of history and structural change, of qualitative variables and choice, of multiplicity of actions and complexity, as Hegel, Marx, Darwin, or Maxwell argued, challenges the positivist certitudes and methods. Yet, it was only with another rupture that positivism was deeply damaged: the defy was suggested from the central corpus of natural sciences and was established by quantum physics. Heisenberg summarized the challenge under ten 'revolutionary propositions' (Lukacs, 1971, 293 f.): (a) the unpredictability of the trajectory and behaviour of particles implies the collapse of traditional positivist determinism, and (b) contradicts the ideal of objectivity;[16] as well as (c) the illusion about definitions; and (d) the illusion about the absolute truthfulness of mathematics; (e) it contradicts the notion of 'factual' truth, considering change; (f) it denies the nature of cause as *causa efficiens*; and (g) reinstates the Heraclitean principle, *panta rei*, everything moves; (h) stresses the importance of relationship and not of the 'essence' of factors; (i) rejects the traditional principle of non-contradiction of Aristotelian logic: in the new concept, *tertium datur*; (j) Cartesian dualism, therefore, collapses.

In other words, method and objects are indistinguishable; no certainty is ever possible. After Popper challenged naturalism, the two remaining dogmas were shot at the same time: universalism and objectivity were severely damaged by these assertions.

2.6. At the boundaries of modern science

The failure of all attempts to support an exclusive demarcation criterion — as 'a necessary and sufficient condition for a belief system to be a science' (Hesse, 1980, 46) — namely the failure of the infirmationist criterion, reopens the epistemological debate. Since *totality*, intelligible as the universe of observational and theoretical statements, cannot be studied according to the principles of exclusive positivist rationality or to the principle of some universal unifying methodology defining immutable and general laws explaining any event in any time dimension (Laplace), a new semantics is central to the agenda of science.

In fact, if verifiability, falsifiability, theoretical or inductive inference are not authoritative, a new pattern of construction must be implemented. It takes the direction of openness to difference, to irreversibility and therefore to time, purposeful action and complex hierarchies and, quite naturally, some of the most important building blocks for such a construction were developed at the boundaries and as criticisms of the previously described corpus of the modern science. This is the case of the transition from logical (Carnap, Hempel, Popper) to historical models (Kuhn, Toulmin, Quine), which implies, consequently, the transition from Euclidean or Cartesian modes of thought — involving reductionism and dualism — to holistic approaches, to the theories of growth of knowledge.

Kuhn, with his *Structure of the Scientific Revolutions*, strongly attacked the positivist paradigm of science. Denying the centrality of the demarcation criterion, and integrating it in 'normal science' ('the conclusions must be logically deducted from socially/scientifically accepted premises', Kuhn, 1989: 330), Kuhn suggested that Popper's falsifiability took the part for the whole, since some rather exceptional events in the history of science — revolutions — were described as a normal and daily method of development of hypotheses (ibid.: 324 f.). Instead, he abandoned the prescriptive epistemology and put forward an historical description of the main stages in the evolution of physics, defining a 'paradigm' as those 'scientific realizations universally recognized which, for a time, provide models of problems and solutions to a scientific community' (Kuhn, 1975: 13). At the origin, the concept was almost literally translated from its original sense of a reference structure for learning languages, but it was later on developed by the author in some distinct ways. In an appendix to his book, Kuhn accepted Masterman's criticism about the ambiguity of his previous use of the concept (Masterman, 1979: 21 f.) and reassessed the notion of paradigm, or 'convergent thought' (Kuhn, 1989: 277) as :

> On one hand, the complete constellation of beliefs, values, techniques and so on, shared by the members of some community. On the other hand, it denotes a sort of element in the constellation of solutions — concrete enigmas which, being used as models or examples, may substitute explicit rules as a normal basis for the solution of enigmas arising from normal science. (Kuhn, 1975: 269)

The first element is sociological, the second indicates some sort of 'disciplinary matrix' (Kuhn, 1989: 358) including symbolic generalizations and models, and is very close to the conception of the metaphoric redescription which will be presented later on. Against the radical anti-historicism of Popper, Kuhn reintroduced the historical comprehension of the evolution of natural sciences. But, nevertheless, his theory does not provide alternative procedures for the validation of hypotheses, except indicating the criterion of agreement with the canon of normal science, and does not deal with the problem of comparison and selection of theories, since the paradigms are considered to be incommensurable.

Lakatos, a critical disciple of Popper, suggested another historicist view and a more sophisticated version of confirmationism. Considering that the 'dogmatic falsificationist' denies the scientific character of some relevant theories and classifies them as metaphysical, that is, as theories not included in the universe of rationality, Lakatos argued for a more general conception: 'One theory can only be said to be scientific if it is intended to express a causal connection' (Lakatos, 1979: 123). In particular, the general use of *ceteris paribus* clauses in science would not make possible falsification tests; furthermore, the refutation of a theory should be made by another theory — 'There is no refutation before the appearance of a better theory' (ibid.: 146) — and not by mere factual tests.

Lakatos was able to present a general model corresponding to the three tasks indicated until now: demarcation, signification (includes cognition) and selection of theories, in the historicist framework. The well-known model describes every theory as including: (a) a *hard core* group of assumptions, supported by auxiliary conditions; (b) a *positive heuristics*, indicating the rules for the development of the auxiliary hypotheses; and (c) a *negative heuristics*, indicating the rules for the refutation of these hypotheses. The evolution of the theory — progressing, enlarging its empirical content, or regressing — is taken as the criterion for selection. So, one theory *i* is preferred to theory *j* whenever $T_i > T_j$, if $C(T_i/E) > C(T_j/E)$ being $P(T_i/E) < P(T_j/E)$, where $T_{i,j}$ are the compared theories, C indicates its empirical content, and T/E is the content given the empirical available evidence, P being the conditional probability. In this framework, the demarcation is merely based on the capacity to create more empirical content, and dramatic confirmations replace infirmation (Giere, 1988: 78). Infirmation eliminates theories, but confirmation develops them.

2.7. A reassessment: the realist-relativist debate

In the previous pages, several theories recuperating a relativist view were presented. This particular aspect is now discussed, since it reinstates the old debate between materialism and idealism and is still today a crucial piece for the construction of a new scientific paradigm.

The first relativist theory to be discussed here was conventionalism. In a very telling metaphor, Duhem referred implicitly to Plato's allegory of the cavern in order to describe the relation of the scientist to reality:

> A slave to positive method, the physicist is like the prisoner of the cave: the knowledge at his disposal allows him to see nothing except a series of shadows in profile on the wall facing him; but he surmises that this theory of silhouettes whose outlines are shadowy is only the image of a series of solid figures, and he asserts the existence of these invisible figures beyond the wall he cannot scale. (Duhem, 1906: 299)

The Cartesian inspiration for his theory was already presented; Descartes or even Spinoza, one of his critics, also considered reality inaccessible to understanding (Meyerson, 1908: 400). Consequently, the theories should be measured by their internal logical consistency, no matter the adequacy to facts and events they may exhibit.

Logical positivists and empiricists used the same approach but from a different point of departure: since theories were considered to be 'language games' of internally connected propositional systems, according to Wittgenstein's early definition of the *Tractatus*, and the ultimate evidence for synthetical claims about the state of the world derived from syntactic objects interpreted through correspondence rules forming a language (logical empiricism), relativism prevails. In this sense, a purely instrumental language should be able to produce univocal determinations, one name corresponding to one meaning. In any case, theories were supposed to be mere predicates, or conjectures, never knowing if they were producing true statements, and the rigour of science was established by a new game of language.

In the same vein, 'postmodern' theories concluded from the quantum uncertainty principle the death of realism and, in consequence, the dispensability of causality. In the extreme position of pragmatism, Rorty rejects epistemology since 'The subject of knowledge is a fiction, an historically determined discursive "fact"'(quoted in Amariglio, 1990: 23), and claims that there is no adequacy between the objects of science and the production of its discourse. In consequence, we are condemned to a total relativist position.

But if truth is theory or discourse-specific, the fragmentary and illusory character of knowledge would be a condition and not a situation of science, and knowledge could not follow any evolutionary process. Furthermore, the meanings emerging from the texts would depend on the subjective interplay of the discursive elements and the purely arbitrary correspondence rules: thus, the ideal of science would finally be the Babel Tower. Relativism is an elaborated form of that old sophism which claimed that no statement or action can be measured or assessed by a controllable procedure in order to define its meaning — and so, it is an inverted form of positivism, or an over-reaction to the impasse of the modern paradigm. Relativism implies that no truth claim is legitimate or definable. In the extreme form, it equates every theoretical statement with an arbitrary narrative,[17] not to be compared with others or selected from others.

In the same sense, relativism denies the accessibility of truth: if no reality is supposed to affect theoretical or observation statements, then no definition of truth is possible, or as Barnes argues, 'But if nothing external determines what concepts are to refer to, then nothing external determines the truth or falsity of verbal statements. If concept application is a matter of contingent judgement, then so too must be the process whereby generalizations are confirmed or refuted' (Barnes, 1982: 30). The argument about quantum physics can be taken in this context. If one states the existence of a multi-level hierarchical organization of reality, the micro indeterminacy is certainly important — and a severe blow to positivism or any form of cultural 'scientism' — but it cannot be taken to imply the rejection of the intelligibility of the mode of determination of the system. In reality, it does not: the orbits of planets and the evolution of macro-realities are approximately determinable and even appeared to follow in a disciplined way the Newtonian canons, which inspired for a long time the methods for the computation of the trajectories of satellites. In other words, the perplexities of knowledge about complexity do not imply that reality be unintelligible or non-existent. Or, as Artigiani puts it,

> Introducing self-referentiality and human creativity leaves scientific information ambiguous and incomplete. But ambiguous and incomplete information is only unrealistic if Nature is presumed to be finished and absolute. ... Nature can then create information as knowledge of its various parts achieves greater signification. But every increase in information changes the whole network of natural relationships. ... A science recognising the incompleteness and ambiguity of its information is, therefore, more 'realistic' than one pretending to greater knowledge of Nature than Nature has of itself. (Artigiani, 1993, 55)

There are then two alternatives in this framework, escaping from relativism and sophism. One is to reform and to reinvigorate the positivist program introducing probability statements in the wake of quantum physics — so that these statements assume the epistemological status of the old certainty claims of the positivist program. The other possible alternative is to take a prudent vision of science: the evolution of knowledge along these complex and holistic structures of determination is therefore to be seen as mapping new and old elements of sense and reference, without the paternalistic protection of a strict demarcation principle and engaging in a necessary reconstruction of science. This is indeed a realist agenda, which accepts that reality exists and is knowable, but that knowledge itself is an evolutionary process, in which characteristic mutations, errors, selection and adaptation are possible.

In this sense, several authors suggested a new and evolutionary conception of the correspondence and coherence principles of truth. Arguing that truth is context-specific and cannot be referred to some final mystical substance of 'objective truth', Bohm claimed that truth corresponded to the evolutionary concept of totality — it is not absolute, but neither is it relative, since in that case it would have no meaning: truth is the understanding of totality (Bohm, 1964: 219-20; also his interview in Buckley and Peat, 1996: 35).

Hesse considered truth as constituted simultaneously by the coherence and the correspondence principles (Hesse, 1974: 4-5), and suggested a theory of social construction of knowledge, denying the convergence towards absolute truth (Arbib and Hesse, 1986: 44, 158). But observational propositions can nevertheless be realistic, since 'Observation language and even perception are theory laden, yet they are rich enough to contain feedback that contradicts the theory' (Arbib and Hesse, 1986: 8).

Maki distinguished between *realism*, an attribute of meta-theories, namely of their semantic and ontological propositions, and *realisticness*, an attribute of concrete propositions, and defined truth as a 'semantic property' of propositions (Maki, 1989: 178-9). In order to assess that property, Maki suggests a specific combination of a *coherence* theory of *justification* ('All beliefs or statements are justified by their relations to other beliefs or statements with which they cohere', justification being indicated by the degree of plausibility in relation to a set of statements, 1993: 27), and, rejecting a coherence theory of truth, supports the correspondence principle. For Maki, this combination should reunite the epistemology of discovery and of justification.

In this sense, realism is the epistemological condition for the definition of the correspondence principle and evolutionism is the theoretical condition to a non-essentialist definition of truth, as an alternative to relativism defining truth as mere correspondence between theoretical propositions. Correspondence is thus defined as the connection between observational and theoretical assertions and reality, and truth may be defined as the meaning of those assertions. It is socially constructed, since it is based on social values and knowledge and not on permanent essences; it is evolutionary, since it propels new inquiries and is challenged by its own feedback controls; it still is truth, since it is responsible for progress in the explanation of reality.

*

Positivism is an ambitious synthesis of modern science: it defines general laws, checks those claims on reality with the deduction of particular instances from those laws, subjects those instances to prediction tests and argues that such a method is universal, objectivist and naturalist. But positivism is neither complete nor coherent: it could not find a demarcation principle in order to discriminate science from metaphysics, nor could it find a signification criterion in order to define the meaning of the observational statements. Furthermore, positivism is based on a mechanistic definition of causality, denying that a cause may simultaneously be an effect. The irruption and epistemological contribution of new sciences, like biology and sociology, proved that several forms of determination should be considered, and not only the asymmetric and deterministic causality in the positivist mood. And the reductionist strategy was simultaneously questioned: the unending quest for the 'ultimon', the ultimate constitutive substance or the secret of matter, amounted to an inconclusive result, since there is always a more elementary particle which can be modelled and the division becomes meaningless (Heisenberg in Buckley and Peat, 1996: 15; Bohm, id.: 50). The three main pillars of positivism — belief in a causal and fully intelligible structure, belief in the predictive capacity of theoretical statements, confidence in mechanistic reductionism — are condemned.

It has been argued here that the failure was inevitable, that Hume's Paradox is not soluble in any way or by any procedure and that positivist science cannot understand complexity and evolution; therefore, a new point of departure is necessary. That is the new task for epistemology: to map the evolution of incomplete, conjectural and ambiguous knowledge about the changing state of the world, through a multi-level hierarchy of determinations and a plurality of principles of explanation. The hierarchical structure supposes an evolutionary auto-catalytic development: the hierarchy is established by the inclusion of several modes of determination and explanation and by a combination of concrete and abstract analysis where abstraction is centred on the determination of real objectal causality at each level, and concrete analysis consists in the combination of the abstract mediations. The place for the determination of truth, in such a system, is practice, and it corresponds to a non-essentialist theory of construction of truth, as the expression of social and scientific activities.

This constitutes therefore a specific cognitive social practice: in such a tradition, a realist strategy is most demanding, since it does not make possible any simplification and does not accept the dominant modes of validation of theories, rejecting 'methodological individualism', since events are not describable as reversible conjunctions of atomistic factors, but as complex determinations in irreversible time. Namely, that realist strategy is strongly opposed to instrumentalism — one of the most influential consequences of the Cartesianism in economics — which implies the acceptance of a theory if the empirical data are 'as if' they would have been generated by the model (Lawson, 1989: 239) and therefore implies that simulation is demonstrative of propositions about reality. Typically, econometrics accepts the closure of the model and the simulation proof, constituting one of the most relevant pieces of the instrumentalist epistemology. Of course, if an evolutionary epistemology is instead used, based on the concept of organic history, that closure and methodological individualism will be rejected. Realism opposes both instrumentalism and relativism.

Epistemology should therefore 'evolve from the consideration of the privileged forms of knowledge through concepts (rationalism) and sense experience (empiricism) to an understanding that it is constituted in cultural and social formations (social theory)' (McCarthy, 1988: 10). The historical nature of the methodology itself is, in general, one concrete link between subjectivity and objectivity, just as social science is, in particular, the abstract link between the subject and the object of the inquiry.

3. Heirs of Positivism

The modern paradigm of sciences incorporated the three main dogmas of positivism — naturalism, objectivity and universalism — and evolved through three different basic stages and conceptual frameworks, namely classical positivism, logical empiricism and infirmationism. The first program established a strong demarcation principle based upon a cognitive criterion — induction and the inference from observational facts to general laws — which was afterwards challenged by logical empiricism. The result was the definition of an inductively based Hypothetico-Deductive Model, adding a signification criterion defined by a deductive logic and a confirmationist procedure. The third program, facing the failure of the logical empiricists to define a universal language for science, returned to the primacy of the demarcation criterion, but the refutation methodology for the progress of science appears to be inapplicable and, in particular, in economics it is incoherent with the basic neoclassical assumptions, since those cannot be specifically tested. The influence of these programs will be now traced in social sciences and, in particular, the puzzling dominance of combined relativist and positivist epistemologies on mainstream economics will be discussed.

Newton and classical positivism influenced the very conception of political economy, since they established the lasting reductionist approach. The *homo economicus*, endowed with the Newtonian laws of motion — locomotion and quantity, no time neither quality — and excluding either cultural propensity or sociability except of course Mandeville's type of egocentrism as an ultimate foundation of social affairs, was presented in economics as the concrete analogue for atoms. During the last one hundred and fifty years, that story was told again and again.

William Petty, friend of Newton and his colleague at the Royal Society, applied his mechanics to economics, replacing arguments by numbers and social relations by agents — this was the foundation of the econometric program (Hayek, 1974: 28; Capra, 1982: 194 f.). Adam Smith, writing just some decades after Newton, reviewed his work around 1750 and considered that it superseded by far the previous Cartesian concepts (Smith, 1758: 189). In *The Principles which Lead and Direct Philosophical Inquiries: Illustrated by the History of*

Astronomy, Adam Smith clearly presents the founding mechanical analogy which was at the core of positivism:

> Systems in many respects resemble machines .... A system is an imaginary machine inverted to connect together in the fancy those movements and effects which are already in reality performed. (Smith, 1758: 116)

Yet, the central piece in Smith's work, *The Wealth of Nations*, is not reducible to this mechanistic paradigm, since it also included descriptive and historical interpretations as part of the reasoning — indeed, it was explanatory and deductive as well as descriptive and inductive. Although fascinated by Newton's work as it was interpreted those days, Smith established a separate research program, defining political economy as a 'moral science'. It was in rupture with that concept that, later in the nineteenth century, economics would be transformed and revolutionized by a whole body of new metaphors and methods.

3.1. The Hypothetico-Deductive Model in action: the foundations of neoclassical economics

Ricardo was the first classical economist to formulate a version of the HDM, combining some of the central features of the positivist paradigm which were already present in Newtonian essentialism — for example, the notion of a fixed, ahistorical essence determining the laws regulating distribution, labour being the substance of the exchange-value — with a deductive procedure. In that sense, Ricardo developed the Cartesian method and was quickly followed by other authors of the late classical tradition: three years after his death, Nassau Senior argued at Oxford that economics should be based upon axiomatics, and that remained ever since the archetype of a pure theory (Hutchison, 1992: 95).

The neoclassical synthesis retained and developed these features, although introducing an important paradigmatic shift while changing the constitutive essence (thereafter the use value and no longer the exchange-value). The Cartesian model for science is in fact at the very heart of the mainstream tradition: the *rational economic man* is the analogue for the mechanism studied by positive science, the rational choice rules are the analogue for the deductive criterion for the scientific inquiry, the rules and axiom about the efficiency of the market correspond to the prescriptive role of epistemology and its certitude, the theory based on individual self-mental valuation is the analogue for the body-mind dualism, just as capital accumulation corresponds to the cumulative character of positive science (Mirowski, 1988b: 55-7, 111-2). The following dominance of the Cartesian standard in mainstream economics is supported by the axiom of equilibrium and its stabilizing properties[18] (Say's Law) and by the rationalistic epistemology:[19] it depicts a mechanical and lifeless society.

In fact, even if classical positivism was not able to sustain its epistemological and inductive claims, the atomist conception was to remain: the synthetic positivist conception of sciences articulated the stable and symmetric Newtonian concept of time and systems with the Cartesian conception of the world as an *automaton*, which could be exhaustively explained by a deterministic causality from which all phenomena could be deduced. A large part of the classical and the entire neoclassical programs in economics were faithful to this paradigm, and the striking paradox survived a long time: envious of the beauty of Newtonian empirical proofs, economics fell in the arms of Descartes. Although claiming to achieve Newtonian certitudes and the aesthetical ideal of the positivist models of proof and prediction, mainstream economics is mostly a product of Cartesianism, based on introspection, general laws derived from it and comprehensive categories of phenomena (capital, labour) to be deductively related in a formal mechanical system. Indeed, Cartesian axiomatics constitutes the epistemological root of mainstream economics, downgrading the status of empirical analyses of data. Stuart Mill and Jevons were the forerunners of those debates and approaches which changed the image and project of economics.

3.1.A. The Babel Tower of early economic epistemology

Stuart Mill was one of the more impressive and contradictory figures of the intellectual landscape of the nineteenth century. An economist, he supported the wage fund theories but collaborated with trade-unions; a humanist, he supported the emancipation of women and the values of solidarity, while adhering to the Malthusian concept of the evolution of population; a socialist, he accepted Say's Law; a Ricardian, he defined the stationary state as the possibility of the cultural development of mankind. These paradoxes were defined at the cross-roads of Newtonian and Comtian influences.

Mill was the first classical author to note the development of sociology and to try to incorporate Comtian positivism into economics. In 1841, he began an intense correspondence with Comte, under whose spell he split with the Benthamite conceptions. Since 'all phenomena without exception are governed by invariable laws' (Mill, 1865: 12), science should be able to discern the essential causation in social movements. As it will be argued later on, this was the basis for his incorporation of Comte's definitions of statics and dynamics and, for the first time as an explicit program, of positivism in economics. But, in spite of a long and close intellectual relation, Mill departed from Comte's political positions rejecting his suggestion for an illuminated despotism, and their rupture became inevitable (Mill, 1873: 201-2).

Newton's influence on Stuart Mill was much more consistent. Reflecting the general consensus of the time, Mill considered that Newtonian astronomy should be the paradigm for social sciences, although it could not be completely translated into the human realm:

> The science of human nature falls short of the standards of exactness now realised in Astronomy; but there is no reason that it should not be as much a science as Tidology is, or as Astronomy was when its calculations had only mastered the main phenomena, but not the perturbations. (Mill, 1843: 846)

The reason for this distinction — the existence of disturbances — is most important, since it opened the Pandora box of the specificity of social sciences. Certainly aware of that, Mill synthesized Newton's and Comte's classical positivist heritage in an extensively but confusedly argued combination of methods and assumptions, forming a new epistemology.

Such an eclecticism may explain the unending discussions about the nature of his methodological contributions. Neville Keynes (1890) presented Mill, as opposed to Ricardo, as the good ambassador of what came to be called the hypothetico-deductive method. While Robbins (1932), Hausman (1992) and Boylan and O'Gorman (1995: 11) described Mill as a typical deductivist, Blaug presented the same Mill as a representative of the last century's inductive vision (Blaug, 1982: 2, 16) and argued that there were two Mills, that of the *System of Logic* and that of the *Principles*. And Hollander suggested an alternative interpretation, stressing instead the unity of Mill's approach, based on a criticism of deduction but also on a 'balanced' articulation on the available methods under the form of the *a priori* epistemology, including an inductive basis, ratiocination on an axiomatic basis and verification (Hollander, 1985: 68-69).

Stuart Mill was indeed one of the forerunners of the positivist methodology and operated a transition from the early classical version to a somewhat more elaborated HDM. Suggesting what he called a 'method of analysis and synthesis', Mill described *analysis* as the process of inquiry about the elements of society, inspired by the individual experience of the scientist, inferring general laws from these elements (Mill, 1844: 65), the synthesis being purely the combination of these laws and effects. The analysis could proceed from the separability of the hypotheses and causal additivity of the effects:[20]

> When an effect depends upon a concurrence of causes, those causes must be studied one at a time, and their laws separately investigated, if we wish, through the causes, to obtain the power of either predicting or controlling the effect; since the law of the effect is compounded of the laws of all the causes which determine it. (ibid.: 53)

As two methods were available to inspect the combination of the causes, one being the *a priori* (induction and ratiocination from personal experience plus deduction from the inferred hypotheses) — in an upwards and downwards evolution combining a variety of specific conclusions — and the other being the *a posteriori* method (induction) — in a pure upwards movement — Mill argued for the first of these methods. 'Reasoning from one assumed hypothesis' should be the best method, since too extensive generalizations are dangerous, and on the other hand the hypotheses could be defined like in geometry (ibid.: 54, 65), as an axiomatic conclusive system. The deductive method should be the procedure for justification, the only conducting to truth, the a posteriori method being 'altogether inefficacious' (ibid.: 58, 62). In this sense, Mill explicitly rejected the Baconian *experimentum crucis* as a rare case (Mill, 1844: 58). In short,

> it is vain to hope that truth can be arrived at ... while we look at the facts in the concrete, clothed in all the complexity with which nature has surrounded them, and endeavour to elicit a general law by a process of induction from a comparison of details; there remains no other method than the a priori one, or that of 'abstract speculation'. (ibid.: 59)

This methodological combination would settle the agenda for the whole next generation.[21] Establishing that the *a posteriori* method was only useful for the verification (justification) and not for the discovery (ibid.: 62) and the development of science, Mill introduced the Cartesian notions and the hypothetico-deductive model, of course without excluding induction from the forms of inquiry, but altogether and conclusively superseded the classical positivist method.

But his solution still depended on two demanding conditions. First, the conjunction, additivity and separability of causes should be unequivocally established. And, second, the disturbances which established the specificity of social sciences should be limited so that the principle of uniformity in nature hold. If both these conditions were met, then general laws could be inferred through the analysis of concordance, difference, causality and simultaneous variations, and the deduction of the explanandum from a general law would be the legitimate explanation of all phenomena. Otherwise, if the plurality of causes, the interdependence of effects or the impossibility of a controlled environment challenged those two conditions, the method could not be applied (Mill, 1843: 452). In his *Autobiography*, Mill accepted that the exceptions to those conditions were the basis for the recourse to one or other of the methods:

> I now saw that a science is deductive or experimental according as, in the province it deals with, the effect of causes when conjoined, are or are not the sums of the effects which the same causes produce when separate. (Mill, 1873: 167)

Both Whewell and Jevons reacted against Mill's *System of Logic* (1843), accused of pure inductivism and of grounding laws in the derivation of particulars from particulars, and argued for a qualified hypothetico-deductive approach. Jevons strongly opposed Mill's conceptions, since he interpreted his epistemology as the vindication of old fashioned empiricism in the Baconian mood (Jevons, 1874: 121) and of deterministic cause-effect chains (ibid.: 222). In a very hostile critique, Jevons supported an extended interpretation of Newton, claiming that 'deductive reasoning must be combined with elaborate experimental verification, that has led to all the great triumphs of scientific research' (ibid.: 507). Jevons recognized, and rightly so, that this was not the literal interpretation of Newton's claims, since the framing of hypotheses was considered to be a crucial step in the scientific work, but he claimed that his concept was closer to the really applied Newtonian practice (ibid.: 583). Nevertheless, there was a curious feature in this argument of Jevons against Mill, since both were in fact engaged in the development of the same program and rejected the pure inductive and experimental method of classical positivism: vindicating Newton, Jevons was in fact contradictorily closer to the Cartesian agenda, except in one topic which will be discussed in the next chapter, the definition of the statistic method. The paradox remained for the whole history of mainstream economics.

Jevons's was the most elaborate early version of HDM. He argued that induction ought to be used in order to infer a general law, but that only probable and uncertain statements could be derived from finite experience (ibid.: 222), and that this process required prior and complementary hypotheses (ibid.: 121, 504). In spite of it, knowledge was possible since it was cumulative and referred to constant features of the universe:

> Happily the Universe in which we dwell is not the result of chance, and where chance seems to work it is our own deficient faculties which prevent us from recognising the operation of Law and Design. In the material framework of this world, substances and forces present themselves in definite and stable combinations. Things are not in perpetual flux, as ancient philosophers held. Element remains element; iron changes not into gold. (ibid.: 2)

This reaffirmation of essentialism and equilibrium is the hard core of mainstream economics. At the same time, it maintained inductivism, as Jevons himself did while considering the logic of statistics: the presumption of constancy in the relations between the atomic and separable elements of the social universe made possible the derivation of functional relations and statistical demonstrations. In other words, his charge against Mill's alleged inductivism was evidently misguided. The early economic epistemologies were born in the shadow of Newton, nourished by Descartes and grown under the light of Laplace. No wonder if they were rather unclear.

3.1.B. Time for syntheses

John Neville Keynes, a generation away from Mill's main methodological text,[22] developed the Cartesian approach in his standard text of neoclassical epistemology. He strongly rejected the inductive method as a cognitive criterion: 'the method of specific experience or direct induction is inadequate to yield more than empirical generalisations of uncertain validity' (Neville Keynes, 1890: 14). Or: 'The need for the a posteriori investigation of the concrete phenomena themselves, at a certain stage of reasoning, remains; but no trust is to be put in an a posteriori method pure and simple. ... In a technical language of logic, the method of direct generalisation is inapplicable, because of plurality of causes and intermixture of effects' (ibid.: 210).

But, as the quotation indicates, he was no longer confident about Mill's separability of hypotheses, and that is why he suggested a more detailed procedure in order to avoid the problem of the interconnection of causes:

> The method in its complete form consists in three steps. It is necessary, first, to determinate what are the principal forces in operation, and the laws in accordance to which they operate. Next comes the purely deductive stage, in which are inferred the consequences that will ensue from the operation of these forces under given conditions. Lastly, by a comparison of what has been inferred with what can be directly observed to occur, an opportunity is afforded for testing the correctness and practical adequacy of the two preceding steps, and for the suggestion of necessary qualification. It will be observed that only one of these three steps — namely, the middle one — is strictly speaking deductive. (ibid.: 216-7).

This is the pattern of a positive, abstract, deductive and hypothetical science, and confirmationism follows directly from its third step (Blaug, 1986a: 840). This still corresponds to the general practice of economists, and has been widely diffused since the end of last century. The development of econometrics is the most well known example of the confirmationist epistemology and a new proof of the paradox of mainstream economics, since its more enthusiastic supporters claim a rigour and fidelity to Newtonian empiricism (including the development of statistical models such as Sims's, presented as purely inductive, hypotheses non fingo), while in reality the implicit hypotheses define an essentially deductive and Cartesian system. As Koopmans put it, the economists define a model under the 'if ... then' qualification and, if the 'findings' survive the confirmation test, the model is confirmed (Koopmans, 1979: 11). Yet, as the next chapters will show, the very conditions of the model and the techniques of its test determine the results of the reasoning and of the experiment.

The main epistemological debate was to be organized around the first and third of Neville Keynes's steps. Since he did not indicate, for the first step, how to 'determine the principal forces in operation' and the laws deriving from them, two extreme versions were possible, and indeed the inductivist account was confronted with the alternative of a purely deductive determination.

The apriorist Austrian school suggested that the information necessary for such a determination was mere personal daily experience and reason, and that axioms derived from reason are true and sufficient for the deduction of a significative set of laws. The central piece for that claim was 1932 Robbins's *Essay on the Nature and Significance of Economic Science*, where he argued against inductivism and Baconian generalization (Robbins, 1932: 74), defining economic analysis as the 'deduction of a set of postulates, being the main facts of experience, almost universal, of all human activity having an economic aspect'[23] (ibid.: 99-100), these being 'qualitative postulates' (ibid.: 114). The 'determination of the forces in operation' was therefore completed by the immediate common experience of human beings, since the economic realities were supposed to be self-evident, and was supported by the cumulative *corpus* of the economic deductions so far developed: in the preface to his work, Robbins argued that the importance of one hundred and fifty years of (deductive) generalizations of economic theory — that is, the *homo economicus* paradigm — was only questioned by ignorant or perverted people (ibid.: xxxviii).

The confidence in this deductive account was such that Robbins claimed that generalizations were more certain in social than in natural sciences, since they were originated in direct personal experience and reason, and not in fallible measurements of the outside world (ibid.: 104-5). A strict positive economic science could consequently emerge without being corrupted by value judgements or normative concerns: 'Equilibrium is just equilibrium' (ibid.: 143; also 24, 131).

The third Neville Keynes's step, the test of hypotheses, was later on considered to be the crucial one. The debate was centred on the infirmationist solution.

3.1.C. The infirmationist query

Hutchison introduced an explicit version of positivism in the 1930s and, at the same time, a first sketch of falsifiability as the sole demarcation criterion (Hutchison, 1938: 19). His own approach was based on four premises: (a) theories must be testable; (b) prediction is the adequate test; (c) therefore, predictive adequacy of theories should be their most important characteristic; and (d) the comparison (and selection) of theories should be organized around concrete tests. Hutchison was not familiar with all the literature about the epistemological turn from classical positivism to logical empiricism, and his stated position was an eclectic combination of several elements: (a) the defence of Keynesianism against Ricardo's method of essentially deductive inferences from simplified assumptions — and thus, against the idea of maximization under certainty and perfect knowledge, the equilibrium model — since a genuine deductive method implies a strong hypothesis about perfect information, useless whenever the explanation of concrete economic phenomena is at stake (ibid.: 2-3; also 1978: 200 f., 211); (b) an early vindication of the Kuhnian type of 'scientific revolutions' (he pointed three: Smith, Jevons, Keynes) as a correct account of the progress of science;[24] (c) a debate against the purely deductive apriorist Austrian school and, finally; (d) a debate with Machlup over empiricism and rationality.

His debate against Robbins and the Austrians was inspired by the rejection of Ricardo's deductive method. As the Austrians took the 'essential facts of experience' as the source of inspiration for the formulation of postulates and the deduction of the theory was based on these postulates — including the central 'rationality postulate', which can hardly be derived from individual concrete experience except in the most trivial fashion — Hutchison criticized the lack of scientific legitimacy of this 'psychological method' and argued for empirical research, for refutation tests as the criterion for demarcation and, thus, as the basis for the social authority of science (Hutchison, 1992: 76). The introspective origin of the basic axioms, non empirically verifiable, was therefore confronted with the requirement for the empirical validation of every postulate.

Machlup, a collaborator of Robbins in the famous 1932 essay and himself educated in Vienna, launched a violent attack against the 'ultra-empiricism' of Hutchison. Machlup argued for an 'indirect verification' principle based on deductive logic: if it was not possible to submit one hypothesis to test or to 'reduce it by direct deduction to an empirically testable proposition', it should be combined with one testable assumption; thus, if the consequence from the combination was tested, and it could only be deduced from the conjunction of both hypotheses, the verification was provided (Machlup, 1956: 199). This was an ingenious way to deal with one of the stated difficulties of infirmationism, but with no convincing practical application. Anyway, the whole debate was reshaped by the development of Popperian epistemology, first adapted to economics by Hutchison himself.

In a late preface to his 1932 essay, in fact a 1981 conference on the subject, Robbins stressed this evolution and his espousal of Popperian falsifiability (Robbins, 1932: xiii-xiv), apparently ignoring that it should imply a wholly different approach and namely the abandonment of the primal role of the introspective method defining the axiomatic — since these principles of psychological observation are not testable in any meaningful sense.[25] At that time, the idea of the test of predictions as the central task for science was already largely dominant in philosophy.

In 1953, Friedman provided the most influential diffusion of an infirmationist version of the HDM in economics.[26] He developed Neville Keynes's methodology and combined it with a logical empiricist account:

> The ultimate goal of a positive science is the development of a 'theory' or 'hypothesis' that yields valid and meaningful (that is, not truistic) predictions about phenomena not yet observed. Such as theory is, in general, a complex intermixture of two elements. In part, it is a 'language' designed to promote 'systematic and organized methods of reasoning' [Marshall's definition]. In part it is a body of substantive hypotheses designed to abstract essential features of complex reality. (Friedman, 1953: 7)

The three central elements of this description — the goal of a positive science, its language system and the substantive hypothesis — summarize the positivist program. As a developed language, economics should 'serve as a filing system for organizing empirical material and facilitating our understanding of it' (ibid.: 7), that is, the purpose of the syntactic system should be to organize and logically validate the inductive inferences. Otherwise, 'Viewed as a body of substantive hypotheses, theory is to be judged by its predictive power for the class of phenomena which it is intended to explain' (ibid.: 8). As in logical empiricism, prediction is considered the normal form of explanation: 'Its task is to provide a system of generalizations that can be used to make correct predictions about the consequences of any change of circumstances' (ibid.: 4, also 39) and again 'the only relevant test of the validity of an hypothesis is comparison of its predictions with experience' (ibid.: 8-9).

But Friedman's position was rather different from the Popperian standard, namely in two essential points. The first is the applicability of pure infirmationism: since it is not possible to create 'controlled experiments', no dramatic and direct evidence is to be expected, and therefore the process of validation is a slow 'weeding out of unsuccessful hypotheses' (ibid.: 10-11). The rejection of the experimentum crucis undermines falsifiability. The second main difference is the instrumentalist consequence from the first one: the important feature of theories is not the application of predictions and refutations, but the fact that their instruments are able to make those predictions, the 'realism' of hypotheses and assumptions being negligible (ibid.: 14-5, 32, 34, 41). As Blaug has put it, the final result is very bizarre: the verification of conclusions without the verification of the hypotheses, assumed to be unrealistic, is not much of a progress (Blaug, 1982: 89; 1986a: 836). Nevertheless, the 'as ... if' unrealistic assumption is a general feature of mainstream economics and indeed it is embodied in the very concept of rationality, as Friedman pointed out in a previous text:

> 'Economic theory' is often taken to be a label for an existing body of doctrine concerned primarily with the allocation of resources among alternative ends and the division of the product among the co-operating resources, and based on the hypothesis that economic events can be 'explained' and 'predicted' by supposing men to behave as if they sought single-mindedly and successfully to pursue their own interests as if their interests were predominantly to maximize the money income they would wring from a hostile environment. (Friedman, 1952: 239)

One striking example is Friedman's discussion about Veblen's 1898 article arguing for 'evolutionary economics' and realism in the formulation of its models: according to Friedman, the criticism of the unrealistic neoclassical program would only be relevant if supported by an alternative model providing better predictions (ibid.: 30-1). The application of the instrumentalist criterion is therefore considered to be the criterion for the legitimacy of instrumentalism.

The necessity of realism of the premises has been one of the main trends in the criticism of neoclassical economics and of Friedman's epistemology. Herbert Simon was one of the authors who pointed to the necessity of ontological realism in the definition of rationality, which includes in the neoclassical version no less than rational economic calculation, constant tastes, independence of decision making, perfect knowledge, perfect certainty and perfect mobility of factors:

> There can no longer be any doubt that the micro assumptions of the neoclassical theory — the assumptions of perfect rationality — are contrary to fact. It is not a question of approximation; they do not even remotely describe the process that human beings use for making decisions in complex situations. (Simon, 1979: 510)

On the other hand, refutation does not provide a demonstrative logic. In fact, if some of the central definitions of neoclassical economics — and, by the way, the same applies to the classical version — are considered, the inapplicability of infirmationism is evidenced. The Law of Diminishing Returns, for instance, defines that if some variable input is applied to some other fixed input, the result will eventually be decreasing additions to the output. Of course, this notion is not falsifiable, because of the ceteris paribus condition which restricts the conclusion. But the Law is decisive in establishing the desirable forms of the short run average and marginal costs curves, and consequently to the determination of the short run equilibrium output of each firm. In the same sense, the Heckscher-Ohlin Theorem in international trade theory was refuted by Leontief's Paradox, but resisted and is still taught in economics. And Ricardo's theory of comparative advantages was refuted by his own example of British-Portuguese trade, but it is still preached at school. In these as in other examples, refutation has no significative role in orthodox economics.

Relativist and instrumentalist epistemologies prevents any meaningful refutation of hypotheses by empirical data. Furthermore, the constitution of the models and theories themselves, and namely of extremal theories describing optimizing behaviour in conservative systems, makes impossible a clear refutation criterion, since auxiliary hypotheses are always present. In fact, demonstrative strategies based on the *ceteris paribus* conditions[27] are never conclusive and merely make possible an infinity of possible combinations of arguments. In this framework, the choice of the hypothesis to be refuted is a matter of ideology, and Popper indicated this quite well when he suggested that the rationality assumption should not be challenged even when negative evidence was present:

> Now, if a theory is tested, and found faulty, then we have always to decide which of its various constituent parts we shall make accountable for its failure. My thesis is that it is sound methodological policy to decide not to make the rationality principle accountable but the rest of the theory .... (Popper, 1983: 362)

The failure of infirmationism is obvious from its general inapplicability and from its arbitrariness when determining the hypothesis to be rejected. Yet, those critiques were wholly ignored by mainstream economic epistemology. In fact, classical positivism founded modern economics and originated the rationality postulate and its atomism, which were taken as the conditions of legitimacy for any relevant science by the next generations of epistemologists and economists. Orthodoxy later on included the HDM program, in a general synthesis which constitutes the dominant modern economic epistemology. It is its 'normal science'.

3.2. The Meccas of economists

The construction of economics as a science has been permanently influenced by a model of epistemology based upon extensive, conclusive and comprehensive laws explaining all phenomena according to general Newtonian-type trajectories in a Cartesian mechanistic and deductive world. This kind of positive knowledge was considered to be the only one compatible with standard scientificity, even if at the very foundation of economic methodology the alert was trumpeted against such a reductionism: 'The *differentia* of economic laws, as contrasted with purely physical laws, consists in the fact that the former imply voluntary human action', wrote Neville Keynes (1890: 216-7). This warning was ignored, and the neoclassical synthesis proceeded to an extensive integration of the concepts of physics into economics, as the legitimate model for description and explanation. In fact, this is the real basis for that bizarre prescription about prediction as a decisive criterion in social sciences.

As it will be shown, this influence was acknowledged and praised by the main authors of the marginalist revolution of the last quarter of the nineteenth century, and it is the constitutive element of neoclassical economics. The energetic metaphor and the biological metaphor suggested as an alternative by Marshall were central pieces in the development of the turn of the century economics.

3.2.A. The marginalist revolution

The marginalist revolution in the 1870s was the single most important paradigmatic shift in the history of economics. It was shaped by the influence of the dominant model of natural sciences, namely by an analogy from physics, taken from the emergence of thermodynamics before the 1840s. In fact, this influence extended over three founding concepts: (a) the definition of human rational behaviour as the maximization of an objective-function under constraints over a stationary field; (b) the metaphor of energetics and, in particular, the law of conservation of energy; and (c) the inclusion of all phenomena under a single denotation, as specific forms of an ultimate form of energy, fully reversible. This argument was developed with minute precision by Mirowski, and the current section just restates his case.

These principles were applied to economics following the law of conservation of energy, established in the 1840s by Mayer, Joule, von Helmholtz and Colding, and when energetics replaced astronomy as the paradigmatic science:

> The energetics movement was a logical extrapolation of some major themes surrounding the original construction of the energy concept: the promise of the reunification of all science; the reification of the fundamental substratum underlying the diversity of phenomena; the expressions of the belief in causal continuity and the determinism of the Laplacean dream; and the prosecution of research by analogy. (Mirowski, 1989b: 56)

Energetics made possible the replacement of the traditional substance theory of value by the new idea of utility to be maximized over a given field. According to the conservation principle, action was a function of position and all motion was fully reversible, creating identity through time in a renewed concept of equilibrium in the movement. The first law of thermodynamics states that all forms of energy are interchangeable and that their summation in a closed system is constant. This is indeed more than an empirical or a deductive law: it is a statement about the ideal behaviour of Nature, 'the mathematical expression of invariation through time, the reification of a stable external world independent of our activity or inquiry' (ibid.: 75).

Mirowski argues at length that this revolution in physics provided the infant neoclassical theory with the epistemological metaphor needed to combine the atomist approach with equilibrium analysis. In this sense, 'neoclassicals did not imitate physics in a desultory or superficial manner; no, they copied their models mostly term by term and symbol by symbol, and said so' (ibid.: 3; also 1988b: 31). The author presents an impressive series of examples:[28] the simultaneity of the discovery of marginal utility, by Jevons, Walras, Menger, and also by Gossen, Edgeworth, Fisher, Pareto, is explained by the common recourse to the energetics metaphor, at that time the most influential current in physics and a model for science in general, as publicized by Spencer.

Gossen explained in 1853, in a rather emphatic way, his adoption of the energetics metaphor:

> I believe that I have succeeded in discovering the force [the central concept of Von Helmholtz's 1847 text being the 'force'], also the law of the effect of this force, that makes possible the coexistence of the human race and that governs inexorably the progress of mankind. And just as the discoveries of Copernicus have made it possible to determine the paths of the planets for any future time, I believe that my discovery enable me to point out to any man with unfailing certainty the path he must follow in order to accomplish the purpose of his life. (Gossen, quoted in Mirowski, 1989b: 193).

Jevons compared the notion of value to that of energy in mechanics, and Edgeworth shared the same frankness:

> An analogy is suggested between the Principles of Greatest Happiness, Utilitarian or Egoistic, which constitutes the first principles of Ethics and Economics, and those Principles of Maximum Energy, which are among the highest generalizations of Physics, and in virtue of which mathematical reasoning is applicable to physical phenomena as complex as human life. (Edgeworth, 1881: V)

Edgeworth also added that 'Pleasure is the concomitant of Energy' (ibid.: 9), and that 'the conception of Man as a pleasure machine may justify and facilitate the employment of mechanical terms and mathematical reasoning in social science' (ibid.: 15). One could not expect clearer declarations.

Irving Fisher, who conceived of hydraulic mechanical models as illustrations for some of his ideas, explained the analogy in detail in his doctoral thesis, where he included a table comparing energetics and economics (Table 3.1, from Fisher, 1893: 85-6). Fisher was very definite about the importance of that map:

> Scarcely a writer on economics omits to make some comparison between economics and mechanics. One speaks of a 'rough correspondence' between the play of "economic forces" and mechanical equilibrium ... . In fact the economist borrows much of his vocabulary from mechanics. Instances are: equilibrium, stability, elasticity, expansion, inflation, contractions, flow, efflux, force, pressure, resistance, reaction, distribution (prices), levels, movement, friction. (Fisher, 1893: 25)

The energetics metaphor was imported into economics through the concept of value and 'utility', therefore concluding a rupture with classical economics. But the incorporation of the concepts and mathematical representations of energetics was not trivial, since the several forms of energy are ontologically comparable and therefore exchangeable, whereas this condition does not apply in economics. Moreover, in economics, the metaphorized conservation of utility-energy was the condition for measurement: it rendered commodities commensurable in the market, formalized the measure and, generally, justified the measure. But the principle of conservation was more a problem than a solution, since it implied the maintenance of the constant total of energy, and a constant sum of the total of expenditure and the total of utility in a closed trading system, although it is obvious that money and utility are ontologically incomparable and therefore non-additive. Of course, without conservation of the summation, the heuristic value of the energetic metaphor is null, since there is no operational maximization principle.

This matter was extensively discussed by Walras's and Pareto's critics, including the mathematician Poincaré and the biologist Volterra. Walras, who defined economics as a physico-mathematico science like mechanics and hydraulics (Walras, 1883: 23), wrote in 1901 a letter to Poincaré asking for his

Table 3.1 Irving Fisher's metaphor from mechanics

Mechanics	Economics
a particle	an individual
space	commodity
force	marginal utility or disutility
work	disutility
energy	utility
work of energy = force × space	utility = marginal utility × commodity
force is a vector	marginal utility is a vector
forces are added by vector addition	marginal utilities are added by vector addition
work and energy are scalars	disutility and utility are scalars
The total energy may be defined as the integral with respect to impelling forces	The total utility enjoyed by the individual is the like integral with respect to marginal utilities
Equilibrium will be where net energy (energy minus work) is maximum: or equilibrium will be where impelling and resisting forces along each axis will be equal	Equilibrium will be where gain (utility minus disutility) is maximum; or equilibrium will be where marginal utility and marginal disutility along each axis will be equal
It total energy is subtracted from total work instead of vice versa the difference is 'potential' and is a minimum	If total utility is subtracted from total disutility instead of vice versa the difference may be called 'loss' and is a minimum

opinion about the *Elements of Pure Political Economy*. Poincaré, although prudent about the matter, answered that he could not accept that 'satisfaction' be measurable, and therefore that he suspected that a function based on that concept would be arbitrary and should be avoided. In other words, Poincaré's argument was that economics could not be compared to mechanics. Furthermore, the specific rationality assumptions of general equilibrium could not be accepted by Poincaré: 'You look at men as infinitely egoistic and infinitely clear-sighted. The first hypothesis may be accepted as a first approximation, but the second one requires perhaps some reservation' (Poincaré, quoted in Ingrao, Israel, 1985: 22, my translation). Volterra discussed the same points with Pareto, arguing that utility and satisfaction were 'non-measurable concepts' (ibid.: 26-7), and as a consequence rejected the whole metaphor:

> The notion of homo economicus which has given rise to much debate and created so many difficulties ... appears so easy to our mechanical scientist that he is taken aback at other people's diffident surprise at this ideal, schematic being. He sees the concept of homo economicus as analogous to those which are familiar to him as the result of long habitual use. ... Lastly, our mechanical scientist sees in the logical process for obtaining the conditions for economic equilibrium the same reasoning he himself uses to establish the principle of virtual work .... (Volterra, quoted in Ingrao and Israel, 1985: 25-6)

These crucial arguments did not receive any convincing answer, or even any answer at all (Mirowski, ibid.: 9, 230-1, 241 f., 250). Nevertheless, the marginalist revolution succeeded in the incorporation of the physical concepts and formalism, and a new economic science emerged from that synthesis by Walras, Jevons,[29] and all the other founding fathers of neoclassical economics.

Their work was followed with unexceeding zeal: Samuelson provided a second revival of the physicist metaphor in his doctoral thesis of 1947, *Foundations of Economic Analysis*. And, more recently, Debreu still argued that theoretical physics is an 'inaccessible ideal' for economics, since no experiment is possible, but that it provides an ideal-type and that mathematical formalization as the condition for the 'scrutiny of logical errors' and for the development of the science (Debreu, 1991: 2). Indeed, that physical metaphor is the foundation for the whole body of theories, models and techniques which dominated economics so far.

In terms of what has been analysed in this chapter, the physical analogy gave a new breath to economics, substantially increasing the sophistication and the mathematization of the discipline and recognizing no need of concessions to plausibility or to realism of hypotheses; it was the protean force behind the positivist epistemology espoused by mainstream economics. Instrumentalism is the heir of this development,[30] and dominated economic theorizing for more than one century.

3.2.B. Marshall: the Mecca of biology

Marshall is sometimes considered as an author influencing the alternative paradigm for economics,[31] even if he was simultaneously one of the forerunners of comparative statics and of the current concept of equilibrium. In fact, his discussion on epistemology is the most telling of all the authors responsible for the paradigmatic shift of the 1870s, and two main topics will be considered now in order to argue that he stood alone as the inspiration for a critical but inconsequent version of the neoclassical program.

The first point concerns the role of the constitutive metaphor of economics, carefully discussed by Marshall in several of his texts. Already in his inaugural lecture at Cambridge, Marshall emphasized the role of the mechanical metaphor, since economics was defined as 'an engine for the discovery of concrete truth, similar to, say, the theory of mechanics' (Marshall, 1885: 159). Later on, he argued that a new metaphor should be considered: his first text dealing with the subject was written in 1898[32] and compared the biological and mechanical analogies. Marshall's argument was a simple one: at an early stage of its formulation, economics must consider supply and demand as 'crude mechanical forces' and derive its knowledge from the study of equilibrium. At such a stage, there is already a possible analogy with biology: 'And here we find a biological analogy to oscillations in the values of commodities or of services about centres which are progressing, or perhaps themselves oscillating in longer periods. The balance, or equilibrium, of demand and supply obtains ever more of this biological tone in the more advanced stages of economics.' And he concluded: 'The Mecca of the economist is economic biology rather than economic dynamics' (Marshall, 1898: 318).

So, the biological analogy was presented as a long-term view of trends of life and decay in order to explain the evolution of forms of equilibrium, namely to study growth. The analogy was always presented, in that article, as a life-cycle metaphor (ibid.: 311), an important one since for some economic problems a simple dynamic solution was considered unattainable (ibid.: 311, 313). Anyway, biology should inspire later stage economics, once the foundations of the theory were established:

> I think that in the later stages of economics better analogies are to be got from biology than from physics; and, consequently, that economic reasoning should start on methods analogous to those of physical statics and should gradually become more biological in tone. (ibid.: 314).

It is not at all clear how such an unilinear evolution could be attained; anyway it implies a continuity of concepts and scientific general procedures, and cannot be taken in any form as an argument for the break with the physical analogy.[33] In his 1898, 1907 and 1910 introductions to the successive editions of the *Principles*, Marshall repeated the same argument with the same words:

> The Mecca of the economist lies in economic biology rather than in economic dynamic. But biological conceptions are more complex than those of mechanics; a volume on Foundations must therefore give a relatively large place to mechanical analogies; and frequent use is made of the term 'equilibrium', which suggests something of a statical analogy. (Marshall, 1907 preface, in 1890: xii)

In a later appendix to the same book, he even presented economics as a branch of biology. But in neither cases was economic biology detailed, allegedly for pedagogic reasons which favoured the teaching of mechanical models as a suitable introduction to economics. In fact, Marshall just pointed out a life-cycle analogy, applied it to economics — simultaneously to the science as such and to one of its topics, the study of growth processes — and afterwards indicated the difficulty of the subject, suggesting the *ceteris paribus* strategy in order to account for situations defined by numerous causes (ibid.: 47-8). Paradoxically, Marshall argued for this choice of the fiction of the representative agent on the basis of the biological analogy (Marshall, 1910 introduction, in 1890: 69-70).

Influenced by the conception of 'evolution' as a social progress along the orthogenetic path (ibid.: vii) and by the positivist tradition, Marshall found no difficulty in combining his mechanical approach with an atomistic biological reference of Spencerian flavour, which was not contradictory with the partial equilibrium methodology and his own adhesion to the marginalist revolution. And, as the next chapter will argue, Marshall's equilibrium concept was not derived from biological metaphors.

The second topic, which distinguished Marshall within the neoclassical tradition, was his proclaimed suspicion about the misleading role of mathematical formalization. In particular in his last years, Marshall argued firmly and repeatedly against the belief in exhaustive formalism:

> I had a growing feeling in the late years of my work that a good mathematical theorem dealing with economic hypotheses was very unlikely to be good economics, and I went more and more on the rules: (1) use mathematics as a shorthand language, rather than an engine of inquiry; (2) keep these until you have one; (3) translate into English; (4) then illustrate by examples that are important in real life; (5) burn the mathematics; (6) if you cannot succeed in (4), burn (3). This last I did often. (Marshall, letter to Bowley, 1906, quoted in Kaldor, 1985: 58-9)

Marshall went even further, arguing that the ceteris paribus clause, so central in partial equilibrium, could be sometimes not even practicable and that formalization could be completely useless:

> In my view, every economic fact, whether or not it is of such a nature as to be expressed in numbers, stands in relation as cause and effect to many other facts; and since it never happens that all of them can be expressed in numbers, the application of exact mathematical methods to those that can is nearly always a waste of time, while in the large majority of cases it is particularly misleading; and the world would have been further on its way forward if the work had never been done at all. (Marshall, ibid.: 59).

This anti-formalist revolt would be clearly expressed by Keynes, under Marshall's influence. This major challenge to the neoclassical research program will be discussed in Part Three of this text.

3.3. The quest for a new map

As part of the positivist paradigm, mainstream economics developed late debates about epistemology, which neither clarified the real issues nor suggested practical solutions to the detected difficulties: an extreme form of scientism is still employed within the general pattern of the physical analogy which inspired the marginalist revolution or a deterministic version of some of the probabilistic concepts of quantum mechanics. At the same time, atomism, objectivity and universalism are still considered indisputable canons of scientificity. The result is by now obvious: the game of unqualified generalizations, the abuse of the *ceteris paribus* protection clauses, the resistance to empirical testing, and the tendency to abstract or to produce unrealist models are lasting characteristics of the mainstream. In fact, economics is positivist by delegation: in spite of its rhetoric, its dominant methods are Cartesian and deductive.

Modern science argued for a constitutive demarcation against metaphysics: yet, Friedman's instrumentalism or Samuelson's operationalism accepted the irrelevance of realistic assumptions and the use of 'predictive statements' or of 'operationally meaningful statements' is metaphysically defined. Such a science is far from the relevant social knowledge and action which was once vindicated by the Enlightenment, and does not even seem to be aware of its blindness. The positivist paradigm has been challenged, and correctly so; it must be replaced.

Three main epistemological problems result from this crisis.

a. The *problem of signification* or, in another form, the problem of the definition of science. The positivist paradigm tried to solve this problem establishing a definitive cognition or a demarcation criterion, or both at the same time. Neither claim was found to be true.

   In particular, the deterministic concept of causality failed to produce authoritative models of explanation, since exogenous causality requires simple and non-reciprocal relations which are generally irrelevant in real economies.[34] As a consequence, economics became the dreamland of paradoxes, where deterministic methodologies and positivism are sustained by their opposite, relativism, and where oaths of fidelity to Newtonian empiricism are formulated in Cartesian language.

   The conventionalist attitude and the confirmationist procedures were challenged by several authors, since signification cannot ignore facts, and induction and deduction are essential tools — but none creates the hypotheses or shapes the theories which make science advance. The solution of the problem of signification requires the definition of a new cognitive map.

b. The *problem of the unit of analysis*. The second major characteristic of the neoclassical research program is its *methodological individualism*, implying that only individuals take decisions or that the economic problems can be modelled assuming that simplicity is a fair representation of complexity.

This postulate raises severe problems, the first being the mathematical tractability of the potential diversity of individuals, even if the strong assumption of rationality defined as the maximizing behaviour creates the fiction of the homogeneous *homo economicus*. But that hypothesis reduces the individual motivations to a common standard and argues that preferences and tastes can be treated as pure passive market realities, an inaccurate contention if the intentional character of economic phenomena is considered. Curiously enough, Walras himself criticized Say's attempts to define economics as a natural science, since he claimed that 'Man is a being gifted with reason and freedom, capable of initiative and progress. In matters of production and distribution of wealth, as in general in every matter of social organization, he may choose between good and evil' (Walras, 1883: 10). But if diversity, change, choice and intention is conceived of in social and economic life, methodological individualism becomes inconsistent and so is perfect rationality. On the other hand, economists know that decision are taken in institutional frameworks and not exclusively by isolated individuals; rationality must therefore be defined so that it includes the logic of institutional motivations and interests of groups.

In the same sense, the Marshallian notion of some epistemic representative and homogeneous agent is irrelevant for practical analysis, although logically equivalent to the *ceteris paribus* condition: equilibrium as a general system of relationship means the total permeability and inter-communicability of the system, but in fact the subsets of economic activity do not really need to interact, since they are homogeneous. In the same sense, the aesthetic concept of perfect information implies that every agent knows everything but learns nothing. As a consequence, economics is constrained to deal with production, innovation and change as some 'black boxes', unexplainable by the theory. Such a theory is empty and irrelevant.

c. The *problem of explanation*. In the neoclassical research program, by influence of the physical metaphor, reversibility of time is assumed. The movement of the particles, or units — economic agents — is determined by universal potential force-fields, inspiring the idea of an universal order. On the other hand, the neoclassical program claimed that this epistemology guarantees a positive science, the postulate of efficiency being independent from any kind of teleological or normative content. But this approach prevents any explanation of real life events, since equilibrium and order are already assumed and the working of science is submitted to that a priori.

A new cognitive map implies alternatively an open and holistic system as the relevant unit of analysis, just as it implies a concept of explanation based on a multiple combination of sense and reference. In this framework, it implies the foundation of a new paradigm, as Robinson[35] suggested, with a strong bias towards realist criteria: 'The methodology is self-conscious. I hold very strongly that the purpose of economic theories should be to try to throw some light on the world that we are living in' (Robinson, 1980: ix).

The general strategy of positivism was to define the cognition criterion as part of the demarcation problem; the alternative, of course, is to abandon the attempts to define some absolute demarcation and to include cognition as part of the general signification problem: we need to know how, but we also want to know why and what for. A substantive shift from the positivist research program is therefore necessary and possible.

4. The Economy of Metaphoric Innovation

In the opening page of *One Hundred Years of Solitude*, Garcia Marquez describes the emptiness of ancient ages: 'The world was so recent that many things had no name and in order to mention them it was necessary to point the finger.' The denotation of things was the vital task to the understanding of the worlds around the human being, and of the human being themself.

But denotation is a difficult and ambitious task. Its process includes interpretation, search, change of sense, errors, comparison, interaction: it is a metaphor of the civilization itself. In particular, the description by the human being of his own feelings was mixed with the most contradictory thoughts: for thousands of years, these were expressed in the external form of several or, finally, of one single divine figure — one of the most striking myths of mankind. The denotation of this creation was thus to include perfection, human appearance and transcendence, immanent exemplary qualities and auto-development features. Jorge Luis Borges described one of these denotation programs through the history of Plato's perfect and uniform figure of the sphere, which was for Xenophon this was the ideal denotation of the single God, the eternal sphere; then the hermetic medieval tradition or the twelfth century French theologian Alanus de Insulis repeated this metaphor: 'God is an intelligible sphere, whose centre is everywhere and whose circumference is nowhere'. The Book of Pantagruel repeated it once again, and so did Dante; many years later on, Pascal discovered again this 'fearful' eternal sphere. 'It may be that universal history is just the history of different intonations given to a handful of metaphors', concluded Borges (1983: 16).

Metaphor — an impertinent denotation process — has been a central inspiration for literature and cultural creation: 'Do not forbid me use of metaphor;/ I could not else express my thoughts at all', wrote Goethe. As Ricoeur argued, 'In service of poetic function, metaphor is that strategy of discourse by which language diverts itself of its function of direct description in order to reach the mythical level where its function of discovery is set free' (Ricoeur, 1977: 247).

Of course, this intervention of the metaphor in the denotation process and in the framing of general narratives was not welcomed by the positivists. One of the most brilliant of them, Voltaire, argued in the name of science against that abuse of freedom. In *Micromégas, A Philosophical History*, Micromégas was a traveller who was discovering the world and blamed his secretary for his colourful and metaphorical descriptions of Nature: 'I do not want to be pleased, I want to be taught', he said (Voltaire, 1752: ii). For Voltaire and the positivists, metaphors are perversions of the ideal of a perfect and unequivocal correspondence between world and word.

But, as the last chapter abundantly proved, metaphors are pervasive in science: the energetic metaphor had a strategic role in the constitution of the marginalist revolution, and Fisher as well as the other founders of the program provided a whole dictionary of analogies from energetics and physics to economics, and the alternative biological metaphor was explored at the same time by Marshall and by Veblen.

In general, the metaphor is part of the associative linguistic processes, as Saussure, Jakobson and Barthes argued (Barthes, 1953: 133), or more recently Sontag (1983: 91), just as it is part of the conceptual systems (Lackoff and Johnson, 1980: 3): economics is no exception.

The argument presented in this chapter is that the same impertinent process of denotation is essential to scientific creativity at all levels, and that metaphoric innovation is not only legitimate but also a crucial part of the growth of knowledge. If this is so, neoclassical economics cannot be charged of its widespread use of metaphor, although it must be criticized by the sense and reference of its own mechanical metaphor.

4.1. Metaphor: word and tone

The metaphor was firstly considered as a figure of speech, and was accordingly studied by rhetoric. In fact, metaphor has a pervasive use in common language and is an essential feature of literature: Aristotle presented his own conception of the metaphor, the signal of genius in language creation, in his *Poetica*. The theory of literature and, generally, hermeneutics, will be used now to establish a classification of metaphors and to discuss its applicability for science, on the basis of previous work by Black, Hesse and Ricoeur.

4.1.A. Substitution metaphors

Consider now the following list of examples:

* Lord is our *shepherd*
* the *head* of the State
* the *arm* of the law
* the *arterial* road
* *falling* in love
* time is *money*

or the following one, as familiar as the previous:

* homo homini *lupus*
* the mind as a *tabula rasa*
* biology is economists' *Mecca*
* consumer *sovereignty*
* human *capital*
* children as '*durable goods*'
* Walras's *auctioneer*
* the *invisible hand*
* transaction *friction*
* the *velocity* of money
* *window* of opportunities
* epigenetic *landscape*
* *putty-clay* in capital market
* market *forces*
* business *cycle*
* *natural* rate of unemployment
* *atoms* of *pleasure*
* *fluidity* of the market
* the genetic *code*
* *game* theory
* money *withering away*
* economic *fluctuation*
* *old* industry
* technological *trajectory*

The first list is taken from daily language or very common cultural references, and it could of course be infinite. It indicates that the normal denotation process is complemented by descriptions or qualifications creating a complex network of references and mapping new connections. This is a standard procedure in literature, and mainly in poetry. The examples are abundant, as the following one: 'Consider, and 'tis easy understanding/ Life is not object, but *the refracted colours*' (Goethe). Obviously, life is not 'refracted colours', but the poet believed that the meaning of his expression was transmitted with more intensity: the metaphor is the carrier of tone.

In the previous lists the metaphorical introduction is indicated by italic characters. The process is immediately communicative: the redescription is accessible to everyone, and it makes possible a linguistic creativeness we all extensively use in daily life, without even noticing it. Furthermore, it is a cumulative process, asking for a new creative activity of the receiver.

The second list has the same characteristics, but it was taken from science, mainly economics but also biology and philosophy, and from as different authors as Hobbes and Carlota Pérez, Edgeworth and Nelson and Winter, Ricardo and Waddington, Marshall and Friedman, Becker and Samuelson, or commonly used technical language. The introduction of a semantic and metaphorical impertinence is also indicated and evidences the importance and generality of the process in scientific definition, classification and communication.

This first form will be called the *substitution metaphor,* as Black did (1962: 31). The metaphorical expression is included as a substitute to the literal expression, such that the metaphorical and literal meanings may be determined in both directions by simple rules of correspondence which represent the cognitive process. Thus, the signification process is stable since dependent on quite evident rules of translation.

Yet, there is a process of learning and communication — the substitution metaphor is not trivial, and it is useful and necessary namely for economic reasons — but it implies no transformation in the meaning of the expression, established in the general frame where it is integrated (the phrase, the narrative). 'Understanding the metaphor is like deciphering a code or unravelling a riddle' (ibid.: 31): the cognitive action is centred at the focus, the word used for the substitution, and does not modify the frame.

4.1.B Comparison metaphors

A new list of twelve metaphors is now taken exclusively from science, in particular from physics.

a. Copernicus considered the Sun at the centre of the planetary system as the King was at the centre of the court or God (the fearful expanding 'sphere' of Pascal) at the centre of the universe (Hesse, 1980: 14). Laplace exemplified the use of analogy and generalized the argument writing that, like the Sun, any star was supposed to be the centre of a planetary system of its own (Laplace, 1812: cxliii).
b. William Harvey's model for the circulation of the blood was inspired by Copernicus's planetary system, the hearth being the analogue for the Sun (Jones, 1982: 5).
c. Descartes described the human body as a machine, and the circulatory system as a clock or automaton (Descartes, 1637: 53, 62).
d. In the Gaia hypothesis, clouds are described as seaweed.
e. The Rutheford-Bohr model was described as a solar system where planets revolve around a Sun, like the hydrogen atom with electrons around more massive nuclei (Wullwick, 1990: 216; Black, 1962: 229).
f. Defining hypotheses for the atomic explanation of the behaviour of the gases, Dalton spent some time using his pure pictorial imagination, drawing several models — 'combining my atoms upon paper' — for the hypothetical geometrical and mechanical evolution of the gases (quoted in Capra, 1982: 54; also Hesse, 1970: 88).
g. The traditional quantitative equation of money, MV=PT, is term-to-term equal to the equation of the state of an ideal gas, 'and has the same status as an irrefutable but useful notion in chemistry as it has in economics' (McCloskey, 1983: 501).
h. Maxwell's notions of fluid flow, heat flow, electric induction, electric current and magnetic field are all describable by similar equations, even if the interpretation differed from case to case (Hesse, 1970: 22, 42).
i. Maxwell's equations for electromagnetic phenomena were used also to explain the transmission of light (Hesse, 1970: 44), and the equations for attraction are transformed into equations for heat: 'We have only to substitute *source of heat* for *centre of attraction*, *flow of heat* for *accelerating effect*

of attraction at any point, and *temperature* for *potential*, and the solution of a problem in attractions is transformed into that of a problem of heat.' (Maxwell, 1890: 157, his emphasis).

j. Huygens used the analogy light-sound to describe the 'waves of light' (Duhem, 1906: 94-5).
k. Malebranche and Young represented monochromatic light with a similar formula to the one representing a sound (ibid.).
l. Ohm did the same with propagation of heat and electricity (ibid.).
m. Juglar described crises as diseases in the economic body, motivated by 'excessive speculation' (Juglar, 1862).

This is a rather eclectic list. The first three items are simple and classical examples of substitution metaphors: their function was essentially that of justification, even if it was a wrong one — the case of the Cartesian description of blood circulation. But this metaphoric extension had a complementary signification: it was also part of a logic of discovery, namely for Copernicus and Harvey. In other words, the metaphors changed the semantic value of the assertions, since they organized a new map of the research.

The three following examples (d-e-f) are clearly cases of search and discovery and, like the following ones, they are metaphors which use active comparisons between two theories from two scientific fields: they will be defined as *comparison metaphors*. Dalton, drawing some possible models, was caling for his whole previous scientific — and metaphysical, as it happens — knowledge, and Bohr used the metaphor as a problem-shifting device: rejecting Maxwell's theory of electromagnetism, his metaphor was organized around a shared language in physics — the Newtonian law of gravitation — in order to operate a transfer of terminology to a different field (Wullwick, 1990: 217), with a comparison of the positive and negative implications of the analogy. These metaphors did not prevent Dalton or Bohr to test the new model independently of the original inspiration.

The last seven examples are new cases of comparison metaphors, extending the comparison between different scientific fields or formulations. In these cases, as the authors indicated, the language — equations, definitions — was transferred and became the potential source of creativeness in the comparison. The cognitive process for comparison metaphors is centred on the relation between the space of the secondary subject, the scientific field where the metaphor is originated, and that of the primary subject, where it is applied. Since the original meaning is established in the secondary subject, the comparison metaphor consists in its application to another field of inquiry under correspondence rules which translate the metaphorical meaning and reorganize the frame in the primary subject. There is a semantic innovation, but it depends upon the specific translation rules and it is one-sided: the production of sense depends upon the change of reference. The heuristic procedure of search and innovation is limited by the correspondence rules, since the metaphor itself consists in the presentation of the underlying similarity.[36]

Another model of the metaphor, in fact the most important one, will be established for those cases in which an active two-sided relation is underway: the interaction metaphor, which is the widely applied metaphoric model for models and scientific explanations. A preliminary discussion of some of the most relevant properties of the metaphor is now in order.

4.1.C. The language innovation of the metaphor

The metaphoric intervention in the primary subject is, as indicated, a very common standard in daily language. It takes place currently as a substitution metaphor, but can also originate some more elaborate metaphors — of comparison — which have been widely used in scientific definitions, in the context of discovery or of justification; both types of metaphor may coexist. These operations can shift, translate and modify the meaning in the primary subject: the creation at the language level also implies a development at the knowledge level, since both are closely related. As Black puts it, 'Indeed, I intend to defend the implausible contention that a metaphorical statement can sometimes generate new knowledge and insight by changing relationships between the things designated' (Black, 1981: 37), that is, by emphasizing, suppressing, reorganizing relevant features of the primary subject (ibid.: 28; also 1962: 44-5).

Positivism, of course, rejected this 'implausible contention'. The objective language of science — the quest of logical empiricism — should be unambiguous, indisputable and clearly describe the meanings of phenomena and the propositions about them, implying testable consequences: positive reality should be literally describable, every word should correspond to an object, and vice-versa. That was indeed the program of young Wittgenstein, of the *Tratactus Logico-Philosophicus*,[37] and it was supported by Russell. This research program for an observational, positive and completely intelligible language was criticized since 1906 by Duhem, but dominated and occupied a large part of the twentieth century scientists' efforts.

Descartes defined the basis for such a program. In a 1629 letter to Marsennes, he stated that 'Order is what is needed: All the thoughts that can come into the human mind must be arranged in an order like the natural order of numbers' (quoted in Mirowski, 1988b: 119). In this pure orderly world, meanings should be unambiguously fixed: metaphoric innovation was excluded by definition. The positivist attack on metaphor was reinforced by Berkeley, supporting literal language and fixed meanings, and thus rejecting deviance in the denotation process; Locke, in his *Essay on Human Understanding*, repeated the refusal of metaphors, since they may 'insinuate wrong ideas' and 'mislead judgement' — these themes have been repeated since then in positivist philosophy, hunting the 'linguistic infection' introduced by metaphors (Horsburgh, 1958: 231, 245). Bertalanffy noted the attack launched by logical positivists against his General Systems Theory, since it established analogies between different systems and considered that isomorphisms in models and analogy were 'potential tools in science' (Bertalanffy, 1962: 8-9). Solow argued recently that methodology should restrict the 'range of permissible modes of argument', stressing that metaphor can be productive but that it can also mislead, namely because the assumptions are not clearly and formally stated and the logical deduction is hidden (Solow, 1988: 32-4).

A different approach was taken by the ex-positivist McCloskey, who fully recognizes the pervasiveness and importance of metaphors: 'Each step in economic reasoning, even the reasoning of the official rhetoric, is metaphoric. The world is said to be "like" a complex model, and its measurements are said to be like the easily measured proxy variable to hand' (McCloskey, 1985: 75). But McCloskey's conclusion is the most paradoxical one: scientific vocabulary is presented as a set of purely literary devices with no specific cognitive content, and metaphor is restricted to literally translatable paraphrases of the substitution type. In this sense, it is rhetoric that establishes the modes of argument, while the pretension of epistemology, which is supposed to legislate about truth, should be rejected (ibid.: 47, 51); as a consequence, economics is defined, as well as all sciences, as a branch of literature (ibid.: 138). This postmodern conclusion reduces the scientific endeavour to conversation and takes a radical relativist view: positivism is not vindicated but it is anyway safeguarded not because it is supposed to be able to exhaustively describe reality, but because reality cannot be assessed in any meaningful way and all science is defined as mere narrative, and therefore all methods are equivalent — in the postmodern night all cats are alike, but you'd better continue with the one you are used to.

In the following sections, the challenge to positivism is organized in two different strategies: at the level of the creation of new languages, and at the level of the epistemic unit as a shift from the explanans to the explanandum.

4.1.C. 1. Language

The argument for metaphoric creativeness was stated by Black in his seminal 1962 book, *Models and Metaphors*. In his approach, a grammar merely based on spatial or translation transfers cannot introduce novelty:

> We can pass from one systematic mode of spatial representation to another by means of rules for transforming co-ordinates, and we can pass from one language to another having the same fact-stating resources by means of rules of translation. But rules for transformation of co-ordinates yield no information about space; and transformation rules for sets of language tell us nothing about the ultimate nature of reality. (Black, 1962: 15)

So, one should abandon the epistemological search for unique and definitive rules of correspondence, and innovation or semantic transformation rules should be looked for. In fact, this is a normal and frequent necessity in science. The catachresis function of the metaphor — the introduction of new terminology where none existed — was emphasized: the 'employment of metaphor serves as nondefinitional mode of reference fixing, which is specially well suited to the introduction of terms referring to kinds whose real essences consist of complex relational properties, rather than features of internal constitution' (Boyd, 1981: 358). Quine suggested that metaphor was particularly useful and needed at these boundaries of science[38], where 'Old idioms are bound to fail [at the 'philosophical fringes of the science'], and only metaphor can begin to limn the new order' (Quine, 1979: 159). For Kuhn, metaphor was the specific creation of adherence of language to the real world: '"metaphor" refers to all those processes in which the juxtaposition either of terms or of concrete examples calls forth a network of similarities which help to determine the way in which language attaches to the world' as part of 'the linguistic machinery of science today' (Kuhn, 1981: 415; 414). In particular, Kuhn stressed this denotative — or catachretic — function of the metaphor.[39]

Generally, the introduction of a metaphor implies the creation of a new language, of a new system of references where innovation at the level of sense replaces the static assumption of the traditional correspondence rules. This process involves, however, a larger domain than that of language, since it requires a violation of the previously accepted linguistic rules.

4.1.C.2. Redescription of the explanandum

In the positivist world, and in particular in the Hypothetico-Deductive Model, the introduction of new vocabularies and languages is a common procedure and indeed it is part of every demonstrative logic. But, the linguistic innovation is confined to the limited space of the explanans, where it constitutes the essence of the explanation.

The HD model sustained the deductibility of the explanandum from the explanans, in such conditions that the explanans should contain at least one law not redundant to the deduction and not falsified by data, and the deduction should culminate in the formulation of predictions: (i) of laws already stated in the explanans having yet unobserved cases; or (ii) of new general laws being deduced, with no change to the fixed set of rules of correspondence. But, for practical purposes, this account is inadequate since: (a) no purely deductive relationships can be established between observational and theoretical statements, since the correspondence rules cannot be derived from the explanans alone, and (b) there is no complete deduction, but at most adequate fitness, depending on the behaviour of the system and not only on the deductive procedures (Hesse, 1980: 120 f.).

In these circumstances, the metaphor provides some useful new insights since it is not tied to the strictly deductive rules and to a fixed set of correspondence terms. On the contrary, it states an interactive system of relations between the primary and the secondary systems and makes possible rational — even if not necessarily true — predictions. This whole operation of explanation through the metaphorical description is located at the level of the explanandum, where observational statements are replaced by theoretical categories — in fact, annihilating the conceptual antinomy between observational and theoretical statements[40] — and these are redescribed through the transfer of the metaphor from the secondary subject (ibid.: 111). In other words, the metaphor not only constructs some of the original premises but also extends the power of deduction.

The cognitive and semantic value of the metaphor flows from these characteristics, unifying the logic of discovery and of justification. The inter-subjective nature of science is assumed, and innovation is considered the object and purpose of the redescription: it is a form of theoretical explanation, although an incomplete one. Its logic is part of its semantics and evolves from conflicts and contradictions — it is a dialectical one.

In this sense, metaphor is as pervasive in science as it is in common language: induction implies the analogy of one particular to another particular instance, and the general relation inferred from these comparisons is of course a metaphor for their relation, as Keynes understood better than any other economist. At the same time, deduction implies a hidden analogy between the known general law and the particular instance which is being affirmed; the Kuhnian 'normal science' develops constructing analogies between what is known and what is not (Barnes, 1982: 49) and Kuhnian revolutions are in this sense 'metaphoric redescriptions of the domain of phenomena' (Arbib and Hesse, 1986: 156).

The precise cognitive implication of metaphor is therefore an open question. Although recognizing the importance and the epistemological function of the metaphor, Schlanger argues that there is no cognitive content in this operation, in spite of providing what she calls an 'index of knowledge', enlarging the intellectual horizons and dislocating the meanings (Schlanger, 1971: 47). On the contrary, it is here assumed that the metaphoric redescription adds new insights and creates new meanings, rather than being limited to highlight the pre-existing similarities between the primary and the secondary subjects: metaphor is part of the innovative creation of hypotheses and has a heuristic function which is essential to scientific endeavour.

4.1.C. 3. Rhetoric and Logic

The interaction metaphor challenges some of the most central scientist canons: the dualism between subject and object, between the explanans and the explanandum, between the meaningless external world and the meaningful world of the scientist and consequently between observational and theoretical statements, where objectivity is firmly based on logical rules of deduction and on the language of inductive inference. The logical status of this challenge will now be stated, in order to define its conditions.

In his *De Poetica*, Aristotle defined metaphor as *a displacement of noun*:

> Metaphor consists in giving a thing a name that belongs to something else; the transference being either from genus to species, or from species to genus, or from species to species, or on grounds of analogy ... . Metaphor, moreover, gives style clearness, charm, and distinction as nothing else can: and it is not a thing whose use can be taught by one man to another. (Aristotle, CW, IX: 1457^{b} 7-18)

Metaphor — a substitutive or comparative metaphor in this case — is thus constructive knowledge. For Aristotle, it is a part of rhetoric[41] and not of logic; indeed, it is considered as the 'sign of genius, since a good metaphor implies an intuitive perception of the similarity in dissimilars' (Aristotle, ibid.: 1459^{a} 4) and, since the explanation is omitted, metaphor works as a riddle (ibid.: 1407^{a} 11-13; 1405^{b} 3-6). In the Aristotelian context, metaphor is a semantic transformation concentrated on the word (ibid.: 1405^{a} 9-10), at the level of the identity of the meaning. Furthermore, metaphors are similes and should 'fairly correspond to the thing signified', that is, no ambiguity should disturb the connection between word and world (ibid.: 1405^{a} 9-10).

But the logic of this transformation is essential and in fact it suggests a larger implication than that of Aristotle: it is not only a deviance in the denomination itself, but also a deviant predicative structure which is at work (Ricoeur, 1977: 66, 143) — the semantic impertinence constitutes the novelty in the primary system, and therefore surpasses the rhetorical search for an effect of sense. In fact, if sense is defined as the internal organization and coherence of the discourse — called until now cognition — and reference as its power to redescribe reality — called until now signification — metaphor is, in general, a creative process in knowledge which involves both sense and reference changes. In the context of rhetoric, metaphor will be considered a trope of resemblance or of dissimilarity, a deviance in denomination, operating at the level of the word; in the context of semantics and semiotics, it is rather defined as an impertinent predication at the level of the phrase. But if the metaphor implies the creation of a radically new semantic pertinence and structure, then it can only be discussed by hermeneutics and logic and its place is in general narrative or discourse.

Classical rhetoric cannot interpret this process, since it does not recognize, besides the effect of sense for persuasion purposes, the production of sense and reference. Instead, the metaphor operates simultaneously as a *denotation* (metaphorization of reference) and as a *connotation* (metaphorization of sense) instrument, and that is why it introduces a new form of logic. The interaction metaphor, which is the most relevant for epistemology, is the result of this dialectical logic.

4.1.D. The creation of metaphors: abduction

Peirce defined abduction as the process of sense and reference innovation at the level of the constitutive hypothesis of a theory:

> Abduction is the process of forming an explanatory hypothesis. It is the only logical operation which introduces a new idea; for induction does nothing but determinate a value, and deduction merely evolves the necessary consequences of a pure hypothesis. (Peirce, 1934: v, also 106)

In this case, science should combine different tools: (1) the definition of laws, by induction; (2) the definition of causality, by 'hypothetical inference', or abduction; and (3) the prediction of consequences, by deduction. Since the truth of a deduction is conditional on the truth of the premises, and induction merely determines magnitudes, abduction is the sole creative procedure to determine new inferences and ideas. In fact, abduction describes an act of insight, a *flash*: indeed, these insights have been fairly commonly described by scientists about their own experience, just like Newton's 'sudden insight', Darwin's reading of Malthus and derivation of Natural Selection or again his analogy for the common descent hypothesis, or Poincaré's 'illumination', or Claude Bernard's 'fecund idea', or Keynes 'grey, fuzzy, woolly monster in one's head'. Bohm summarized the task of scientific research with the metaphor of a dark room, scarcely illuminated by candles, until a sudden insight gives a picture of the totality: 'In what is generally described as a very sudden process, a "click" or a "flash", one grasps the basic principle of the circle, which is to say, one sees it as a totality' (Bohm, 1964: 215).

This theme has been widely discussed in epistemology, history and sociology of sciences, originating some descriptive concepts without any established autonomous logical content. Some striking exceptions are Quine's 'ontological imputation' concept (Lawson, 1989: 68), Kuhn's presentation of the formation of theories as 'imaginative statements' (Kuhn, 1989: 338), Popper's definition of the 'meta-scientific knowledge' prior to the test (Popper, 1982: 242) and, somewhat more surprisingly, even Friedman's conclusion in his essay on positive economics:

> The construction of an hypothesis is a creative act of inspiration, intuition, invention; its essence is the vision of something new in familiar material. The process must be discussed in psychological, not logical categories; studied in autobiographies and biographies, not treatises on scientific method; and promoted by maxim and example, not syllogism or theorem. (Friedman, 1953: 43)

Even if the argument is overstated and excludes the logical and epistemological dimensions, creation and imagination are rightly considered as the main tools and tasks for science. In fact, abduction is nothing but the process of selection of metaphors. In particular, a network model considers the legitimacy of the three processes of growth of knowledge — induction, deduction, abduction — and such a combination presents the viable alternative to Hume's Paradox: 'induction without abduction is blind, abduction without induction is empty' (Apel, 1968: 89).

Abduction and metaphoric innovation are responsible for revolutionary insights in science. The very evolution of Darwinian evolutionism depended on such types of processes, since its scientific status changed when it was able to incorporate Mendel's laws, which were novel in the context of the traditional and inconclusive experiments with hybridization by the botanists:[42] heredity was abductively hypothesized, and it transformed botanics; then its metaphoric retranscription into biology defined the neo-Darwinian synthesis. In astronomy, Kepler's discovery of the orbit of Mars is an example of such a process.[43] Abduction is nothing more than a new name for the metaphorical choice of hypotheses, which is essential in the constrution of every theory.

In economics as well as in social sciences, this formulation of new hypotheses is decisive: in his homage to Schumpeter, Frisch, one of the fathers of the econometric revolution and certainly one of the most influential mathematical economists in the neoclassical program, stated that 'no amount of mathematical technicality, however refined, can ever replace intuition, this inexplicable function which takes place in the brain of a great intellect' (Frisch, 1951: 9). He was right.

4.2. Analogy: the flight of the metaphor

The logic of the metaphor has been widely described under the name of analogy. In fact, the progressive heuristic role of analogy is much less disputed in epistemology than it is the more general role of metaphor, which is sometimes feared as the opening the door to metaphysical assertions, alien to positive science. Some conclusive examples indicate now this attitude of general acceptance of the analogy as part of science, and illuminate the metaphorical content of those analogical operations:

a. Hegel argued in his *Logic* that analogy was decisive in experimental sciences:

> In the experimental sciences Analogy deservedly occupies a high place, and has led to results of the highest importance. Analogy is the instinct or reason, creating an anticipation that this or that characteristic, which experience has discovered, has its roots in the inner nature or kind of an object, and arguing on the faith of that anticipation. Analogy it should be added may be superficial or it may be thorough. (Hegel, 1830: 254)

b. Jevons, writing in 1874, argued for the same sort of analogies:

> whoever wishes to acquire a deep acquaintance with Nature must observe that there are analogies which connect whole branches of science in a parallel manner, and enable us to infer of one class of phenomena what we know of another. (Jevons, 1874: 631)

c. Maxwell argued that analogy of particulars with generals could be compared to explanation:

> When a certain phenomenon is susceptible of being described as an example of a general principle applicable to other phenomena, this phenomenon may be said to be explained. (quoted in Meyerson, 1908: 92)

d. Duhem stressed the same point:

> The history of physics shows us that the search for analogies between two distinct categories of phenomena has perhaps been the surest and most fruitful method of all the procedures put into play in the construction of physical theories. ... Analogies consist in bringing together two abstract systems; either one of them already known serves to help us guess the form of the other not yet known, or both being formulated, they clarify each other. (Duhem, 1906: 95-6)

e. Poincaré argued that analogy is a general procedure for science and gives the example of gases and star systems:

> And when sciences have no direct connection, they highlight one another by analogy. (...) Gases are, to a certain point, the image of the Milky Way and those facts, which appeared to be uninteresting except for physicists, will soon open some new and unexpected horizons to Astronomy. (Poincaré, 1908: 330, my translation)

or also

> In this way through Carnot's second law fresh analogies are revealed to us, which may often be followed in detail; ohmic resistance resembles the viscosity of liquids; hysteresis would resemble rather the friction of solids. In all cases, friction would appear to the type which the most various irreversible phenomena copy, and this kinship is real and profound. (Poincaré, 1903: 151)

f. Gelderen, one of the forerunners of the study on long waves of economic development, indicated that analogy with the shorter business cycle was the starting point for the research:[44]

> there exists an analogy between the longer periods of production development and the on average ten-years cycle .... It is exactly this analogy, we feel, that should be the starting point for investigating the price increase .... (Gelderen, 1913: 45)

g. Penrose, arguing against the biological analogies for economics, nevertheless states the epistemological relevance of the procedure:

> The purpose of logical reasoning in which we consciously and systematically apply the explanation of one series of events to another very different series of events is to help us better to understand the nature of the latter, which presumably is less well understood than the former. ... Analogies of this sort are not only useful but almost indispensable to human thought. (Penrose, 1952: 806-7)

h. Meyerson, himself a positivist, compared the constructive role of analogy with mechanics:

> But as to the process itself of analogical reasoning, it must be clearly understood that it is still more indestructible, if that is possible, than mechanism, for by it alone can we approach reality. Whatever we do we are always obliged to suppose — at least momentarily — that nature proceeds, as does our reasoning. The errors of Descartes and the Natur-Philosophen, as also of Comte, consisted solely in using analogy, not to formulate assumptions to verify, but for apoditic affirmations. (Meyerson, 1908: 416)

The list, of course, is not exhaustive. However, the eight cases have one characteristic in common: they all exclusively refer to the logic of discovery, or at most to the logic of persuasion, under the forms of inspiration, extension of thought, or combination of scientifically founded assertions and arguments. In this sense, analogy is very generally accepted as a useful tool of thought: Descartes used the same logic,[45] in spite of his appeal for legitimate reasoning to be expressed in numbers, which would exclude other forms of analysis; Einstein's work is an example of the general use of this procedure (Pessis-Pasternak, 1993: 207).

But the same authors denied the relevance of metaphor for the logic of justification, that is, its demonstrative function. Kant[46] argued that no demonstration could be supported by analogy, Pearson[47] that it was just part of creative imagination, Duhem[48] insisted upon this 'sensory' limited aspect of analogy, Poincaré[49] that metaphors 'should be no more interdict than to the poet' but that they were restricted to 'indifferent hypotheses', Bertalanffy[50] considered them as 'scientifically worthless' as far as explanation is concerned, Maynard-Smith[51] excludes the explanatory capacity of analogies and Gervet[52] adds that analogical reasoning can be the premise for some 'ideological contamination' of science. These warnings are fully justified; yet, they miss something: the analogy is not logically demonstrative, but is still part of the demonstration and is not limited to the abductive definition of the initial hypotheses.

Of course, the whole positivist tradition implies that explanation must exclude analogies from demonstration, since analogies are not acceptable among the premises in the deduction of general laws. In such a conventional view of the metaphor, two boundaries are built. The first one imposes restricting analogy to the laws connecting the properties of the analogues: that is what Duhem intended by the causal analogy — the case of Maxwell's analogies between equations for different systems remains the standard example. The second limits the analogical operation to the logic of discovery, implying the heroic assumption of some comparability between systems, so that no explicit retranscription takes place once the analogue is transported to the primary system; in other words, the analogy is only legitimate if it is a passive operation between completely isomorphic systems. Otherwise, an interaction metaphor's transformation in the primary system would undermine the conventionalist view. The interaction model of metaphor, the third type of metaphor to be considered and the most pertinent for science, challenges this approach; it will be summarized in the next section.

4.3. The interaction model of the metaphor

Even in the very restricted sense of the inductive construction of a theory, there is a redescription operation going on:[53] 'where the theory is regarded as justifying inductive inference from the evidence to further predictions, then the theory must be taken to be an assertion, for the empirical instances it covers, of a previously perceived analogy between those instances' (Hesse, 1974: 221-2). Each model implies in any case an implicit analogy between the modelled theory and the proposition about the phenomena to be explained, the explanandum: a model is a metaphor from the space of theoretical statements applied to the space of the observational statements, and the rigour of such a metaphorization indicates its adequacy as a model.

Thus, there is always an operation of redescription going on and the analogy is not limited to the context of discovery. Analogy will therefore be defined as the operation — or the logic of the metaphorical operation — of the redescription which is neither confined to a separate field (under the untenable positivist assumption of the epistemic distinction between discovery and justification) since it is a permanent feature of all scientific inquiry which is progressive à la Lakatos and, in this sense, it is not a recourse restricted to persuasion, nor should it be considered as an all-inclusive category of all figures of speech.[54]

As an operation of transportation, analogy is an agent of transformation: it establishes a bi-directional link, an interaction. This interaction model of the metaphor was developed by Black (1962, 1981), Hesse (1966, 1974, 1980) and Ricoeur (1977, 1979) and it includes: (a) the interaction between the secondary predicate and the primary subject; (b) the cognitive and signification value of the statement defined by this interaction; (c) the production of new information, namely of sense and reference; and (d) the inexhaustibility of the cognitive appropriation of the metaphor through paraphrase, the literal retranscription into the primary subject. It defines a 'tensional truth' (Ricoeur, 1977: 243, 313), which corresponds to the construction theory of truth presented earlier in this book: the new meaning is created by the tension developed between the focus and the frame, the function and the identity, the literal and the metaphorical interpretation.

The interaction metaphor creates sense as it implies a new pattern of implications within the primary subject, and accordingly also changes the secondary subject where it is originated, since the metaphor extends its implications. Unlike the comparison metaphor, which develops a literal description from the secondary subject, the interaction metaphor establishes its referent in the primary system and constitutes a powerful tool for explanation (Hesse, 1980: 120 f.). Namely, it can organize explanation through the distinction between positive and negative analogies, the negative ones indicating the domains where the secondary subject shows a causal structure non-applicable to the primary analogue (Hesse, 1970: 24). The general use of models in science is a particular and illustrating form of interaction metaphors.

4.4. Models

Models, rigorous, definite, formal models, are presented as a major achievement of the positivist paradigm and are even adored as the synonymous of modernity in science. The formalization of knowledge, the reductionist approach transforming facts into numbers and equations, the mathematical reification of social or natural relations, the use of proxy variables and protective ceteris paribus clauses, all that paraphernalia is considered to be the indisputable pattern of scientificity. Considered itself as a complex model, the world is then reduced to limited simpler models, and it is argued that the test of the assumptions about each of these sub-worlds is conclusive and a decisive criterion to distinguish science from metaphysics. This conventional theory of models will be checked against a metaphorical model of models.

4.4.A. Standard formulations of models

In the standard HDM, the models, explainable in a theory, include a set of behavioural assumptions (in economics it could be the demanded quantity as a function of price, Q=f(P), for instance), a set of simplifying assumptions (for example, Q=a+bP) and a set of assumed parametric specifications obtained from inductive verification (for example, b=-4.2), all the variables being completely defined and quantifiable. The predictive consequences of the model, under the form of statements like 'b*= 6' should be considered for testing. This test is then used to confirm the model: the generation of hypotheses and the confirmation of predictions structure its correspondence with reality. The model is a functional metaphor of reality.

But the model is also a metaphor from a theory. In that sense, Tarski defined the model in a more general deductive plan: the model is that deductive portion of a theory that includes the satisfaction of every theorem derived from the axiom system, 'A possible realization in which all valid sentences of a theory T are satisfied is called a model of T' (Tarski, 1953: 11); the model is a model for the axiom system of the theory.[55] The 'primitive statements' or axioms determine the deduction of all other theorems and therefore their logical relations. As a consequence, the model itself is like a predicative statement, a list of assumptions, containing no claim about reality (Hausman, 1992: 25). This kind of model is considered for conceptual exploration rather than for confirmation of theories, and this is why the functional forms came to dominate the positivist exploration of models.

In other words, all models have in common the positivist dilemmas and are part of the general problem of justification by induction: they share the difficulty of justifying the inference of hypotheses by means of the models themselves, or of developing a deductive system from given postulates in order to confirm it against factual evidence. Either as arguments from theories or as representations of reality, all models are metaphorical if not strictly rhetorical, and of course economics provides some of the most obvious examples: in the 'production function', under a general form $Y=A(t)f(K,L)$, we have a metonymy, something is associated to a thing that stands as its symbol (K,L), and a synecdoche, taking the part for the whole, $A(t)$ (McCloskey, 1985: 83 f.). If this is so, the metaphoric inspection of models provides a powerful tool for their discussion and logical development.

4.4.B. The error of oblique transfer

Hesse suggests the following definition for a realist model[56]: 'A model is intended as a factual description if it exhibits a positive analogy in all respects hitherto tested, and if it has surplus content which is in principle capable of test' (Hesse, 1970: 27). The negative analogy indicates the properties of the secondary subject not existent in the primary one, closing several directions of extension, while the positive analogy suggests a possible path for the development of the inquiry, and the neutral analogy implies the possibility of predictions, in Hesse's account. Thus, the validation of the analogy and the use of the model depend upon the extension of the positive analogy compared to the negative one. One of Hesse's examples is the following table used for the comparison of properties of sound and light (Table 4.1).

There are one-to-one horizontal relations between the two analogues, and each set of properties exhibits internal vertical causal links, which are compared. In the horizontal relation, two types of analogies can be suggested: formal analogies, if the corresponding terms are situated in the same relative position in the same causal networks, or material analogies, models that can

Table 4.1 *Hesse's comparison between sound and light*

	sound	light
↑	echoes	reflection
causal	loudness	brightness
relations	pitch	colour
↓	detected by ear	detected by eye
	propagated in air	propagated in 'ether'
	← similarity relations →	

provide the formulation of predictions if there are observable general similarities (ibid.: 68). The vertical relations are logical causal relations defined by probability, necessary conditions, or deduction from a hypothesis; there can be structural or functional analogies, whose validity depends upon the scientific acceptability of the causal connections (ibid.: 84-5).

The necessary and sufficient conditions for the formulation of a model based on the analogy are therefore, according to Hesse, that: (a) the horizontal relations be relations of similarity (identity or difference) between the sets of characters; (b) the vertical relations be causal and of the same kind; (c) the essential and causal properties of the model be not part of the negative analogy (ibid.: 86-7). These are the logical simultaneous conditions establishing a model, and they define the rules for justification and explanation and for the equivalent negative and positive heuristic.[57] The definition of new theoretical terms in this model does not avoid the requirement for external justification by facts, although it develops the semantic associations by analogy: in this sense, metaphorization is a cognitive operation, not only aiming at revealing hidden isomorphisms but also at the creation of new meanings and abductive hypotheses throughout all the inquiry, that is, it is cognitive because it defines a new heuristic.[58]

That interpretation highlights the importance and pervasiveness of what will be called the *error of oblique transfer*: the illegitimate transposition of propositions defined in the space of the horizontal relations to the space of the vertical relations, causing an excessive inference which ignores the negative analogy. Current examples are the juxtaposition of formal and substantial signification, or of similarity and causality. And the inductive paradox may be reinterpreted on the basis of the obliquity error, since induction is typically based on the exclusive existence of positive analogies between the successive instances, and inductive inference will only avoid the false transposition under the strict and implausible conditions of the constancy of the system of determination and of the environment: Hume and his scepticism are fully justified.

4.4.C. The dictionary theory

The 'dictionary theory' was formulated by Campbell in 1920 as a conventionalist approach to the definition of analogies as models and as a synthesis between inductive and deductive procedures. Campbell defined a scientific theory as including a set of hypotheses — some axioms plus some non-confirmable deduced theorems —, a dictionary for the hypotheses and an analogy to another system with previously well-established laws. This logical structure was supposed to organize a theory as a 'linguistic system' relating the concepts by a series of rules of interpretation (Campbell, 1920: 122; 1921: 96). Models were defined as metaphors whose vocabulary differs from that of the modelled theory.

Eberle, in this context, defines the theory as a set of postulates and its vocabulary as the class of names, predicates, and connections which occur in the sentences of that theory. The interpretation of the vocabulary is then determined by the universe of the discourse, the semantic function of the interpretation rules and the forms of assignment to predicates and relation expressions (Eberle, 1970: 220). The dictionary theory is formulated for the case of isomorphic theories, in terms of horizontal and vertical relations. Identical structures and correlation are thus assumed, and simple translation rules define the working of the metaphor. But the dictionary theory suggests that there is an idiomatic innovation, since two vocabularies are then adequate for operations in the primary subject, that introduced from the secondary subject and that originally resident in the primary subject.Accordingly, the interaction metaphor implies the creation of a new language.

Some characteristic examples were already indicated: in the conceptual architecture of the marginalist revolution, the table by Irving Fisher is typically a dictionary of isomorphisms between economics of equilibrium (the primary subject) and energetic physics (the secondary subject). But this dictionary is limited to the domain of formal metaphorization since it does not allow for an a posteriori verification in relation to reality; the whole operation is strictly deductive and therefore the cognitive process is tautological and limited, and it is dangerously vulnerable to the error of the oblique transfer.

The critique of the neoclassical metaphor is not that it is a metaphor, but rather that it is self-sufficient in its own rhetoric: it is mere self-persuasion, and very successful it has been.

4.4.D. Models as metaphors

Even accepting the importance of metaphors in the constitution of models, it still may be argued that the class of models is more vast than that of metaphors. Toulmin does so:

> Certainly, it is the suggestion, and systematic deployability, that makes a good model

> more than a simple metaphor. When, for instance, we say that someone's eyes swept the horizon, the ancient model of vision as the action of antennae from the eye is preserved in our speech as a metaphor; but when we talk of light travelling our figure of speech is more than a metaphor. (Toulmin, 1953: 39)

This is a rather paradoxical statement. In both cases, we have in fact metaphors: 'the eyes swept the horizon', and 'light travelling' are very simple metaphors. The single difference is the fact that for the first metaphor there are only horizontal relations of similarity and it does not constitute a valid model, as we now know that the secondary subject, the theory of vision as action of antennae, is wrong and we can infer from it no causal relations; while the second exhibits vertical causal besides horizontal relations. In general, the class of models coincides with that of metaphors: 'Every metaphor is the tip of a submerged model' (Black, 1981: 31). Every model is a metaphor of some relation, every metaphor is a model of some entity.

The global picture of models as network metaphors can be drawn, considering the three dimensions already indicated: the production of sense, dominant in the substitutions and comparison metaphors, the production of reference, relevant in the interaction metaphor, and the explanatory redescription of the explanandum as a consequence of both. A network of different modes of explanation is organized, preserving the realism of the model: the inductive logic is reintroduced as a heuristic device, the abductive insight is added to the heuristic.

A synthetic map summarizes these conclusions. Let us consider the two co-ordinates: the production of sense (S, cognition) and reference (R, signification). Now, if a new dimension is added to the map, namely the scientific redescription of the explanandum as a specific explanatory purpose, we have an effect of translation:

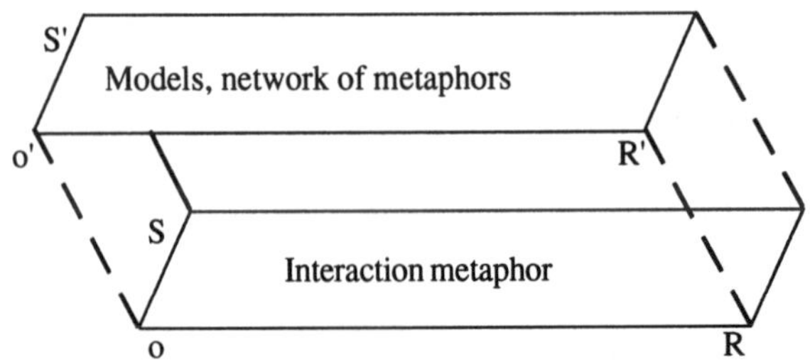

Figure 4.1 Interaction metaphors and network models

where R' and S' are respectively the scientific extension and scientific innovation, the surface R'oS' defining the space of models or network metaphors, that is, the interaction metaphors which establish a multiple connection between several domains of organized knowledge.

4.5. Metaphors in economics, again

The classification of metaphors and the network model of models will now be applied to the two main fields of inquiry referred until now, the analogies from physics and those defined in statistics. Some new methodological problems will be presented and discussed, before dealing with the biological metaphors in economics and the evolutionary paradigm in the next chapter. The role of metaphor in discovery and even in justification was asserted by several authors writing on economic methodology, since the very beginning of political economy. Adam Smith argued that persuasion was essential in trading and that social sciences were part of rhetoric, but it was Bentham who took a larger view of the linguistic processes involved in scientific development, arguing that analogy was part of the 'rules of invention':

> 5. For means and instruments, employ analogy. Analogias undique indagato.
> 6. In your look-out for analogies, for surveying that quarter of the field of thought and action to which the art in question belongs, employ the logical ladders, the ladders made of nest of aggregates placed in logical subalternation. In analogiarum indagatione scalis logicis utere. (Bentham, 1843: 276)

Furthermore, Bentham argued that analogy may be used in induction as well as in deduction, giving the examples of chemistry and mechanics (ibid.: 278). At the same time, Stuart Mill studied and discussed the role of analogy and metaphor with great care. He has been misinterpreted as rejecting metaphors (for example, Mini, 1974: 55 f.) since he pointed out that they 'assume the proposition which they ought to prove' (Mill, 1843: 377-8). In fact, Mill argued that no proof is inherent to the analogical setting of hypotheses and that analogical reasoning is 'imperfect' or 'incomplete' induction (ibid.: 554, 560, 1101). Yet, he added that the metaphor may outline the demonstration: 'For an apt metaphor, though it cannot prove, often suggests the proof. ... A metaphor, then, is not to be considered as an argument, but as an assertion that an argument exists' (ibid.: 800-1). According to Mill, analogy as a general case of metaphor was to be used in every inductive study on similar and adjacent circumstances. Although it constituted no proof, it defined the orientation of the inquiry:

> It is hardly necessary to add that, however considerable this probability may be, no competent inquirer into nature will rest satisfied with it when a complete induction is attainable; but will consider the analogy as a mere guide-post, pointing out the direction in which more rigorous investigations should be prosecuted. It is in the last respect that considerations of analogy have the highest scientific value. (ibid.: 559)

Although taking in general a very critical and hostile attitude towards Mill, Jevons adopted the same basic position, stressing like Bentham that analogy could lead to discovery, and added that it provided powerful means of relating different scientific fields:

> Whoever wishes to acquire a deep acquaintance with Nature must observe that there are analogies, which connect whole branches of science in a parallel manner, and enable us to infer of one class of phenomena what we know of another. It has thus happened on several occasions that the discovery of an unsuspected analogy between two branches of knowledge has been the starting point for a rapid course of discovery. ... The analogy, once pointed out, leads us to discover regions of one science yet underdeveloped, to which the key is furnished by the corresponding truths in the other science. An interchange of aid most wonderful in its results may thus take place, and at the same time the mind raises to a higher generalization, and a more comprehensive view of nature. (Jevons, 1874: 631)

According to Jevons, analogy is part of the language and the logical processes to be used in economics like deduction and induction, generalization or classification, and is part of the inferences about relations between objects (ibid.: 1, 11, 627-8). In fact, Jevons, like Bentham and Mill and so many of the scientists of their times, accepted that metaphor and analogical reasoning are part of the progress of science and could in some way provide a complementary support to induction. And since economics and social sciences in general have been considered not to be as developed as natural sciences, the incorporation of insights from those related areas of inquiry has been a constant feature of their progress. The two cases to be considered now, probabilistic inference in statistical models for chronological series and metaphors taken from physics, prove this point on the pervasiveness of the metaphoric transfer in the constitution of a science.

4.5.A. Probabilistic inference: the axiom of sufficient reason

The central reason for Jevons's dismissal of Mill was the latter's alleged inductivist bias. But, in spite of that, Jevons accepted a major exception in his anti-inductive crusade, since he drew heavily from Laplace, a faithful Baconian (the 'true method', Laplace, 1812: cxl) and the father of inductive inference in statistics: Jevons was indeed one of the first scientists to develop thoroughly the new methods of statistical inference in economics. In this framework, he studied the previous work of Laplace, and quoted him in order to argue for the three conditions of inference: ontological continuity, causal stability and chronological reversibility. For the first, iron does not change into gold, as Jevons wrote, and under such an anti-alchemist certainty a substantive continuity could be asserted. For the second condition, Jevons argued that analogy itself was based on a probability assumption and invoked Laplace's authority for that argument:

> Analogy is founded on the probability that similar things have causes of the same kind, and produce the same effects. The more perfect the similarity, the greater is the probability. (Laplace, 1812: cxli; this passage was also quoted by Jevons, 1874: 597)

In other words, Laplace established the very strict condition that relations of similarity coincide in probability with the relations of causality, so that the analogy works. Furthermore, he considered probability to be defined by (observable) similarity and not by (unobservable) causality. But since causality was defined by logical or situational contiguity, similarity could always be taken as a good index of its deep structure;[59] moreover, the stability of causation was commonly assumed as a necessary condition for scientific inference,[60] making possible the induction on the basis of analogy (Laplace, ibid.: cxxxviii-cxxxix; also Hegel, 1830: 252).

The third condition follows from this one. Structural stability is assumed to be a permanent feature of the observable entity; otherwise, probability could account for the passing of time or for the appearance of new instances under the same condition:

> Inference but unfolds the hidden meaning of our observations, and *the theory of probability shows how far we go beyond our data in assuming that new specimens will resemble the old ones*, or that the future may be regarded as proceeding uniformly with the past. (Jevons, 1874: 219, his emphasis)

The three conditions empower the scientist with the extreme capacities of the Laplacean demon, as Jevons recognized quite clearly:

> We may safely accept as a satisfactory scientific hypothesis the doctrine so grandly put forth by Laplace, who asserted that a perfect knowledge of what was to happen thenceforth and forever after. Scientific inference is impossible, unless we may regard the present as the outcome of what is past, and the cause of what is to come. To the perfect intelligence nothing is uncertain. (Jevons, 1874: 738-9)

These three bodies of assumptions provided the tools for the development of statistical inference: Jevons, like Laplace, assumed that the causal relations are to be ignored — under the hypothesis that there is a high probability of coincidence of similarity relations with causal relations — and that similarity is epistemologically sufficient in order to define the proceedings of inference. As a consequence, the procedure of statistical inference in chronological series accepted the operational concept of identity between the future and the past, that 'the future may be regarded as proceeding uniformly with the past' (ibid.). A large part of conventional statistics is based on this precise assumption, named by Laplace the 'axiom of sufficient reason': 'Current events have with precedent ones a connection based on the evident principle, that one thing cannot commence without a cause producing it' (Laplace, ibid.: vi); as furthermore the causal connection should be interpreted as logical contiguity, the condition of structural causal stability was transformed in the condition for the scientific inquiry.

Sceptic as always, Hume had pointed out that induction would be demonstrative only under the restriction of no change in nature and in time, and that this was quite untenable:

> For all inferences from experience suppose, as their foundation, that the future will resemble the past, and that similar powers will be conjoined with similar sensible qualities. If there be any suspicion that the course of nature may change, and that the past may be no rule for the future, all experience becomes useless, and can give rise to no inference or conclusion. It is impossible, therefore, that any arguments from experience can prove this resemblance of the past to the future. (Hume, 1748: 37-38; also 1740: 119)

The divorce of 'science' — defined by analogy to natural science and physics in particular — with history can be traced back to this moment of acceptance of the principle of causal stability in order to explain events under a general law. But the acceptance of this whole catalogue of the obliquity errors was nevertheless necessary in order to respect the positivist framework. Karl Pearson, a distinguished statistician, took the very extreme position of rejecting both Mill's and Jevons's interpretations, since their work was supposed to be too much infected by the 'pure field of conception', and vindicated Laplace as the true Baconian master (Pearson, 1892: 33 n.): statistics, a truly inductive branch of science, should follow his lead and this method should unify all sciences (ibid.: 16). But of course, these extreme claims about inductive inference are only coherent if some reversibility of time, the most extreme form of causal stability, is assumed. Pearson willingly did so, claiming that a colleague of Maxwell's demon could very well travel backward and forward in history, provided the necessary conditions; so could statistics, given the reversibility of natural processes (ibid.: 343-4).

The consequences are devastating, since time collapses into some eternal and indistinguishable present ruled by similarity between the successive instances. The inductive logic of statistical inference is nevertheless re-enthroned, Hume's Paradox is once again hastily ignored and the HD model suggests than that the derived laws — such as functional relations confirmed by the statistical test — represent the accurate model for reality. Indeed, the self-confirmation and self-correcting virtues of the model are certain by definition, and each new instance will adequate the statistical result and provide a new vector of parameters for the same model and under the same law.

But the whole operation stands or falls on the acceptability of the founding claim, that similarity relations are the counterpart of causal relations. This is logically untenable since structural relations obviously change with time if the economy does not describe some sort of aseptic and eternal automaton. The axiom of sufficient reason is an insufficient reason for the justification of the metaphor.

The introduction of the obliquity error annihilates the power of the metaphor and requires no more and no less than the death of the subject, that time does not have a constructive role, that economic relations are permanent and do not suffer any structural change, that the similitude relations do not cease to correspond to causality relations and, in short, that nothing changes in the economy through time. In this framework, the functional model may be perfect although being meaningless.

4.5.B. Keynes and the principle of unlimited independent variety

Keynes's early interest in the logic of mathematics and in philosophy motivated his first major contribution, the *Treatise on Probability* (published in 1921, after fourteen years of preparation), which discussed thoroughly the problem of statistical inference. The book criticized the assumptions and the conclusions from inductive correlation on two grounds. The generalization of laws from induction supposed, first, the atomistic character of reality and, second, the composition of the constitutive elements, the 'legal atoms', as part of a permanent structure:

> They [the 'material laws'] appear to assume something much more like what mathematicians call the principle of the superposition of small effects, or, as I prefer to call it, in this connection, the atomic character of natural law. The system of the material universe must consist, if this assumption is warranted, of bodies which we may term (...) legal atoms, such that each of them exercises its own separate, independent, and invariable effect, a change of the total state being compounded of a number of separate changes each of which is solely due to a separate portion of the preceding state. ... Each atom can according to this theory, be treated as a separate cause and does not enter into different organic combinations in each of which it is regulated by different laws. (Keynes, TP: 276-7)

This atomistic nature of society was implicitly accepted by Mill, so that 'his methods and arguments would fail immediately, if we were to suppose that phenomena of infinite complexity, due to an infinite number of independent elements, were in question, or if an infinite plurality of causes had to be allowed for' (ibid.: 302) — complexity versus simplicity.

To this principle of the 'legal atomic' nature of society and to the assumption of the composition of forces which derives from it, Keynes called the 'principle of limited independent variety' (ibid.: 301), and it is the epistemological premise of inductive correlation. Of course, since causal stability and structural permanence are assumed, the frequency of confirmatory instances is relevant and should uncover the law regulating the phenomena: the analogy from instance to instance is assumed, the general inference is possible and asymptotically approximates certainty. Alchemy, claimed Keynes: 'No other formula than Bernoulli's Law of Great Numbers in the alchemy of logic has exerted more astonishing powers. For it has established the existence of God from the premise of total ignorance' (ibid.: 89), and this principle of stability of statistical frequencies is obviously dependent on the metaphysics of causal rigidity: 'Poisson seems to claim that, in the whole field of chance and variable occurrences, there really exists, amidst the apparent disorder, a discoverable system. Constant causes are always at work and assert themselves in the long run, so that each class of event does eventually occur in a definite proportion of cases' (ibid.: 366) — complexity reduced to simplicity.

The argument, a brilliant anticipation of the contemporaneous critique of the Law of the Large Numbers when applied to social phenomena which are dominated by structural instability and mutation, led Keynes to oppose the frequencist approach to probability and the excessive claims of inductive correlation. Instead, he suggested two methods to develop the statistical research: first, the researcher should investigate the solidity of the positive analogy, namely using Lexis's technique for checking the stability of frequencies among sub-series (ibid.: 428); second and essentially, he should develop the negative analogy, so that the domain of the phenomena to be explained be investigated in its own peculiarities (ibid.: 206).

This battery of critiques evidences the generality and deepness of the methods of analogy, both in the domain of discovery and in the domain of inductive inference and justification, and points to the three rules of the logic of metaphoric redescription which should be used: forbid the oblique transfer, control the positive analogy, increase the negative analogy. Otherwise, if the 'legal atomistic' properties are accepted and reality is representable by the mechanic analogon ruled by the demon of Laplace, then these rules can be safely ignored and nothing is impossible to the perfect intelligence which knows everything, since there is no time, no change and therefore nothing to be learnt.

4.5.C. Physics against physics

Positivism made possible an extraordinary development of law-like assertions, based on generalized analogies based on known properties of nature. Attraction, movement at a distance, atomism and reductionism established astronomy and then mechanics as the reference sciences and constituted the potential inspirations of the analogies for other sciences.

In economics, the extensive use of analogies of this origin is rather impressive. Physiocrats defined the economy as a Newtonian universe. Say used an early and implicit concept of conservation of energy in his famous law, and so did Smith in his substance theory of value. Gossen, Jevons, Fisher, Walras, Edgeworth, Pareto, almost all of the founding fathers of marginalism used the concepts, the symbols, the mathematical techniques and the philosophy of energetic physics and of the first Law of Thermodynamics, in what constitutes the most relevant and most influential case of incorporation of a whole body of metaphors in economics. And the mechanical models became ever since a major source of metaphoric inspiration: the steam engine for the representation of loanable funds (Tugan-Baranowsky), the pendulum (Yule, Fisher, Frisch), the rocking-horse (Wicksell, Frisch), the list is immense. Of course, causality in the original physical sense of a strict determination of a time and sequential path from cause to effect was considered by positivist sociology and economics as the very subject of the inquiry, and therefore the physical analogy was supposed to intervene in the strongest possible way.

Nevertheless, this approach is particularly misleading and inaccurate for analogies developed in social sciences, where relations of similarity are mostly irrelevant and causality refers to plural and ever changing structures of determination. In short, economists and social scientists defined scientism according to a conception of physical analogies which was inadequate in the social sciences and which was soon to be rejected by physics itself throughout the twentieth century. In this sense, Prigogine and Stengers argue that:

> we can hardly avoid stating that the way in which biological and social evolution has traditionally been interpreted represents a particularly unfortunate use of the concepts and methods borrowed from physics — unfortunate because the area of physics where these concepts and methods are valid was very restricted, and thus the analogies between them and the social or economic phenomena are completely unjustified. The foremost example of this is the paradigm of optimization. It is obvious that the management of human society as well as the action of selective pressures tends to optimize some aspects of behaviour or modes of connection, but to consider optimization as the key to understanding how population and individuals survive is to risk confusing cause and effects. Optimization models thus ignore both the possibility of radical transformation — that is, transformations that change the definition of a problem and thus the kind of solution sought — and the inertial constraints that may eventually force a system into a disastrous way of functioning. (Prigogine and Stengers, 1985: 207)

The program of physical analogies for social sciences collapsed, as far as it depended on the positivist paradigm and reinforced its claims for scientism, objectivity and universalism. Conscious of the failure of the equilibrium and the metaphor of the First Law, many scientists looked for alternative inspiration from biology.

5. Time for Evolution

The previous chapters discussed the necessity of replacing the positivist paradigm as a general reference for sciences, and the last one presented a relevant building-block for that task, the construction of network models using analogy in order to create new hypotheses and to broaden the space of the explanation. Indeed, the use of metaphors is current in economics, as proved by the incorporation of physical metaphors since the marginalist revolution and by the excessive indulgence in relation to the obliquity error in statistical inference. And since the crucial mistake of those analogies lies in the ignorance of the specific dimension of time, possible alternatives have been looked for in the development of evolutionary metaphors, in order to construct a post-positivist paradigm. The initial inspiration for that shift was Darwin's work.

Darwin's 1859 *The Origin of the Species* constituted a Kuhnian-type scientific revolution, since it implied the reintroduction of historicity in non-teleological systems. Challenging the limitations of 'normal science' — the essentialist and creationist accounts in biology, as expressed by the Archbishop James Usher, who declared that all organisms were created the Sunday 23 October of 4004 BC, at nine o'clock sharp — Darwin reorganized simultaneously the hard core hypotheses and the heuristics of the discipline: it was a successful major attack against naturalism and universalism. Consequently, evolutionism influenced several currents in social sciences, and correctly so, as a constitutive metaphor for an epistemological break with mechanicism and the physical metaphors, and for new research programs. These constitutive metaphors are now discussed.

5.1. Darwin's use of metaphor

A summary of Darwin's methods was previously presented, stressing the use of metaphors. Three cases illustrate now his recourse to metaphorical innovation.

The first is the most evident, since Darwin himself stated its influence: the metaphor taken from Malthus's essay on population.[61] In his notebooks, Darwin explained how the reading of Malthus helped him to configure his new theory; in *The Origin of Species*, it was presented as a general method: 'It [the 'principle of geometrical increase'] is the doctrine of Malthus applied with manifold force to the whole animal and vegetable kingdoms' (Darwin, 1859: 117). Simultaneously, the co-founder of the theory of evolution, Wallace, read and was influenced by Malthus in exactly the same sense.

In fact, arguing against Condorcet's naturalistic ideal of progress through successive stages[62] and claim for the unlimited perfectibility of men, Malthus defined human beings as submitted to the same laws as other animals (Malthus, 1798: 225) and to a general tendency to conflict for natural resources and food (ibid.: 72, 124, 238). But the conclusions were quite opposite to Darwin's, since Darwin indicated the continuity of evolution, and Malthus's pessimistic and catastrophic conclusion denied it. Nevertheless, as far as the conception of the theory of evolution was concerned, the metaphoric insight from Malthus was essential to Darwin in order to develop a new hypothesis, even if the choice of methodologies and the conceptual framework was independently developed.

The second example is the concept of the 'survival of the fittest', introduced after much resistance by Darwin as an analogy to Spencer's evolutionism, which was a unilinear version of a teleology of progress. Hayek praised this idea and came to the radical conclusion that 'in many respects Darwinism is the culmination of a development which Mandeville more than any other man has started' (Hayek, 1978: 265). The vindication of the evolutionist legitimacy of the Panglossian praise of capitalism or of the Fable of the Bees can eventually be very useful for ideological purposes, but has no basis in Darwin's work, where social co-operation and learning is stressed instead. This topic will be discussed later on, as a critique of sociobiology.

Anyway, the point is at the heart of some of the most disputed interpretations of Darwin's method. Peirce[63] and Marx[64] sustained that *The Origin* was an extension — a metaphor — of the savage competition under British capitalism of the nineteenth century type, that was certainly an important influence on Darwin (Maynard Smith, 1993: 43). But the existence of such an obvious influence does not imply necessarily any kind of Social Darwinism, as that argued for by Spencer: not only did Darwin resist the incorporation of this metaphor[65] — he only accepted it in the sixth edition, probably as a concession to his own cultural environment — but so did Wallace, always hostile to that idea. In fact, Wallace even wrote to Darwin suggesting the non-applicability of the Malthusian conflict to human evolution, given the spread of voluntary forms of co-operation which denied the law (Young, 1985: 48 f.).

A third example of a metaphor employed by Darwin was the formulation of some of the central hypotheses: the analogy from artificial selection to natural selection, and the metaphor of common descent. The analogy should establish a space of probabilities:

> I believe that animals have descended from at most only four or five progenitors, and plants from an equal or lesser number. Analogy would lead me one step further,

> namely to the belief that all animals and plants have descended from some one prototype. But analogy may be a deceitful guide. Nevertheless, all living things have much in common, in their chemical composition, their germinal vesicles, their cellular structures, and their laws of growth and reproduction. ... Therefore I should infer from analogy that probably all the organic beings which have ever lived on this earth have descended from some one primordial form, into which life was first breathed. (Darwin, 1859: 454-5)

Analogy and metaphoric innovation are therefore legitimate forms of creation of hypotheses, wherever deduction or induction cannot reach. And an even larger metaphoric intervention is important to science, according to Darwin (ibid.: 116), as it is used to create new interpretations where logical deduction does not allow for any conclusion or to formulate hypotheses not based on known facts (ibid.: 263, 455), and it is a permanent action in the process of ostension and semantic incorporation (ibid.: 456).

These examples are no less than the core of the theory responsible for the scientific revolution in biology and natural history.

5.2. Evolutionism

Organicism was certainly previous to Darwin's work. By the end of the eighteenth century and in the nineteenth century, the organic analogon was considered as an archetype of rationality and was closely related to the dominant vision of the universe since it indicated the presence of an essential harmony. The dominant conception consequently implied that social sciences should part of the natural sciences since both refer to the same structure of organization, Newtonian planets and Smithian markets behaving in essentially identical or comparable ways. In fact, the idea of internal regulation, of the harmony of the clock — the *machina machinarum* of Newton and of Descartes — was widespread in both natural and social sciences, and represented this first organic principle. Organicism was born as an interpretation of the closed systems such as the perfect sphere of Pascal; instead, Darwin introduced the new dimension of the open systems.

In this sense, the organic metaphor was not clearly emancipated before Darwin, since it was still part of the dominant naturalistic paradigm. The metaphor of the tree,[66] in order to describe the State functions (Hegel) or that of the State as an organism[67] (Rousseau), was frequently used in a purely rhetoric sense; at the same time, social sciences were pervaded by ambiguous metaphors, as Hobbes's Leviathan — an animal, but also a machine — or Saint Simon's or Comte's concepts of society as an organized mechanism. Instead, Kant presented a holistic concept of formative energy distinguishing between organisms and machines, and so did Marx: organic totality and social intentionality were considered as forming the specific difference between human organization and those inanimate things constituting mechanical wholes.

Evolutionism, the general conception of natural history inspired by Darwinian biology, was established as an autonomous research program in the second half of the nineteenth century, liberated from the mechanical metaphors. Its major achievements are not described here, since several authors have presented a detailed account of such a paradigmatic change (Scoon, 1968; Maynard Smith, 1972, 1993; Waddington, 1975; Mayr, 1982; Hull, 1988; Jantsch, 1980; Brooks and Wiley, 1986; Mani, 1991; Faber and Proops, 1991; Depew, and Weber, 1996). In summary, Darwin's conception included the following elements:

a. The concept of production of variability, namely from the following four origins: (a1) inheritance; (a2) reversion; (a3) use and disuse; (a4) direct effect of environment. But variation was dominantly produced as a process independent from selective advantage.
b. The hypothesis of intervention of natural selection on variation, as the 'predominant power' preserving the favourable variations in a very slow process of historical change:[68]

> Owing to this struggle for life, any variation, however slight and from whatever cause proceeding, if it be in any degree profitable to an individual of any species, in its infinitely complex relations to other organic beings and to external nature, will tend to the preservation of that individual, and will generally be inherited by its offspring. The offspring, also, will thus have a better chance of surviving... . I have called this principle, by which each slight variation, if useful, is preserved, by the term of Natural Selection, in order to mark its relation to man's power of selection. (Darwin, 1859: 115)

or

> Of all these causes of change I am convinced that the accumulative action of Selection, whether applied methodically and more quickly, or unconsciously and more slowly, but more efficiently, is by far the predominant power. (ibid.: 100; also 169, 203-4, 231)

c. The theory of common descent with modifications through natural selection (ibid.: 435).

From (a) and (b), the lineages were considered to be transformed by a long process of replication and interaction with the environment; from (c), they were considered to have a common tree.

This theory constituted a major change in biology. By the time, the essentialist account (the natural species were considered fixed essential elements) and creationism (the doctrine of divine intervention in the formation of the species as stated by the Bible), were the dominant ideological features. Lamarck, some decades before Darwin, presented a first challenge to those views, based on his own scientific research about the development of species, stating that the gradual evolution was determined by two main factors: (a) the innate tendency of species to evolve towards increasing complex structures under the direct action of the environment; and (b) the mechanism of the inheritance of acquired characters; by (b1) directly imposed environmental changes (for example, effects of light or heat; and (b2) inheritance of functionally produced modifications (Lamarck, 1809: xxxiv), those being 'laws of nature' always verified by observation.

A classical positivist, Lamarck strongly believed that *natura non facit saltum* and that the animals were arranged by some 'natural order' according to their complexity, with human beings as the final step of the ladder of development. His laws of evolution were accepted by Darwin, although he did not follow his predecessor as far as to argue for a natural scale of animals or for some finalist evolution.

This difference and the similarities between these authors is in some ways obscured by the current and dominant version of the Neo-Darwinian synthesis (for example, Cohen and Stewart, 1995: 108). But it is interesting to reconsider now that difference, since it was related to one of the most widespread evolutionist metaphors in social sciences. In fact, Darwin rejected the finalist approach by Lamarck, who represented quite well the typical end-of-eighteenth century rationalist anthropomorphic vision. Nevertheless, Darwin did not reject the idea of inheritance of acquired characters as a relevant form of production of variation, and even in his last works, namely in a 1874 reface, maintained it:

> I may take this opportunity of remarking that my critics frequently assume that I attribute all changes of corporal structure and mental power exclusively to the natural selection of such variations as are often called spontaneous; whereas, even in the first edition of 'The Origin of the Species', I distinctly stated that great weight must be attributed to the inherited effects of use and disuse, with respect both to the body and mind. I also attributed some amount of modification to the direct and prolonged action of changed conditions of life. (Darwin, preface to *The Descent of Man*; 1871: viii)

Darwin argued for the importance of inheritance of acquired characters in such different domains as domesticated and savage animals,[69] plants,[70] and human beings.[71] But he considered nevertheless the natural selection process as the central one,[72] and it is plausible that he insisted on the inheritance of acquired characters simply because he did not know the genetic mechanism for the production of variation, and so his system of explanation for causality was clearly deficient. Another reason for the acceptance of the Lamarckian mechanism was obvious in Darwin's polemics with Lord Kelvin about the time scale needed for evolution in the framework of the entropic process: in the sixth edition of *The Origin*, Darwin accepted that the Earth was not old enough for the slow evolutionary process his theory accounted for, so that another influential process should be at work to explain the new species, and that was the inheritance of acquired characters (Depew and Weber, 1996: 460). But it is evident that Darwin was very prudent about this explanation and indeed he rejected acquisition as the single or the most important factor, since he was convinced that both his grandfather Erasmus Darwin, a forerunner of the theory of acquisition, and Lamarck missed something essential: as he told in his *Autobiography*, his reading of both did not 'produce any effect on me' (Darwin, 1876: 49), and in a letter to Lyell, written by the time of the publication of *The Origin*, he restated that he did not incorporate a single fact or idea from Lamarck's work (ibid.: 153). The missing elements were mutation and selection.

Even if Darwin did not openly reject the Lamarckian process of acquisition of characters by use and disuse and of its inheritance, there were still two major differences between both scientists on this question: (a) for Darwin, variation was the result of a complex and changing causal process, and therefore indeterminate; and (b) Darwin decoupled the processes of mutation and selection as autonomous variables, unlike Lamarck who coupled them (Hull, 1988: 455). The epistemological distinction between both processes is the hallmark of Darwinism: mutation constituted the pool of variation, while natural selection originated order and direction in the evolutionary process.

Afterwards, the Neo-Darwinist synthesis transformed this position in a definitive conclusion, Weissmann's 'central dogma' which firmly rejected Lamarckism: there is no possible influence of the phenotype over the genotype of an individual or, in other words, the germ line is independent of soma, the proteins do not transmit information back to DNA (Waddington, 1975: 251; Stenseth, 1985: 55 f.; Maynard Smith, 1993: 2, 79) — there can be no inheritance of acquired characters in any case. Microbiology has essentially confirmed this result, for the moment,[73] although a developmental view emphasizes the fact that DNA itself evolves historically and that it is not a stored immutable program: 'The central dogma of molecular biology, according to which information flows solely from DNA to RNA to protein, seemed to underwrite the isolation of evolutionary from developmental biology. It has become clear, however, that "reverse information flow" does in fact occur' (Depew and Weber, 1996: 396).

This very paradoxical debate opens new problems, as far as the metaphors for social sciences are concerned.

5.3. Some general metaphors from evolutionist biology

Two general evolutionist metaphors are now presented, both derived from Lamarckian biology and the Central Dogma debate, and both having important implications for social scientists: the metaphor inspired by Lamarck for the evolution of cultural processes and social behaviour as forms of acquisition of inherited characters, and the selectionist and optimization view of sociobiology. Both families of metaphors are here rejected on grounds of the obliquity error.

5.3.A. The Lamarckian cultural process

Social or cultural evolution is commonly metaphorized as a Lamarckian process by many evolutionist authors, namely economists (Hayek, 1988: 25; Clark and Juma, 1987: 40; Mani, 1991: 36, 55-6; Faber and Proops, 1991: 58; Goonatilake, 1990: 40; Saviotti and Metcalfe, 1991: 12-3; Hodgson, 1993: 234; Nelson, 1995: 54). Claiming that 'The biological evolution is Darwinian; it does not transmit acquired characters. Tradition, on the contrary, is definitely Lamarckian, i.e., it transmits only acquired characters' (Georgescu-Roegen, 1971: 359), some announce consequently their 'espousal of Lamarckism' (Nelson and Winter, 1982: 11). Gould, a palaeontologist, used the same argument in order to justify the opposite conclusion, that the biological metaphor is inadequate for social sciences (Gould, 1987: 18).

The metaphor suggests a first terminological problem. As argued before, both Darwin and Lamarck were 'Lamarckian' concerning the acceptance of the existence of a process of inheritance of acquired characters, and it is inaccurate to argue for an opposition between their views in that precise matter, although their general theories were quite different, namely concerning the role and the explanatory capacity of acquisition. But the conception of inheritance of acquired characteristics was shared by both authors, and the genetic proof of the origins of variation — the notion of a single causal random process of generation of mutations and the Mendelian random conjunction of chromosomes — was only established by the Neo-Darwinist synthesis later on. Only Wallace, unlike Darwin, was very clearly against the Lamarckian concept of inheritance of acquired characteristics:

> The hypothesis of Lamarck — that progressive changes in species have been produced by the attempts of animals to increase the development of their own organs, and thus modify their structure and habits — has been repeatedly and easily refuted by all writers on the subject of varieties and species, and it seems to have been considered that when this was done the whole question has been finally settled; but this view here developed renders such an hypothesis quite unnecessary, by showing that similar results must be produced by the action of principles constantly at work in nature. (Wallace, 1858: 112)

As a consequence, it is meaningless to present the cultural process as Lamarckian as opposed to a supposedly Darwinian natural process: in the best of the cases, one can only argue that Lamarckism was, on this topic, opposed to Wallace's version of the theory. But the most relevant problem is the definition of Lamarckism itself: not only did Lamarck defend a 'behaviourist' intentionality in use and disuse — so that species could adapt to environment, as supposedly in the case of the giraffe's neck — but he also pointed out that such an intentionality was part of a general design to perfection in animal development. To define cultural evolution as Lamarckian implies necessarily the simultaneous importation of both theses and their stated causal relations, besides the invoked relations of similarity. And even if, to oversimplify, one reduces Lamarckism to adaptive evolution, the metaphor still has no viable interpretation. At most, it could be considered as a comparison metaphor, stressing some elements of similarity in horizontal relations: but those who argue for that metaphor should readily accept that such a similarity — a fictitious one, since it refers to non-existing processes in biology — is the less interesting of its qualities, since the difference between socio-cultural evolution and biological evolution is precisely what is at stake. On the other hand, no causal relations can be compared since, in the literal sense, Lamarckism would imply for the metaphorized primary subject that cultural traditions and beliefs should become programmed into the genes or, in a more general sense, that cultural ideas and social organization are some sort of analogues to the genes, even if it is common notice that they do not have a genetic function, since they are permanently transformed in society.

The metaphor is therefore a trivial one: it only states that social and cultural ideas are essential to the human organization, as the genes are essential to the body. In any case, the metaphor does not add any new element to what we already know, that is, that social organization and behaviour develops and adapts itself in response to internal and external stimuli, in a cumulative and complex way. To stress intentionality, one should not use the Lamarckian metaphor, since purposeful action is confused in this framework with teleological determinism, which is an inaccurate description of cultural evolution. If the intention of the metaphor is to argue for evolutionism in the cultural realm, then the final result is to undermine it. If, on the contrary, the intention is to legitimate the inquiry in the light of evolutionary biology and of a model which is alternative to positivism, then it is absurd to look for references in that body of theories which does not lead to any progressive research program.

On the contrary, scientists should consider the processes of social learning, like cultural production and transmission of information, as rather distinct and indeed as an alternative to biological transmission: 'This gives man a second evolutionary system superimposed on top of the biological one, and functioning by means of a different system of information transmission' (Waddington, 1975: 288) — no single entity carries the information, which is not produced at one single point of time but it is continuously changed, the modes of 'para-genetic transmission' are multiple and, last but not least, human beings learn how to learn. Moreover, in biological evolution there are divergences of lineage without any possible subsequent reunification, while this is the norm in cultural lineages and a major source of variation. The specificity of the social realm is, finally, what leads to the rejection of the Lamarckian metaphor for cultural evolution.

5.3.B. Social Darwinism and Sociobiology

Even before the publication of *The Origin of the Species*, Spencer published his own Malthusian interpretation applied to biology and to human society, developing the idea of the struggle for the 'survival of the fittest'. This kind of theory was from then on developed by the Social-Darwinist approach. It implies a Lamarckian type of finalism for the process of natural selection in biological evolution, and it was easily adapted to social theories and attitudes: Spencer was a very influent publicist, and the nineteenth century idea of progress — the Panglossian confidence in capitalism — bears his mark.[74]

Social Darwinism evolved afterwards in two different paths: on the one hand, there was the eugenic politics defined by Darwin's cousin, Galton (and later on supported by Keynes); on the other hand, there was its right-wing version, such as that of Karl Pearson and which took an open political stance with the adhesion of Marinetti, the Italian futurist who became a fascist supporter, with Konrad Lorenz's appeal for the elimination of 'morally inferior beings' (Sacarrão, 1989: 299), and with several other arguments for racism or holocaust.

There is no support for those theories in Darwin's work: his books are clear statements against racism or any form of discrimination. He explicitly and emphatically stated in 1859 that co-operative behaviour was part of the process of natural selection:

> I ... use the term 'struggle for existence' in a large and metaphorical sense, *including dependence of one being to another*, and including (which is more important) not only the life of the individual, but *success in leaving progeny.* (Darwin, 1859: 116, my emphasis)

So, not only co-operative behaviour was considered, but also the struggle for existence was indirectly developed through the number of offspring. In the later *The Descent of Man*, a very ambiguous book marked by a strong influence of Galton's eugenics,[75] which he did never fully endorse, Darwin still presented social history as a long and intentional movement against natural selection:

> We, civilised men, on the other hand, do our outmost to check the process of elimination; we build asylum for the imbecile, the maimed, and the sick; we institute poor-laws; and our medical men exert their utmost skill to save the life of everyone to the last moment. (Darwin, 1871: 205-6)

Huxley, Darwin's favourite disciple, emphatically rejected eugenism. In the same mood, Tort suggested recently the concept of a *Darwinian reversive effect of civilization.* The reversive effect had been stated by Darwin in a very limited scope: some domesticated animals, changed by functional adaptation, could revert to the previous forms and lose the acquired changes (Darwin, 1859: 77, 203-4). Mani quotes the example of the peppered moth in some industrial regions of England in the eighteenth century: the growth of melanic forms corresponded to the increase in pollution, since the lighter forms suffered a high degree of predation, being more visible. Some decades later on, with the decrease of the pollution due to the change of combustible, the reverse evolution took place — in other words, phenotypic reverse evolution can be produced if the variant genotypes are not lost in the population (Mani, 1991: 35). But Tort means much more: he implies the historical selection of anti-selective processes by means of the social instincts of human beings, generalizing co-operative action through the creation of health and education systems or other means (Tort, 1985b: 176). In this sense, the degree of civilization can be measured by the impact of the reversive effect on natural selection processes.

Darwin's notion of social instincts included the definition of human ethics as something more than the result of the process of natural selection, as deliberate choice between alternatives including the open behavioural program developed by education and social learning in each human being's life (Mayr, 1988: 88). Extending this notion to socio-cultural evolution, Tort stresses the essential differences between biological and social evolution.

Sociobiology, following Social-Darwinism, generalized instead the argument for a biological determination of the genetic system on human and social behaviour.[76] Defined as 'the systematic study of the biological basis of all social behavior' (Wilson, 1975: 4), Sociobiology explains the social status quo by a genetic determinism: 'A key question of human biology is whether there exists a genetic predisposition to enter certain classes and to play certain roles. Circumstances can be easily conceived of in which such genetic differentiation might occur' (ibid.: 554). In particular, sociobiology associates a global maximum principle to the concept of Natural Selection, which had been defined by Darwin as a local principle of minimum change or, as Dobzhansky put it, as a trend towards tolerable fitness, not towards an optimum; moreover, the natural selection principle rules animal behaviour and its functions (maintenance, satisfaction, survival) but not the whole social human behaviour, as Bertalanffy pointed out many years ago (Bertalanffy, 1962: 17). In this sense, rationality for Darwin meant adequacy and adaptedness and not maximization (Toulmin, 1972: 327).

Sociobiological arguments are not based on Darwinism, and in fact they are an extreme genetic version of earlier Social Darwinism.[77] On the contrary, Darwin argued that human social evolution influences and dominates the natural selection process through deliberate choices.

5.3.C. The allegoric use of biological references

The biologist Maynard Smith suggested two possible types of analogy for social sciences. On the one hand, he considered a category of analogies based on isomorphic models, in which parallel causality relations are involved and which would eventually be developed in order to create new predictions. On the other hand, he considered another category of analogies between non-isomorphic systems and which could be heuristically useful, since it helps us 'to think about unfamiliar things',[78] even if they are not demonstrative.

His examples of different modes of causality suggest the inapplicability of evolution theories to social sciences. According to Maynard Smith, we have the following Weissmann's model of heredity (Maynard Smith, 1972: 38 f., adapted):

➔E➔E➔E➔
↓ ↓
A A
↑ ↑
➔G➔G➔G➔

N: E, environment; A, adult, phenotype; G, genotype

Figure 5.1 *Weissmann's model of evolution*

The opposed model of historical evolution would be:

standard historical model	idealistic version	economic deterministic version
➔E➔E➔E➔ ↓ ↓ ↑ ↑ ➔C➔C➔C➔	E E ↑ ↑ ➔C➔C➔	➔E➔➔E ↓ ↓ C C

Figure 5.2 *Historical models of evolution (C: habits, ideas)*

According to Maynard Smith, in none of these versions can a model of historical behaviour be compared to the Neo-Darwinian model of evolution through heredity, from the point of view of causality and general relations between the variables. Nevertheless, there is an important body of literature in economics which has been developed on the basis of the evolutionary metaphor: that inspiration is productive, justified and indeed fundamental since referred to the necessity of new models of causality, although equivocal and contradictory when attempting to uncritically transfer the specificity of those models to the social realm.

One of the persuasive arguments was that of Veblen. Veblen's defence of the evolutionary character of science was directly inspired by considerations against mechanistic causality. In particular, Veblen criticized the cause-effect chain accepted by earlier natural scientists and by classical economists as the standard definition for a law in order to 'exercize some sort of coercive surveillance over the sequence of the events' (Veblen, 1898: 378). Therefore, the difference between evolutionary and pre-evolutionary sciences is one of philosophical or methodological attitudes, 'a difference in the basis of valuation of the facts for the scientific purpose' (ibid.: 376). The teleological character of economic action and its cumulative and reciprocal forms of determination and causation suggested a genetic account of science, and 'The evolutionist point of view, therefore, leaves no place for the formulation of natural laws in terms of definitive normality, whether in economics or in any other branch of inquiry' (ibid.: 392).

This plea was intended to leave no space for the mechanical and deterministic metaphors, which looked for the analogues for maximization and 'legal atomicism' in biology and in economics, in order to legislate about the desired equilibrium properties of the models. In fact, if the analysis of the genetic organization and development of society is the main characteristic of evolutionary economics, then the sole function of the biological metaphor is a very general redescription of the determination structures, in order to define a structural-genetic methodology. It is a meta-metaphor, an allegory, whose function is to reorganize the content of the discipline against the traditional meta-metaphor from energetics, as derived from positivist scientism and universalism. Therefore it would be an useless scientific effort to try to give to the evolutionist metaphor a concrete content for the interpretation or formation of hypotheses at a lower level, since no vertical relation of causality can be translated from biology into economics. The Lamarckian metaphor for cultural evolution and the sociobiological maximization metaphor are some of the examples of such failures of efforts to legitimate new concepts in economics, which either do not have any practical or clarifying content, or are misleading.

Yet, to indicate this specific wrong use of metaphors does not imply their rejection as a pertinent procedure. Of course, this allegorical use of biological processes has been most useful: it was the primary tool for the emancipation of theories and scientific disciplines from positivism. But the power of metaphoric redescriptions should be stated in a prudent way, in order to make possible new, practical and creative conjectures. Metaphors are a tool for creative thought and should not be a mere rhetorical argument for the legitimacy of a new formalism.

In fact, the differences between the two analogues — the biological (the secondary subject) and the economic system (the primary subject) — do not permit metaphors of substitution or of comparison. These differences at the central level of the genetic and information system are obviously: (a) the existence of dominant adaptive systems in society, namely of some feedback processes which are absent in natural evolution; (b) the speed of transformation of the units of transmission of information in society, as compared to the genetic system, so that the rate of evolution of the general historical system is much slower in biology than in society; (c) the genetic systemic information is transmitted through direct sexual forms of reproduction, whereas the information in society is transmitted through multiple, indirect and simultaneous forms; (d) lineages in biological evolution do not recombine, while recombination is the normal procedure for creation of novelty in social information. Still, there are several possible epistemological comparisons, such as the basic indeterminateness of the systems, which do not allow for meaningful general predictions: in biology as in economics, prediction of future evolution implies at least the ability to fully predict mutations as well as to fix a constant environment so that adaptation may follow a single path — and those requirements are contradictory and impossible to meet, either each one or both at the same time.

As a consequence, concrete biological metaphors can be used only at a local level, where similarity and causal connections can be indicated and established as the root for the metaphorization. Some examples of a meta-metaphor of self-organization in the social as well as in the organic realms are discussed in the next sections.

5.4. Economic metaphors derived from evolutionism

More than fifty years after the appeal by Veblen, or of Marshall's prophetical suggestion of a new way to Mecca, no significant use of the biological metaphor or of evolutionist methods impressed the scientific community. One considerable achievement was Schumpeter's work on cycles and the historical interpretation of capitalism, but it was a very ambiguously stated rupture with the economic mainstream and it did not abandon its physical concept of equilibrium. The landscape of evolutionary economics has not been very crowded.

The reason for this general situation was not, of course, the lack of motivation — since the neoclassical synthesis was in great difficulties and under a severe attack from Keynes and others from the first decades of this century — but the real difficulty of the question. Some examples of evolutionist metaphors confirm these difficulties.

5.4.A. Organic totality: Marx

Simultaneously and independently from Darwin, Marx and Engels developed an evolutionary theory applied to history and, in particular, one of the first general critiques of positivist economics. This conception may be summarized by the following main features: (i) social relations (for example, capital, labour) are organic wholes; (ii) a social organism comprehends distinct coherent and significative structures, which have their own regulating features; (iii) dynamic processes of change are the basis of evolutionary history; (iv) dialectics is the method that apprehends and represents those contradictions and multiple determinations.

Both Marx and (mainly) Engels studied Darwin's writings and praised his theory, considering that it was a convincing although eventually unintended critique of British laissez-faire:[79]

> Darwin did not know what a bitter satire he wrote on mankind and specially on his country man, when he did show that free competition, the struggle for existence, which the economists celebrate as the highest historical achievement, is the normal state of the animal kingdom. (Engels, 1882: 28, my translation)

Although Engels supported a Darwin-Lamarckian version of the inheritance of acquired characteristics, he correctly summarized Darwin's thesis pointing out that inheritance was a secondary feature of evolution (ibid.: 335). On the other hand, he stressed that this theory included an essential and non-deterministic concept of an internal relation between necessity and contingency, a considerable break from the traditional antinomy established by positivism, and a concept that Darwin shared with Marxian dialectics.

The organic metaphor was again dominant in Marx's main texts. He also used in the first volume of *Capital* some other metaphors, inspired by different sciences, such as chemistry (for example, the continuity and difference of fluids and crystals) and geometry (metaphors of distance, space, area) as well as a concept of weight in order to distinguish his position from the substance-essentialist versions of the labour theory. But the biological and organic metaphors were decisive (the concepts of metabolism, metamorphosis, embodiment, incarnation, etc.), and society was described as an organism. These concepts of organic totality and evolution were a crucial part of the theory of history defined by Marx.

In the preface to the second German edition of the first volume of the *Capital*, Marx lengthily transcribed and commented on a critique just published in a daily of St. Petersburg, 'The European Messager'. The passage stressed the organic method of Marx, which is no other than the dialectical method:

> The one thing which is of moment to Marx, is to find the law of the phenomena with whose investigations he is concerned; and not only is that law of moment to him, which governs these phenomena, in so far as they have a definite form and mutual connection within a given historical period. Of still greater moment to him is the law of their variation, of their development, that is, of their transition from one form into another, from one series of connections into a different one including the 'necessity of successive determinate orders of social conditions'. ... Marx treats the social movement as a process of natural history, governed by laws not only independent of human will, consciousness and intelligence, but rather, on the contrary, determining that will, consciousness and intelligence. ... But it will be said, the general laws of economic life are one and the same, no matter whether they are applied to the present or the past. This Marx directly denies. According to him, each historical period has laws of its own ... . In a word, economic life offers us a phenomenon analogous to the history of evolution in other branches of biology. The old economists misunderstood the nature of economic laws when they likened them to the laws of physics and chemistry. A more thorough analysis of phenomena shows that social organisms differ among

> themselves as fundamentally as plants or animals. ... the scientific value of such an inquiry lies in the disclosing of the special laws that regulate the origin, existence, development, death of a given social organism and its replacement by another and higher one. And it is this value that, in point of fact, Marx's book has. (quoted in Marx, 1873: xxvii-xxix)

And Marx added immediately: 'Whilst the writer pictures what he takes to be actually my method, in this striking and (as far as concerns my own application of it) generous way, what else is he picturing but the dialectical method?' (ibid.: xxix). No other passage in his work was so explicit about this essential trend: for Marx, the concept of evolution was indistinguishable from his whole philosophical approach. In other words, a meta-method was applied as well to biology as to economics: dialectical logic.

5.4.B. The maximization or viability analysis: the metaphor of natural selection

The dominance of the mechanistic metaphors and in particular those derived from the maximization principles in energetics implied for economics the definition of a new agenda and a new set of tools adapted to those problems. The examples discussed in this section are samples from a much larger population and represent some of the attempts to refine the optimization procedures.

In 1950, Alchian suggested dealing with the problems detected in normal science by 'reverting to a Marshallian type of analysis combined with the essentials of Darwinian evolutionary natural selection' (Alchian, 1950: 213 n.). According to Alchian, the introduction of the realistic assumption of uncertainty implied the impossibility of a simple profit-maximization hypothesis for the behaviour of firms, since in that case profits were no longer unambiguously determined by the system, and the research would finally be limited to some restricted knowledge about the distribution of potential outcomes of expected profits (ibid.: 212). As the very concept of an optimum distribution of probability is ambiguous, Alchian suggested another concept in order to characterize the firm's behaviour: the 'realized positive profits' should be the criterion for viability and therefore for success, since positive profits imply relative superiority, and firms with no profits or losses would have to disappear.

In order to reinforce the Darwinist analogy, Alchian introduced chance — 'fortuitous circumstances' — as a method for achieving success, along with trial-and-error, imitation or other adaptive behaviours. In this context, mutation implied the creation of a new type of organization, and the process of selection indicated the probability of its survival, or its viability (ibid.: 220). So, the maximization hypothesis was replaced by a competitive system in which agents just tried to obtain profits: 'The suggested approach embodies the principle of biological evolution and natural selection by interpreting the economic system as an adoptive mechanism which chooses among exploratory actions generated by the adaptive pursuit of success or profits' (ibid.: 211).

In this case, the determination proceeds at two levels: (a) the environment adopts the innovative-competitive firm (positive profits); (b) the firm adapts to a competitive behaviour by (b1) chance-mutation transformations and by (b2) imitation. The problem for the biological metaphor, of course, is that (b1) and (b2) do not explain innovation: mutation is a random process, neither explainable nor explanatory for social evolution, and imitation is a strategy of follower and not genuinely creative behaviour; furthermore, their effect on the environment is unpredictable. On the other hand, we are back to Social-Darwinism, since a viability hypothesis requires that evolution increases and selects efficiency, although this is not necessarily the case in biology and it is obviously not always the case in economics. This point was taken by Penrose, who criticized the analogy stating that there is no analogue for genetic heredity, since imitation cannot fulfil the same role in social evolution (Penrose, 1952: 814); Alchian answered that the analogy was 'merely expository' and non-demonstrative (Alchian, 1953: 601).

The following table summarizes Alchian's analogy from biology:

Table 5.3 *Alchian's dictionary*

Biology	Economics
organisms	firms
genes	imitation
mutation	innovation
natural selection	positive profits

This kind of orthodox attempt to incorporate Darwinism was developed again by Friedman, on the basis of a similar dictionary in his famous 1953 essay. The traditional hypothesis of 'maximization of returns' was there considered as the consequence of a general natural selection process: 'The process of "natural selection" thus helps to validate the hypothesis — or, rather, given natural selection, the acceptance of the hypothesis can be based largely on the judgement that it summarizes appropriately the conditions for survival' (Friedman, 1953: 22). This natural selection metaphor was thus consistent with a physical and positivist meta-metaphor presented by the author: 'In short, positive economics is, or can be, an "objective" science, in precisely the same sense as any of the physical sciences' (ibid.: 4), since he added the assumption of perfect knowledge. Once more, this biological analogy is meaningless, since the metaphorization is inconsistent in the domain of the secondary subject.

Some contemporary economists go even further. Presenting inversely biology as a branch of economy — therefore the maximization principle is extended to the whole realm of nature and society — Becker (1976: 817 f.) indicate the following metaphor:

Table 5.4 *Becker's dictionary*

	Biology	**Economics**
selector	nature	preferences
factor of production	nutrients	inputs
information	genotype	technology
	organisms	agents

The maximization is of course possible since there is no uncertainty or change in this bio-economic system. Now, these examples suggest a further conclusion, since they all deal with economics as a system of interactions between independent units, adopting the atomist or reductionist approach. But biology, just like economics, denies this pure atomist approach: this is why the natural selection metaphor is not interpretable as optimization, and its economic meaning collapses. This was indeed the point made by Marshall in his *Principles*, where he stated that 'struggle for survival' may imply the dominance of those methods best fitted to prosper in the environment, but not necessarily of those best fitted to benefit the environment (Marshall, 1890: 596-7): in that case, even the welfare properties of individual maximization are implausible.

5.4.C. The genotypic evolution metaphor

The conditions for the incorporation of evolutionary analogies at a local level can then be indicated as: (a) the definition of a genetic system for the organization and transmission of the information, and also (b) the definition of the population and its forms of production and reproduction of such a system. A new series of examples is taken from economic models built on biological analogies and addressing the first of these tasks.

Boulding criticized the 1898 article by Veblen as defining some sort of 'celestial mechanism of society' under stable parameters. The criticism was unfair, but it is true that Veblen did not present a developed alternative model for evolutionary science. Boulding tried to outline such a model: 'My own definition of evolution is that it consists of ongoing ecological interaction, of populations of species of all kinds which affect each other, under conditions of constantly changing parameters' (Boulding, 1981: 23; 1991: 10). Social, physical and biological species were included under the same definition. Positive feedbacks in economic evolution — such as Smith's example of the division of labour depending on the extension of the market, and reciprocally (ibid.: 12) — were presented as a proof of the generality of the metaphor.

Boulding presented the following metaphor for economics (ibid.: 24-5, 30):

Table 5.5 Boulding's dictionary

Biology	**Society, economy**
Genotype	Knowledge, information structure
Phenotype	economic goods
Selection	ecological interaction
Mutation	change of parameters (for example, new genetic organization, evolution of fashion)
Population	stock of existing commodity
* Birth	* production
* Death	* consumption
(no analogue)	price system

In this metaphor, two different and non-interacting systems of information are considered: the genotypic and the price systems; in consequence, production is separated from the conditions of production. And, of course, neither mutation nor selection are of a Darwinian type, and no vertical comparison can be established for the analogues; instead, mutation was defined in the saltationist way, as a punctuated evolution rather than as a Darwinist process (Boulding, 1991: 13). In spite of those differences, Boulding concluded in a very confident way: 'Thus, the process by which a fertilized egg becomes a chicken is not essentially different from the process by which knowledge in the minds of the automobile company members is transformed into an automobile' (ibid.: 25: also 1976: 11).

Here is again the same difficulty as we found before for the excessively general metaphors. Indeed, for substantive reasons, the biological production of the chicken and the social production of the automobile are quite incomparable: the social artifacts are multiparental, the genetic information is not included in the offspring but in other objects (the manual of operations for constructing or driving the automobile), social subjects are permanently creative and self-generators of information and not mere gene transmitters, and dialectical conflict dominates in society, as Boulding was forced to accept (Boulding, 1981: 25-6, 34, 45).

The problem is that a firm is not an organism, in the sense that it does not organize and develop information like an organism, it does not grow and change like an organism, and its economic and social relations cannot be modelled as the submission of the organism either to some genetic determination, or to the environmental natural selection process, still less to both at the same time. In other words, this metaphor fails just because it is neither useful nor possible to indicate a genetic system for the generation and management of social information — unless in the complex framework of the global evolution of the population. The genetic characteristic of the economies, which is its information system, cannot be metaphorized as in neo-Darwinian biology in singular organisms or atomistic factors.

Nelson and Winter presented another and more influential metaphor, which goes in the same direction of the viability metaphors, but in a better way. They assume that firms are profit motivated, but that there is no possible maximization over a well-defined set of exogenously defined choice set (Nelson and Winter, 1982: 4). So, firms are considered profit-satisficing and not profit-maximizing (Winter, 1971: 245). The metaphor is organized around the idea of some structural genetic endowment of the firm, indicated as 'the process by which traits of organizations, including those traits underlying the ability to produce output and make profits, are transmitted through time' (Nelson and Winter, 1982: 9). The decision rules of the firm and its routines and knowledge are the counterpart of the genetically transmitted information (ibid.: 14). So, we have a Lamarckian process — the acquired characters are transmitted to the offspring — where routines determine the long term behaviour of the firm: traits, which are passive and phenotypical in a Darwinist account, are transformed here in the analogue for the genotype and in the active and dominant elements.

Even excluding teleological evolution from these 'Lamarckian' concepts, the analogy is still a very limited one. 'The selection mechanism here clearly is analogous to the natural selection of genotypes with differential net reproduction rates in biological evolution theory' (ibid.: 17), but the natural selection process does not operate over a phenotype exclusively determined by the genotype and its stochastic evolution, since the phenotype itself is also able to incorporate new information. So, the phenotype and the genotype have the same epistemic nature in these cases and the Darwinian inspiration is dissolved.

The authors understand the nature of the difficulty — the absence of a clearly defined relation between individuals and population — since one of them previously stated: 'Thus, while decision rules themselves are the economic counterpart of genetic inheritance, the failure-stimulated search process apparently has no analogue in biological evolution — it would correspond to a mechanism that automatically generates a burst of mutations when they are needed' (Winter, 1971: 245). This analogue cannot in fact be found at the level of the single organism.

These models are restricted to a metaphoric redescription based on the genotype evolution, which in biology is unpredictable, determined by chance mutation, and non-purposeful, characteristics quite inadequate to model the vertical causal relations in the economic world. The solution was to go back to Lamarck in order to get rid of this genetically bounded system. Here is how Nelson summarized his own metaphor from biology (Nelson, 1995: 68-9):

Table 5.6 Nelson and Winter's dictionary

Biology	Economics
fitness	profitability
genes	routines
phenotypes	firms

Besides the previous argument against the Lamarckian metaphor, these dictionaries indicate a consistent difficulty: traits, routines or other information structures are not stable, do not mutate unpredictably and are not organized as a genetic code — the metaphor ignores the secondary subject and, therefore, is a mere procedure of persuasion established in the framework of the primary subject.

Faber and Proops dealt with these problems and presented the following map for the metaphor (Faber and Proops, 1991: 69 f.):

Table 5.7 Faber and Proop's dictionary

Biology	Economics
Genotype	stock of available techniques (including technology, legal system, preferences of the agents, institutions)
Phenotype	Market system (employed techniques, social system of distribution, type of capital goods, quantities/prices)
genotypic evolution	invention
phenotypic evolution	innovation

Again, there is no clear distinction of primacy and determination between the genotype and the phenotype: each one can be considered essential and definitional, according to the point of view; from the historical account, the phenotype is in this case the primary system. The authors recognize this difficulty, since the evolution of the genotype is quicker than the transformation of the phenotype (ibid.: 76, 68, 74). But, contrary to the previous models, they indicate the interactive economic system as the unit of analysis: 'we define the unit of selection to be the entire set of interacting economic agents and their institutions and artifacts' (ibid.: 76). The coevolution of the system — conceived of as an ecological system — defines the appropriate unit of analysis in economics.

The same direction was taken by the last metaphor to be considered here, Hodgson's model, which has two very distinctive features: (a) it indicates purposeful action and intentionality as the relevant characteristics of the system of transmission of information; and (b) it takes a biological hierarchical view of selection, like Bertalanffy or Waddington, considering the coevolution of the whole system of species (Hodgson, 1993: 108, 229). Instead of a simple system of replication by chance (mutation) and necessity (natural selection), there is a hierarchy of institutions and modes of transmitting information, including the chreodic type of institutional evolution, whose cumulative character can generate adaptive crisis (ibid.: 126, 258 f). Since in economics something more than the Darwinian type of stochastic variation is involved, Hodgson argues that the essential feature is the creation of variation and innovation; in that case, a non-deterministic (or a non-Spencerian) evolution metaphor is needed to account for the irreversibility processes, the short and the long term dimensions, the quantity and quality changes, the situations of non-equilibrium and equilibrium, and the possibility of non-optimizing behaviours (ibid.: 23, 38). Veblen's metaphor is thus defended in this context (ibid.: 124):

Table 5.8 *Veblen and Hodgson's dictionary*

Biology	**Economics**
Unit of selection for variation	Institutions (replicators of information)
Principles of continuity and heredity	Habits and instincts (including technology)
Natural Selection	number of offspring of better adapted organisms, extinction of organizations

The author develops a large effort to emphasize this metaphor, against the mainstream and also against Schumpeter (who is presented as denying the evolutionary metaphor altogether) and Marx (who is presented as arguing that selection entails the withering away of variation). Avoiding some of the main pitfalls in metaphorical redescription, this type of hierarchical system is simultaneously more controversial and more explanatory than the traditional ones, even if the unit chosen, institutions, is hardly tractable in economic formal models. More explanatory, since institutions refer to those forms of organization which are specific of human societies and, therefore, that distinguish it in the space of evolution; more controversial, since it implies the abandonment of the mechanical properties, tools and world-vision which were responsible for the previous development of economics. In short, this dictionary moves beyond the strict limits of the biological inspiration.

5.4.D. Population and organisms in evolution

The balance sheet of the precise metaphors taken from biology is rather poor: for most of the cases, the metaphors amounted to extreme versions of the obliquity error, not only translating similarity into causality but also implying wrong relations of causality.

This was to be expected. The heuristic importance of evolutionary thought is predominantly related to the reconceptualization of economics as a whole and to the inclusion of the constructive dimension of time, and not to disciplinary dictionaries. This does not preclude the use of evolutionary metaphors; indeed, the differences the two fields of inquiry are substantive but so are some of the theoretical reasons to reject positivism and mechanicism in both sciences. Like the principle of maximization, the corollary of rationality — that individual maximization behaviour generates maximum social welfare — is not accurate either in biology or in economics.

That departure from the atomist perspective was already present in Darwin's work and it did become a central issue in the evolution of evolutionary thought. The Neo-Darwinist program was in fact a first answer to the puzzle, with the Central Dogma assuming the individualistic approach. But several biologists, like Waddington, argued that at the level of the population all characters are in some sense acquired, since natural selection is also a process at work from the point of view of the population as a whole, through some kind of *genetic assimilation*:

> I have found myself impelled to envisage evolution as dependent on processes which affect phenotypes and have only secondary repercussions on frequencies of genotype. This is perhaps a more profound change than the others, since it demands radical alterations in some of the most deeply ingrained biological dogmas: (1) it points out that, although an 'acquired character' developed by an individual is not inherited by its individual offspring, a character acquired by a population subject to selection will tend to be inherited by the offspring population, if it is useful; (2) it argues that genotypes, which influence behaviour, thus have an effect on the nature of the selective pressures on the phenotype to which they give rise; (3) it introduces into the theory an inescapable indeterminism quite different in nature from quantum indeterminacy, but almost as intractable, since identical phenotypes may have different genotypes, and identical genotype may give rise to different phenotypes. (Waddington, 1975: v-vi)

Then, at the level of population, there is one adaptive mechanism compatible with the process of interaction with the environment, which is natural selection: the adoption of an adaptive response to the environmental change depends on the selection of genetically controlled positive reactivity of the organisms present in the population (ibid.: 21-2, 39). It is not a Lamarckian process, since the genetic endowment of the organisms is decisive and there is no inheritance of characters individually acquired, but it is not an atomistic neo-Darwinian process either, since those organisms better prepared to respond to the environmental change will have more chances to survive and will shape the next generation of the entire population. This indicates two different processes of acquisition of characters, following Waddington:[80] from mutations, independent from the environmental stimuli, random processes at the organismic level, but also from genetic assimilation, canalized processes at the ecological level. These two processes of selection form a doubly indeterminate and combined system of evolution. In short, methodological individualism is incompatible with evolutionary biology.

Waddington suggested also that this process at the level of the population — that is, for a generation of phenotypes — is influenced by four distinct systems: the genetic system and natural selection as in the Neo-Darwinian biology, but also the exploitive (the niches of animal action and the choice of habitats) and the epigenetic systems (ibid.: 57 f.; Waddington, 1975: 15; also Jantsch, 1980: 55 f.; Mayr, 1982: 320 f.; Maynard Smith, 1993: 320 f.). The genetic assimilation, that influence of environmental activity on the process of storage and organization of genetic information of a whole population, proceeds through the *chreodic canalization of development*, that is, the development along lines fixed by the process-oriented evolution guiding ontogeny. This chreodic process explains the development of the different organs, without intermediate forms, from cells which are rather similar once in the genotype — and may be metaphorized in economics as processes of path-dependency and coevolution.

The random mutation explanation is conserved, but inserted into a general framework of hierarchical causes and effects in a network system. This is indeed a striking example of the earlier stated model of multiple and combined structural-genetic causation system. It also emphasizes the importance of the level of the population in order to understand the multiplicity of causality and to construct an explanation. As far as the introduction of the evolutionist metaphor in economics is concerned, this strongly suggests the importance of considering a larger unit of analysis than the single — abstract — representative firm, taking instead the level of the industry or the general movements of coevolutionary economic variables. It is obviously outside the scope of this book to judge about the biological interpretations and polemics just indicated; but the argument is nevertheless necessary and sufficient in order to reject the vindication of that metaphysical claim based on the comparison between the complexity of organic life and the simplistic maximization behaviour postulated for fictitious entities.

The Fable of the Bees is just a fable.

5.5. The self-organization paradigm

A paradigmatic shift is necessary in order to replace positivism. That was suggested more than half a century ago, when Smuts extended Darwinism to an 'ultimate synthetic, ordering, organising regulative activity in the universe' (Smuts, 1926: 317). This principle was holism: 'Evolution is nothing but the gradual development and stratification of a progressive series of wholes, stretching from the inorganic beginnings to the highest levels of spiritual creation' (ibid.: 1). Even ignoring the prophet's tone, this position was incompatible with the definition of a singular system of causality or of a single explanatory factor.

Furthermore, the holistic approach contradicts the Cartesian dualism and positivism, since the system is defined by the fact that significant connections among parts are always referred to the whole. Bertalanffy defined the 'fundamental problems of life' as being order, organization, wholeness and self-regulation, and presented organisms and organizations as hierarchies.[81] In spite of his rather mechanistic version — he stated the possibility of establishing general laws, exact and deducible within a theory, in the same sense as in a theory of physics (Bertalanffy, 1952: 19, 151, 171) — the holistic approach still suggested a change of metaphorical reference. In later writings, Bertalanffy developed a more general vision of self-organized systems opposed to the simplistic view of unorganized complexity. Self-organized systems are characterized by the progressive differentiation up to higher levels of complexity, are open systems and import negentropy, and are driven by feedback processes which correspond to the holistic feature of the system (Bertalanffy, 1962: 2, 5-6). This corresponds quite closely to Jantsch's and Prigogine's later syntheses.

Organicism or holism should not be defined in a trivial way — the relation of everything with everything — so that they do not become useless concepts. Alternatively, holism must be defined as the epistemic reference for self-evolution, or self-organized development, therefore compatible with the genetic analysis. In that case, the dominant feature of a holistic inquiry is the choice of the unit of analysis including the totality of parts: this is the switch of metaphors which is going on in biology, from reductionism to holism, from the notion of survival to that of autopoeisis, from inheritance to emergence. Necessity and contingency: complexity, in one word.

In that sense, the development of thermodynamics provided a new interpretation for organic and inorganic phenomena overcoming this rigid antinomy, and suggested a synthesis including both social and economic processes: 'evolution is an axiomatic consequence of organismic information and cohesion systems obeying to the second law of thermodynamics' (Brooks and Wiley, 1986: ix), although their specific differences cannot be ignored since 'On the human level irreversibility is a more fundamental concept, which for us is unseparable from the meaning of our existence' (Prigogine and Stengers, 1985: 298). In that framework, Darwinism is a model of irreversible transformations, through the active system of genotypic change and the single autocatalytic mechanism of DNA replication, the phenotype being the dissipative system of the organism.

The self-organization, self-referentiality and irreversibility of these open systems one of the building blocks of the new paradigm. Consequently, the extension of the Second Law of Thermodynamics to the domain of open systems annihilates the metaphor of equilibrium — a standard feature of closed systems — and suggests that the creation of new forms of order occurs mainly in situations out of equilibrium (Prigogine, 1989: 397). Furthermore, as Schrodinger emphasized in 1944, negentropy is created by the organism so that survival and evolution be possible. In summary, complexity is defined as the general characteristic of self-organized systems: it is a common property of those universes which exhibit nonlinear dynamics and, a fortiori, of those social and human systems that have learning capabilities. This implies unpredictability (Smale, 1980: 106) and highlights the distinctive nature of the historical processes, a new level of complexity to be added to those of quantum uncertainty, of genotypic random mutation, or of the indeterminacy of human agency.

The recent works of developmental genetics added new insights to this approach. Waddington, Stephen Jay Gould, Lewontin, Stuart Kauffman and Brian Goodwin criticized positivism and the reductionist strategy, stressing that there is an irreducible complexity in nature: the organism is part of the building of the environment, and self-organization contains natural selection, which drives evolution. In this sense, life is an emergent collective property (Kauffman, 1993: 340), and 'much of the order in organisms may not be the result of selection at all, but of the spontaneous order of self-organized systems' (Kauffman, 1995: 25). More radically, this implies that the traditional antinomy phenotype-genotype may even be partial and probably misleading: 'With autocatalytic sets, there is no separation between genotype and phenotype. The system serves as its own genome' (ibid.: 73).

This world recognizes itself as complexity.

6. Conclusion of Part One: Pertinent Impertinence

In his *Ignorant Philosopher*, Voltaire, a master of positivist rationalism, did not hide his surprise since 'it should be very singular that all nature, all the planets should obey eternal laws, and that there should be a little animal, five feet high, who in contempt of these laws could act as he pleased, solely according to his caprice' (Voltaire, in Olson, 1971: 303). In fact, the history of science in the last hundred years is simultaneously the history of the search for and also of the revolt against these eternal laws. Both attitudes were illustrated and discussed in this Part One: the positivist conformism reigned for a long time, whereas the change of paradigm is being based on the suspicion against that world of omniscience and equilibrium — and there is evidence indicating that this change is underway. Such a caprice is not recent: the Earth lost its place as the centre of the universe, the Human Being lost his uniqueness in the animal kingdom, and even the Self is found to be divisible and divided. There are indeed many more things between the Earth and the Moon than dreams in our vain philosophies ...

The positivist paradigm, inspiring Voltaire's cosmological compliance, was here described through three main characteristics: naturalism, universalism and objectivism. Its implications were traced in the history of science, from the early classical positivism and its Baconian inductive and experimental principle, to the mature positivism of logical empiricism, and to the final form of Popper's infirmationism. All these research programs were generally based on the dogmas of positivism — even if the later one challenged and abandoned the radical naturalist definition — and tried to solve the same epistemological problems: the problem of demarcation of science against metaphysics, the problem of cognition and the problem of signification. Their solutions were criticized: classical positivism was challenged by Hume's Paradox and its scepticism, and the confidence in the demarcation principle based on inductive cognition had to be abandoned; logical empiricism, although developing a new domain of signification, tried to solve the problem of induction with the introduction of a new language in the space of the explanans, and no such language could be found or unambiguously defined; and, finally, infirmationism did not make possible a complete demonstrative logic or a general applicable and useful criterion of demarcation as a necessary condition for science. In consequence, the first question of the Introduction — on the need to replace the positivist paradigm — was affirmatively answered. Science is not to be defined by its demarcation against other forms of knowledge, since those boundaries are historically variable, but instead by its progressiveness and search: science must be simultaneously a metaphysics of suspicion — and not of belief (for instance, Santos, 1989) — and a heuristic of organization.

It was then suggested a new affirmative answer to the second question: metaphor is a powerful and necessary tool for creation of new hypotheses, that is, for the introduction of semantic impertinence and redescription of the explanandum. Indeed, this process of generation of innovations is more relevant for science than the quest for a complete and demonstrative inductive logic, doomed to fail. Hume's Paradox was not solved and cannot be solved.

The successive failures in that long pilgrimage for the solution of the problem of induction revealed anyway a deeper problem: the incapacity of encapsulating reality in the framework of the mechanistic descriptions of the universe. Simplicity, independent and atomistic events and factors — or, at most, unorganized complexity — cannot account for emergent phenomena, for organized interactions, for nonlinear relations, for institutions or purposeful action. Complexity and change cannot be captured by that science dedicated to the metaphor of the clockwork in the universe. If time is to be considered, not as a simple parameter, a space coordinate or yet another variable, but as real evolution, then a paradigmatic shift is fully justified.

This new approach is based on a genetic and structural conception of causality recognizing several levels of specific causality: an example is Darwinian biology, where a natural selection process operates on non-adaptive genotypic random mutations at the level of the single organism (a closed program), and a process of genetic assimilation and ecological selection and interaction operates at the level of the populations (an open program). As a consequence, the natural selection process can in fact select anti-selective processes: this is the reversive Darwinian effect presented by Tort. In this sense, the introduction of the combined dimensions of choice and irreversibility, of contingency and necessity, flows from the vindication of historicity by the metaphor borrowed from evolutionist biology.

To several modes of determination correspond several modes of evolution and of time: just as in natural history we have the paleontological, the life cycle, the genetic or the micro-organismic times, in society we have several different modes of events and development of relations and conflicts — besides the biological one. Each of them introduces a form of causality: *strict cause-effect relations* (as in some physical or astronomic processes), *irreversible and indeterminate developments* (as in other physical and chemical processes, establishing the uniqueness of events), *functional relations* (as in some biological processes) and *intentional causal relations* (as in the majority of social processes). Several of these modes of causality can include distinct forms of determination: relation (interaction), generality (inclusion), necessity (implication), purposefulness (choice).

In particular, coevolution in time implies the existence of populations and hierarchies, of systems of interactions and specifically of these different modes of determination. In open systems, like the social or economic ones where the fundamental processes are irreversible and therefore historical, this conception strengthens the rejection of the orthodox models, which assume strictly deterministic causality in closed universes. For the new paradigm of complexity, the relevant questions are those naturally ignored by positivism: the creation of change and disequilibrium are the essential topics.

The use of biological metaphors was discussed in this context. The evident advantage of the evolutionist metaphor was stated — inspired by the Darwinian type of hierarchical and historical evolutionary models — even if it has only been successfully used as a general allegory against positivism, and not as a concrete workable metaphor. In fact, it was confronted with another allegory: the machine as the prototype of physical and social relations, implying a conservation principle and a maximization rule — the archetype of the economic interpretation which inspired the marginalist revolution and the neoclassical synthesis.

Such a mechanistic approach is rather influential, even in biology. In fact, a specific type of biological analogies, based on viability or a supposed maximization principle of natural selection, was developed in economics (Friedman, Alchian) after the same interpretation had been stated by Social Darwinism and then incorporated by Sociobiology. It was argued that there is no maximization principle in biology, 'Never say higher or lower', as Darwin warned in his Notebooks (Depew and Weber, 1996: 136): the intrinsic drive towards complexity is distinct from the Spencerian idea of unilinear progress. On the other hand, all kinds of orthogenetic versions — as Lamarck's — are inaccurate: selection works on the creation of variation as a random process at the organismic level and as a coevolutionary process at the population level. The maximization metaphors were rejected, both in the secondary and in the primary subject.

A Lamarckian metaphor for cultural or social evolution was also strongly rejected: even if one avoids the finalist trend in Lamarckian evolutionism, there is no meaningful way to metaphorize purposeful human action as the environmental pressure on the adoption of functional changes in the phenotype, changes which are then programmed into the genotype. Furthermore, genotype and phenotype do not have clear and epistemologically distinct analogues in social systems (Boulding, Faber and Proops).

Other metaphors were reviewed (Nelson and Winter, Hodgson, Veblen), which provide some useful insights for the creation of new hypotheses, such as the chreodic canalization of development for the 'technological trajectories', if those are deprived of their Newtonian deterministic content. These are local interaction metaphors, corresponding to the two conditions indicated for the validity of biological analogies: they state an analogue for the genetic system of creation, storage and evolution of information, and they indicate an unit of analysis including the hierarchical causal determinations supposedly involved. If this criterion is applied in economics, that unit is the population of firms, the industry or the economy as a whole, and as a consequence the single, abstract and Marshallian representative agent should be rejected as some useless fiction.

The introduction of metaphors is useful and necessary: it is a normal part of the process of growth of knowledge and, as shown, it is a general and permanent way of generating new hypotheses — what Peirce named abduction and Schumpeter called a pre-cognitive vision. The metaphor of self-organizing systems, dissipative and far-from-equilibrium systems is a generalization from biology: self-evolution is the analogue for mutation, self-sustained and auto-catalytic development is the analogue for heredity, and self-reference is the analogue for selection. In the framework of this meta-metaphor, *selection, creation of variety* and the *process of coordination of development* are the core concepts for evolutionary social sciences, thus replacing the physicist metaphor of equilibrium, maximization and conservation.

Finally, the third question of the Introduction — is the evolutionary metaphor an adequate tool for the reconstruction of economics? — is positively answered, although prudently. This evolutionary metaphor is useful and necessary, but it is also limited: the analysis of social systems must consider that their more distinctive characteristic in relation to biological systems is the presence of positive and controlled feedbacks (and this control is the subject of the social struggles). In other terms, self-organization and complexity in social systems are qualitatively determined by human choice: after all, the small animal whose free will so much disturbed Voltaire still has a say about its universe, be it a pertinent impertinence.

Newton's powerful synthesis was decisive for the emergence of modern science. Closed systems, deterministic, reversible and atomistic factors could be dissected, studied and encapsulated in general laws: *organized simplicity* ruled the universe. Darwin suggested another vision, the drive *from simplicity to complexity* as he put in the last phrase of *The Origin*:

> There is grandeur in this view of life, with its several powers, having been originally breathed into new forms or into one; and that, whilst this planet has gone cycling on according to the fixed law of gravity, from so simple a beginning endless forms most beautiful and most wonderful have been, and are being evolved. (Darwin, 1859: 459-60)

Newton was vindicated, but there was also something else: a new form of order, life. Boltzmann and Maxwell, shortly afterwards, suggested a new development, the first generation of indeterminism: *disorganized complexity*, namely thermodynamics as the study of the probability of arrays of molecular motion. The synthesis emerged as the Neo-Darwinian program and the Central Dogma: evolution was explained as mutation plus natural selection and that dogma lasted for generations. The current paradigmatic transition challenges that result and its epistemology, suggesting a second wave of indeterminism: evolution is the result of *organized complexity*, complexity and order. This is the theme for the next parts of this book.

*

Toulmin once suggested an analogy between the construction of physical theories and drawing maps — although complaining that no analogue exists in maps for the laws of nature (Toulmin, 1953: 105 f.). Evolutionary economics emancipates from that difficulty: its new cognitive map does not depend upon known certainties or straight roads — it is a process of search, a heuristic program for the discovery of time dimensions and changing processes, in which the scientist is himself a part of the mutation and the result is normative and purposeful rather than positive and neutral.

This process of evolutionist thought is at the very centre of the creation of a new historicist current in economics.

Notes of Part One

1 In fact, Newton's philosophical polemics against his predecessor Descartes or his contemporary Leibniz are very helpful to map the evolution of modern science. Leibniz accused him of ignorance of the divine task of science: to know the stability of universe from the permanent action of God, science should ideally be omniscience. On the contrary, Clarke, speaking for Newton, argued that God should come back to his creation only from time to time in order to correct it, and that the world was determined by general trajectories. God should be something like the watch-maker of Kepler's letter, and the field for human inquiry was completely opened. In other words, Newton considered 'forces' and Leibniz discussed the 'conservation principle' in a closed system (Prigogine, 1989: 397; Stengers and Schlanger, 1991: 144).

2 Newton's *Opticks* argued this point and, nevertheless, also his attachment to the inductive method: 'As in mathematics, so in natural philosophy, the investigation of difficult things by the method of analysis, ought ever to precede the method of composition. This analysis consists in making experiments and observations, and in drawing general conclusions from them by induction, and admitting of no objections against the conclusions but such as are taken from experiments, or other certain truths. For hypotheses are not to be regarded in experimental philosophy. And although the arguing from experiments and observations by induction be no demonstration of general conclusion, yet it is the best way of arguing which the nature of things admits of, and may be looked upon as so much the stronger, by how much the induction is more general. And if no

exception occur from phenomena, the conclusion may be pronounced generally. But if at any time afterwards any exception shall occur from experiments, it may then begin to be pronounced with such exceptions as occur. By this way of analysis we may proceed from compounds to ingredients, and from motions to forces producing them; and, in general, from effects to their causes, and from particular causes to more general ones, the argument end in the most general. This is the method of analysis: and the synthesis consists in assuming the causes discovered, and established as principles, and by them explaining the phenomena proceeding from them, and proving the explanations' (Newton, 1704: 404-5). Even under such limitations, Newton was confident enough to establish general trajectories in the universe, whose basic characteristics are lawfulness, determinism and reversibility (Prigogine and Stengers, 1985: 60).

3 The same can be said about the following generations of positivist scientists. Kelvin stated that 'I never satisfy myself until I can make a mechanical model of a thing' (quoted in Meyerson, 1908: 92). Kelvin's confidence in positivist science was so large that he suggested to close down the Patent Office, since all major discoveries would be already registered.

4 In a letter to Coates (1713), Newton presented a new meaning of hypotheses which was not covered by his rejection of the concept: the laws of motion as the conclusions derived from induction, in order to explain phenomena, or 'the difficulty you mention which lies in these words, "since every attraction is mutual", is removed by considering that, as in geometry, the word "hypothesis" is not taken in so large a sense as to include the axioms and postulates; so, in experimental philosophy, it is not taken in so large a sense as to include the first principles or axioms, which I call the laws of motion. These principles are deduced from phenomena and made general by induction, which is the highest evidence that a proposition can have in this philosophy. And the word "hypothesis" is here used by me to signify such a proposition as is not a phenomenon nor deduced from any phenomenon, but assumed or supposed — without any experimental proof' (Newton, 1713, in 1953: 6).

5 Campbell presented a case of the inductive proof's failure: until the discovery of Australia, all the known swans were white, and any new inductive evidence would verify the statement 'all swans are white'. But some Australian swans were found to be black: 'These examples seem to prove that a large number of favourable instances, even if without exceptions, is neither sufficient nor necessary condition to establish a law' (Campbell, 1921: 63).

6 'No object ever discovers, by the qualities which appear to the senses, either the causes which produced it, or the effects which will arise from it; nor can our reason, unassisted by experience, ever draw any inference concerning real existence and matter of fact' (Hume, 1748: 27-8).

7 In the first pages of his *Discours de la Méthode*, Descartes stated that the book presented a 'fable' with certain examples worthy of imitation' (Descartes, 1637: 5). In other words, it is a metaphor. In spite of that assumption of the rhetoric character of the book, the author decided to publish it anonymously, since he feared the reaction of the Church against that metaphor.

8 'All science is certain, evident knowledge. We reject all knowledge which is merely probable and judge that only those things should be believed which are perfectly known and about which there can be no doubts' (Descartes, quoted in Capra, 1982: 42). The primacy of thought over existence meant in fact the elevation of human reason to the supreme position.

9 'I suppose that the [man's] body is nothing but a statue or a machine made of earth' (Descartes, quoted in Meyerson, 1908: 63). In the same direction, Hobbes presented the human body as an 'engine' moved by 'appetite' and 'aversion', pleasure and pain (Katouzian, 1980: 19); the echo of such a conception in marginalist economics is very obvious.

10 See the example of evolutionist Biology, as Bertalanffy put it: 'A lover of paradox could easily say that the main objection to selection theory is that it cannot be disproved. With a good theory, it must be possible to indicate an experimentum crucis, the negative

outcome of which would refute it. ... In the case of selection theory, however, it appears impossible to indicate any biological phenomenon that would plainly refute it' (Bertalanffy, 1952: 89; also Toulmin, 1982: 145). Much after Bertalanffy's text, a 'lover of paradox', Popper, criticized Darwinism as being not an 'empirical theory' and a mere 'logical truism' (Popper, 1972: 69). Furthermore, Darwinism was presented as a 'metaphysical research program', since it was not testable (Popper, 1974: 167, 168). The impossibility of generating predictions implied its failure as an explanation: 'it does not really predict the evolution of variety. It therefore cannot really *explain* it' (ibid.: 171, original italic).

11 In other words, we can never have the desired situation where $\{H \longrightarrow M \longrightarrow P; \sim P\} \longrightarrow \sim H$, but we have instead the undesired situation $\{H \longrightarrow M_o \longrightarrow P; \sim P\} \longrightarrow \sim M_o$, where H is the set of hypotheses to be tested, $\sim H$ is the negation of H, M the model, P the prediction, $A_{j,p}$ the auxiliary conditions, and $M_o = H_i.A_j$ is the combination of a particular set of hypotheses with some auxiliary conditions, the other notation being as usual. Even when $\sim M_o$, some other $M_n = H_i.A_p$ can be true.

12 But, to the distress of Blaug, Darwin consistently refused to make predictions.

13 Darwin's Notebooks show that he read Smith's *The Theory of Moral Sentiments* in 1838-1839; at the same time, he summarized *The Wealth of Nations*. The emergence of order from chaotic interactions and of diversity rather than disharmony from the social complexity certainly influenced Darwin's vision. Of course, the Malthusian influence was in the opposite sense, since he suggested that potential disorder could arise from social complexity. 'Population thinking' can be interpreted following these two opposite directions, as in fact it happened afterwards (Spencer, Hayek and others suggesting evolution to be the creation of some 'spontaneous order', while Malthus and others suggesting a 'spontaneous disorder'). In the same sense of Smith, the influence of Babbage, who attended the same intellectual circles as Darwin and studied the effects of specialization and division of labour, should not be ignored in this first biology-economics metaphoric interaction.

14 Until Darwin, this analogy was used in the opposite direction, denying natural selection. This was the case of Leyell (Ruse, 1982: 46).

15 In spite of this suggestions by Bunge, in this book the common terminology is followed: all forms of causality are included in a general category of determination, and in particular the 'deterministic causality' is defined according to the standard positivist model, which is simply called 'causality' by Bunge.

16 Heisenberg: 'We cannot completely objectify the result of an observation, we cannot describe what "happens" between [one] observation and the next ... , any statement about what "actually happened" is a statement in terms of the classical concepts and — because of the thermodynamic and of the uncertainty relations — by its very nature incomplete with respect of the details of the atomic events involved' (quoted in Lukacs, 1971: 294).

17 In spite of the general reference by the pragmatic authors, this was not Peirce's position. For Peirce: 'There are Real Things, whose characters are entirely independent of our opinion about them; these Reals affect our senses according to regular laws and though our sensations are as different as are our relations to the objects, yet, by taking advantage of the laws of perception, we can ascertain by reasoning how things really and truly are' (Peirce, quoted in Ayer, 1968: 30; also Apel, 1981: 191).

18 'The mere circumstance of the creation of one product immediately opens a vent for other products' (Say, 1803, I: 167).

19 Rationalism defined in the restricted version of neoclassical economics has a self-confirmation virtue: 'Its highest aspiration is to create a "well-rounded" system, or "model", whose qualities will be symmetry, internal consistency, simplicity, economy of axioms, and "elegance". Forced by its epistemology to contemplate itself, the rationalistic mind acquires a narcissistic quality' (Mini, 1974: 63).

20 This concept of additivity of causes was sometimes defended by Mill for whatever complex situation: 'however complex the phenomena, all their consequences and coexistences result from the laws of their separate elements. The effect produced, in

social phenomena, by any complex set of circumstances, amounts precisely to the sum of the effects of the circumstances taken singly' (Mill, CW, VIII: 895).

21 Mill certainly shared some of Hume's ambiguity about the critique of induction, since it remained for him the ultimate source of knowledge, although not establishing a demonstrative logic. In his last work, the 1873 *Autobiography*, Mill presented his early work on method, the *System of Logic* (1844) as a critique of Whewell and the 'German doctrines' of the a priori method (Mill, 1873: 214) and a defence of induction, all knowledge emanating from experience. But, as proved in the current text, his position was rather more sophisticated and included those conditions for which only deductive analysis could overcome the inseparability of effects.

22 The text was published in 1890, the same year as Marshall's *Principles*. These dates indicate the late epistemological development of economics: the main reflections about its method were published when the neoclassical synthesis was already on its way. This is a rather triking fact since economics claims to be a science thanks to its method, and not to the demarcation of its object, as sciences generally do. Schumpeter considered Neville Keynes's text to be 'one of the best methodologies for economic sciences of all times' and the 'best guide' (Schumpeter, 1990: 356)

23 The 'praxeologic' aspect of Robbins's or the Austrians' definition and approach are not discussed in this book since they were commented on elsewhere (Louçã, 1987). For the general discussion about the epistemological shigts, see Louçã (1994).

24 Ward supports this view of the 'Keynesian revolution' as a Kuhnian change of paradigm after the verification of severe anomalies in the 'normal science' of the 30s (Ward, 1972: 34). He is certainly correct, but the case also shows the coexistence of several paradigms and even the possibility of their recombination, since the neoclassical synthesis partially reintegrated Keynesianism. That is not conceivable in Kuhn's terms.

25 Nevertheless and unlike Popper, Robbins did not present falsifiability as the single criterion for demarcation (Hutchison, 1992: 74).

26 Friedman never quoted or referred to Popper in his 1953 essay, even if it was clearly influenced by the infirmationist debate in epistemology. But his trivial 'Popperianism' was not faithful to the original, and Blaug rightly and bitterly complained about the effect of this version among economists.

27 'To speak of biological or social evolution in a non evolutionary environment is a contradiction in terms. Ceteris paribus — the indispensable ingredient of every physical law — is a poison to any science concerned with evolutionary phenomena' (Georgescu-Roegen, 1971: 203).

28 Mirowski's arguments provoked a lively debate. For the arguments against this account of the marginalist revolution, see Blaug (1992), Walker (1991) and Hoover (1991). Blaug argues that the historical evidence is wrong, but does not present any case. Walker argues that the historical description does not constitute evidence against the actual program of neoclassical economy.

29 McKenzie argued that the inspiration for Walras's incorporation of the energetic metaphor was a *Treatise on Statics* by Poinsot (McKenzie, 1989: 3). Walras was educated as an engineer and Jevons as a physicist before coming into economics, and both were certainly aware of the most important analytical tools of the profession by that time. Although some mainstream historians and epistemologists, like Blaug, condemned this argument about the origins of the marginalist revolution, the point is today generally accepted (also Keynes, for instance in vol. X: 262; Shackle, 1968: 248). See for instance the editorial of the first issue of *Econometrica*, where Frisch stated that the purpose of the new discipline was 'to promote studies that aim at a unification of the theoretical qualitative and then the empirical-quantitative approach to economic problems and that are penetrated by constructive and rigorous thinking similar to that which has come to dominate in the natural sciences' (Frisch, 1933: 1). Arrow and Intriligator explicitly described the early period of the neoclassical program as defined by the physical metaphor: 'The early period 1838-1947 of mathematical economics was one in which economics borrowed methodologies from the physical sciences and related mathematics to develop a formal theory based largely on calculus' (Arrow and Intriligator, 1981: 1).

30 Recently, some major supporters of the neoclassical synthesis have sharply criticized the mathematical and abstract drive of theories and hypotheses without any realism. In 1966 and then in his presidential address to the AEA, Leontief sharply criticized that 'theoretical structure based on so scarce and superficial fact foundations' (Leontief, 1966: 54). As Samuelson puts it: 'There is really nothing more pathetic than to have an economist or a retired engineer try to force analogies between the concepts of physics and the concepts of economics. How many dreary papers have I had to referee in which the author is looking for something which corresponds to entropy or one or another form of energy. Nonsensical laws, such as the law of conservation of purchasing power, represent spurious social science imitations of the important physical laws of the conservation of energy' (Samuelson, 1972: 254). Given that the author was one of the founding fathers of the modern synthesis in economics and that he extensively developed the notion of maximization under constraints, one should consider the importance of the statement. But, of course, the origin of this drive is at the very foundations of neoclassical economics.

31 This is the case of Saviotti and Metcalfe (1991: 2). But Clark and Juma (1987: 49 f.) and Witt (1993: xiii) present the opposite view, which is taken here: Marshall did not use the energetic metaphor (Mirowski, 1988b: 18), but his reference to the biological metaphor was inconsequent.

32 1898 was also the year of the publication of Veblen's article 'Why is Economics not an Evolutionary Science?'; the coincidences highlights the fact that an alternative version of the role of both metaphors was possible and indeed it was argued for.

33 Mirowski argues that this article was intended to answer to a strong pressure from biologists against economics. In fact, following Whewell, Galton argued in 1877 for ousting Section F (Statistics and Political Economy) of the British Association for the Advancement of Science, as these scientists did not proceed in a 'scientific manner' (Mirowski, 1989b: 262). Marshall's article should therefore be considered a piece in a legitimization strategy. Even if the intention was true, the difference of dates (1877 to 1898) and the posterior persistence of Marshall in the same theme, indicates that it was a genuine puzzle for the author.

34 Hicks argued that this traditional concept of exogenous cause is a limitation to economics: 'It must indeed be conceded that the abundance of exogenous elements in economics is no cause for congratulation; it is an indication of the modesty of its scientific status, if indeed it is a scientific status, which is all that economics can hope to achieve' (Hicks, 1979: 22).

35 Robinson argued that economics must take important insights from other sciences and even from metaphysics: 'Metaphysical propositions also provide a quarry from which hypotheses can be drawn. They do not belong to the realm of science and yet they are necessary to it. Without them we would not know what is it that we want to know' (Robinson, 1962: 3). She was right. Marshall also stressed the role of common sense: 'economists must be greedy of facts, but facts by themselves teach nothing. ... Economic science is but the working of common sense aided by appliances of organized analysis and general reasoning, and drawing inferences from particular facts' (Marshall, 1890: 32). In the same direction, also Santos (1987, 1989).

36 This means, as it will be argued later on, that these metaphors have a very limited application in science, since they consist in literal interpretations across the primary and secondary subjects. Peirce, who studied metaphor, reduced it to this single level, since he included metaphor in the theory of signs, as an icon, 'a sign which refers to the object that it denotes merely by virtue of the characters of its own', that is, only using similarity. Even more metaphorically, he defined the metaphorical function as 'to represent the representative character of a representamen by representing a parallelism in something else' (Peirce, in Ayer, 1968: 100-1). That type of metaphor is part of literature but not an essential part of innovation in science.

37 Later on, Wittgenstein changed his mind and argued that words were related to use and not unambiguously related to facts or objects, as in the *Tratactus*.

38 For Quine, the use of metaphor produces a constant linguistic growth in human knowledge: 'If the crux of metaphor is creative extension through analogy, then we have forged a

metaphor at each succeeding application of the early word or phrase' — an argument for the metaphorical character of Marquez's process of denotation. And in a specifically metaphorical statement, Quine asserts the scientific relevance of the process for discovery: 'The neatly worked inner stretches of science are an open space in the tropical jungle, created by clearing tropes away' (Quine, 1979: 160).

39 'I take metaphor to be essentially a higher-level version of the process by which ostension enters into the establishment of reference ['epistemic access'] in natural-kind terms' (Kuhn, 1981: 413-4). See the extremely successful example of ostension, which is provided by the naming of 'chaos' theory, which became by itself a vehicle of diffusion of the idea.

40 But it maintains an empirical basis, since cognition and verification is still realist: 'In this account, theoretical concepts are introduced by analogy with the observational concepts constituting the natural descriptive language. Scientific language is therefore seen as a dynamic system which constantly grows by metaphorical extension of natural language, and which also changes with changing theory and with reinterpretation of some of the concepts of the natural language itself. In this way an empirical basis of science is maintained, without the inflexibility of earlier views, where the observation language was assumed to be given and to be stable' (Hesse, 1974: 4-5).

41 Adam Smith also considered all discursive relations as the realm of rhetoric, and included economics in rhetoric in his work, *Lectures on Rhetoric and Belles Lettres*, written in 1748-9.

42 In fact, Mendel 'started his experiments with a novel and ingenious abstraction. He broke with the conceptions prevailing up to then' (Bertalanffy, 1952: 69). His whole work represented a successful example of abduction, more than of scientific rigour.

43 'This example [the discovery of the orbit of Mars by Kepler] is instructive in two different ways. It shows that the inductivist model of scientific theories as mere generalizations of observed facts may be inadequate even with respect to theories which remain at the observational level: at the very best, the observations have to be selected: we need some hypotheses to tell us what to observe and under what conditions. But it also shows that if we take the view that the testing of theories consists in an endeavour to refute them, we must remember that the refutation of a theory is not an end in itself, but rather a means to obtain a better theory' (Apel, 1981: 88-9).

44 Kondratiev also acknowledged the role of metaphor from one science to another: 'To transport a concept from one science to another is not arguable if it is fruitful. If it is fruitful, it is justified; there are no other, and there can be no other criteria to decide about that question' (Kondratiev, 1992: 14).

45 In his *Dioptrique*, Descartes compared the change of direction of a light ray coming from air and entering water with the path of a ball passing from one physical medium to a more resistant one (Duhem, 1906: 34). Kant, for instance, criticized Descartes' use of metaphor, when comparing beasts to machines (Kant, 1790: 399 n.).

46 'We can indeed think one of two dissimilar things, even in the very point of their dissimilarity, in accordance with the analogy of the other; but we cannot, from that wherein they are dissimilar, conclude from the one to the other by analogy' (Kant, 1790: 399-400).

47 Pearson, 1892: 21, 37.

48 'The appeal to analogy forms in many cases a valuable means of investigation or test, but it is well not to exaggerate its power; if at this point the words 'proof by analogy' are uttered, it is well to determine their meaning exactly and not to confuse such a proof with a genuine logical demonstration. An analogy is felt rather than concluded' (Duhem, 1906: 301-2).

49 Poincaré, 1903: 142.

50 Bertalanffy, a (non positivist) biologist, separated analogies — 'superficial similarities in phenomena that correspond neither in the factors operating nor in the laws applying to them' — and explanation (Bertalanffy, 1952: 200). He did not follow Darwin in this respect.

51 Analogies 'may give us new ideas, although they cannot in the nature of the things

prove that these ideas are correct' (Maynard Smith, 1972: 38; also 45).

52 Analogy is presented as an 'index' to enlighten reason, but not as a valid proof. Its intervention is at the moment of discovery, afterwards being an obstacle to science (Gervet, 1985: 83).

53 The author, Mary Hesse, previously presented metaphor as a degeneration of an analogy (Hesse, 1954: 144-5), before coming to the conclusion of the importance of the process of metaphoric redescription of the explanandum, which includes analogy. Her work was one of the crucial early contributions to the study of metaphors in science.

54 That is McCloskey's position: divided according to the criteria of 'explicitness' and 'extent', all figures of speech, simile, metaphor, allegory, symbol, are included in analogy (McCloskey, 1983: 505). It is the exactly opposite procedure which has been followed here.

55 'If a relation R is reflexive and has the property O in a class K, we say that K and R together form a model or a realization of the axiom system of our theory or, simply, that they satisfy the axioms. Our model of the axiom system is formed, for instance, by the class of the segments and the relation of congruence, that is, the things denoted by the primitive terms; of course, this model also satisfies all the theorems deduced from the axioms ... . However, this particular model does not play any privileged role in the construction of the theory. On the contrary, on the basis of universal logical laws like I' and II' we arrive at the general conclusion that any model of the axiom system satisfies all theorems deduced from these axioms. In view of this fact, a model of the axiom system of our theory is also referred to as a model of the theory itself' (Tarski, 1965: 123).

56 Of course, models are not necessarily realist. But only scientific interpretations of reality are here considered, having some empirical properties: 'Scientific models are a prototype, philosophically speaking, for imaginative creations or schemes based on natural language and experience, but they go beyond it by metaphorical extension to construct symbolic worlds that may or may not adequately represent certain aspects of the empirical world' (Arbib and Hesse, 1986: 160-1).

57 'Analogical arguments from models have been formulated in terms of characters which are independently observable but not experimentally separable. In connection with the justification of analogical argument, a distinction has been made between the *logical* problem of justifying inferences from similarities and the *causal* problem of deciding whether the type of vertical relation implied in the analogy is acceptable as causal for either or both of the analogues' (Hesse, ibid.: 100, her emphasis).

58 'The function of analogical models is not directly cognitive but essentially heuristic: it must ... enlarge the space of the tractable' (Stengers and Schlanger, 1991: 86).

59 Morgan proves that Jevons's theory of sunspots causing the business cycles was based on the similarity of the length of both types of cycle (Morgan, 1990: 24-5): it is an example of the obliquity error. Juglar developed simultaneously his own theory of the cycle, based on an organic metaphor.

60 And, consequently, for the analogy itself: 'Analogy (in a qualitative signification) is the identity of the relation between reasons and consequences (causes and effects), so far as it is to be found, notwithstanding the specific difference of the things or those properties in them which contain the reason for like consequences (i.e., considered apart from this relation)' (Kant, 1790: 399 n.).

61 Malthus's influence was nevertheless contested. Schumpeter denied it, and so did Hayek. But Schumpeter did not know Darwin's Notebooks, where he presented the reading of Malthus as an illuminating event, and Hayek did (Jones, 1989: 418; Hodgson, 1993: 55, 63, 229n). Of course, the argument about Malthus's influence was also explicit in *The Origin*.

62 This was also Lamarck's position, and it was a general idea of the Enlightenment period.

63 'The "Origin of Species" of Darwin merely extends politico-economic views of progress to the entire realm of animal and vegetable life' (Peirce, 1934: 196).

64 'Darwin was led by the struggle for life in English society — the competition of all with all, *belum omnium contra omnes* [Hobbes] — to discover competition to [word missing]

as the ruling law of "bestial" and vegetative life. The Darwinists, conversely, consider this a conclusive reason for human society never to emancipate itself from its bestiality' (Marx, letter to the Lafargues, 1869, in Marx, Engels, 1988, CW, v.43: 216-7; also a previous 1862 letter to Engels, v.41: 381). The argument was wrongly taken by some Marxists to imply the rejection of the theory of evolution. In fact, Marx read Darwin and considered *The Origin* as a new departure for science and for his own thought: 'this is the book which, in the field of natural history, provides the basis for our views', and 'Darwin's work is most important and suits my purpose in that it provides a basis in natural science for the historical class struggle. ... Despite all shortcomings, it is here that, for the first time, 'teleology' in natural science is not only delt a mortal blow but its rational meaning is empirically explained' (Marx, letters to Engels, 1860, and Lassalle, 1861; ibid., v.41: 232, 245). The correspondence by Marx to Darwin (Colp, 1982: 461 f.), Engels's preface to the *Communist Manifesto* and, overall, Engels's speech at the funeral of Marx indicated this long term debt to Darwin. The dogma of the incompatibility of evolutionism or genetics with the dialectical method, which led to the scientific fraud of Lyssenkism, was produced much later.

65 Darwin was very hostile to Spencer, as he showed in his *Autobiography*, and not only for personal reasons (he was 'extremely egoistical', Darwin, 1876: 108), but mostly for scientific mistrust: 'His deductive manner of treating every subject is wholly opposed to my frame of mind. His conclusions never convince me ... . His fundamental generalizations (which have been compared in importance by some persons with Newton's laws!) — which I daresay may be very valuable under a philosophical point of view, are of such a nature that they do not seem to me to be of any strictly scientific use. They partake more of the nature of definitions than of laws of nature. They do not aid one in predicting what will happen in any particular case. Anyhow they have not been of any use to me' (ibid.: 109).

66 This was indeed a very common metaphor for the early accounts of evolution. Schumpeter addressed a violent attack against trivial evolutionism based on the analogy of the tree in an implicit critique of the Marshallian school, and of Keynes in particular (see Part Three).

67 Rousseau wrote in his entry on 'Political Economy' for the *Encyclopédie*: 'public finances are the blood that a wise economy, working as the heart, distributes through the whole body as food and life' (Rousseau, 1755: 36).

68 Stengers and Schlanger interpret Darwin as a 'dramatic discontinuist' since he stressed the role of mutations, as opposed to Lamarck (Stengersm and Schlanger, 1991: 112). This is certainly not accurate. Darwin explicitly used the 'natura non facit saltum' motto, and his theory corresponded quite well to it: that 'canon' which 'every fresh addition to our knowledge tends to make more strictly correct' was the basis of Darwin's natural history (Darwin, 1859: 445).

69 'Any variation which is not inherited is unimportant for us. But the number and diversity of inheritable deviations of structure, both those of slight and those of considerable physiological importance, is endless' (Darwin, 1859: 75, also 72). He presented several examples: the wings of domesticated and savage ducks, the Baltic shells, some dwarfed plants in the Alps, the fur of northern animals (ibid.: 74, 101). The same was applied to savage animals (ibid.: 130, 135, 175; 1871: 928).

70 Darwin, 1859: 74.

71 Darwin, 1871: 49-50, 74.

72 'I had two distinct objects in view: firstly, to show that species had not been separately created, and secondly, that natural selection had been the chief agent of change, though largely aided by the inherited effects of habit, and slightly by the direct action of the surrounding conditions' (Darwin, 1871: 92).

73 The general rule of the genetic character of the transmission of the characters does not exclude all other forms of transmission of information independent of nuclei acids. Maynard Smith presents some examples of what geneticists call a 'cytoplasmatic inheritance', implying the replication of changes in an organism being transmitted to the offspring: the developmental flexibility of the locust female is partially

inheritable, and some experiments on the slipper animalcule Paramecium establish forms of inheritance of acquired characters independent of nuclei acids (Maynard Smith, 1993: 80 f.).

74 Spencer argued for instance against the British system of poor relief in the name of Darwinism (Nasmyth, 1971: 136). Many followed in his foot steps.

75 Darwin explicitly accepted Galton's views about the existence of 'inferior members' of a society (Darwin, 1871: 929, 945) but did not adopt his political conclusions: he only suggested the voluntary restriction of marriage for those 'inferior members' (ibid.: 205-206) and the incentive to the 'best' to leave the largest number of offspring (ibid.: 945). Darwin accepted a policy for social protection, in spite of the selective disadvantages which he presented in a rather unpleasant way: 'Thus the weak members of civilised societies propagate their kind. No one who has attended to the breeding of domestic animals will doubt that this must be highly injurious to the race of man' (ibid.: 206).

76 As Wilson presents it: 'It is part of the conventional wisdom that virtually all cultural variation is phenotypic rather than genetic in origin. This view has gained support from the ease with which certain aspects of culture can be altered in the space of a simple generation ... . The extreme orthodox view of environmentalism goes further, holding that in effect there is no genetic variance in the transmission of culture. In other words, the capacity for culture is transmitted by a single human genotype. Dobzhansky stated this hypothesis as follows: "Culture is not inherited through genes, it is acquired by learning from other human beings ... . In a sense, human genes have surrended their primacy in human evolution to an entirely new, non biological or superorganic agent, culture." ... The very opposite could be true' (Wilson, 1975: 550).

77 This extreme genetic determinism denies historical reality. Maynard Smith (1972: 36) presents the case of the Arabic civilization which, a thousand years ago, would make us suppose — according to sociobiological standards — that the Arabs were better fitted to science than Europeans. Since the decay of this civilization — for whatever ecological reasons, or social conflicts and historical defeats — one can now argue just the other way round, for the fitness of the Europeans. But how could genes change so dramatically in so short a period from the point of view of genetic history?

78 'Now, the use of analogies is this way is widespread in biology. It is not necessary that two systems be isomorphic, or that an exact mathematical description be given of either of them, provided that they have something in common in their behaviour. Examples of such analogies are the "psycho hydraulic" model of the brain evolved by Lorenz, or the comparison by Waddington of a developing organism with a ball rolling down a "genetic landscape"' (Maynard Smith, 1972: 37-8).

79 Later on, Keynes shared the same interpretation: the Darwinian survival of the fittest is 'a vast generalization of the Ricardian economics' (Keynes, IX: 276).

80 'I have suggested that evolution operates not so much by the differential reproduction of chance variations of the phenotype, but by selection of genotypes which endow their possessors with the capacity to react adaptively with their surroundings' (Waddington, 1975: 36).

81 'A living organism is a hierarchical order of open systems which maintain itself in the exchange of components by virtue of its system conditions' (Bertalanffy, 1952: 129).

PART TWO

The Rocking Horse

It is demonstrable, said Master Pangloss, that things cannot be otherwise than they are; for as all things have been created for some end, they must necessarily be created for the best end. Observe, for instance, the nose is formed to bear spectacles, therefore we wear spectacles. The legs are visibly designed for stockings, accordingly we wear stockings .... It is not enough therefore to say that everything is right, we should say every thing is in the best state it probably could be.

Voltaire, 1759: 230

7. Introduction to Part Two: Simplicity from Complexity

Let's go back for a moment to turbulence. The most influential theories in economic literature, the canons of orthodoxy, selected a particular problem in social reality — equilibrium, the intrinsic drive to the changeless state of rest — for ideological but also for very practical reasons: it provided a justification for the emerging market relation and simultaneously it allowed for quantification and computation, identified as legitimate knowledge corresponding to the positivist criterion. In the Cartesian tradition, intelligibility requires immutability: like the first theories of hydrodynamics, neoclassical economics just denied the existence of turbulence.

Walras argued clearly for such a solution, with his powerful metaphor of the lake: the market permanently tends to equilibrium, which is permanently disturbed like the level of the lake is disturbed by the wind, 'the water always looking for equilibrium, without ever attaining it. ... Just as the lake is in some days deeply agitated by tempests, so is the market violently agitated by crises, which are sudden and general perturbations of equilibrium' (Walras, 1883: 207-8, my translation). This metaphor inspired a transition from the typical mechanical world of causal regularities — the metaphor of the clock, dominant in the seventeenth and eighteenth centuries — to the next generation of metaphors, since irregularities were conceived of under the condition of being generated by external events impinging on the stabilizing mechanism.

The 'normal' state of affairs or the natural form of evolution were still identified with the tendency of the system towards a rest point, but cycles may enter the picture through 'winds' and 'tempests'. Indeed, the pervasiveness and the impact of the business cycles in industrial capitalism transformed this problem into a major question for the emerging science of economics.

This Part Two deals with the building blocks of the orthodox analyses of cycles and growth, which dominated the profession for more than one century and inspired new techniques and models which are influential until our days. Like in hydrodynamics, this transition supposed the description of cycles as independent modes of oscillation created by factors alien to the natural equilibration of the systems. The rocking horse — the most impressive of the late mechanical metaphors — provided the epitome, the model and the procedures for that explanation.

The next pages will deal with the cyclical social processes as indicated by the fluctuations of macro variables and consequently develop the previous methodological insights in the more concrete field of analysis of social and historical reality. The next chapter presents some of the available taxonomies of business cycles theories, and discusses their common paradigmatic foundation, that organizing metaphor of the rocking horse — with its exogenous impulses and the endogenous propagation system which is supposed to represent the real economy —, the cornerstone of the Frischian paradigm for cycles. This specification was essential to make viable the econometric revolution and to the successful incorporation of the concept of randomness into mainstream economics: as in fluid dynamics, equilibrium was conceived in the first generation of theories as a state of rest, while this second generation added energy under the form of random and independent movements of particles whose correlation, again, created order. This was a quite different approach and, yet, equilibrium was maintained as the centre of gravitation of the theory. The following chapter develops these concepts in the context of three central problems for the theory of fluctuations: the concept of equilibrium, the definition of the nature of the explanatory variables, and the articulation of the different modes and times of fluctuation (trend and cycle). Mitchell, Keynes and Kaldor provided some of the most powerful attacks on the orthodox concepts of equilibrium and certainty, and their critiques are reviewed. In particular, it is proved that the rocking-horse metaphor is indeed a particular case of the mechanical analogies which defined economics since the marginalist revolution. Frisch, Samuelson, Hicks and so many of the most influential authors of this century understood the crucial importance of that physical metaphor, and said so: the notions of statics, dynamics, stability and a fortiori equilibrium cannot be understood if that origin is ignored.

Since Walras, the general equilibrium program has been defined by the simultaneous requirements of existence, uniqueness and global stability, making possible optimality and comparison between different equilibria. But the stability condition, in particular, presents major difficulties that the ingenious concept of the *tatônnement* process was unable to cope with. This dead end required a substantial change: since the 1930s Hicks and Samuelson redefined the mathematical formalism of the program, and Debreu chose to neglect the stability condition, while concluding with Sonnenschein and Mantel that those requirements could not be simultaneously met in the traditional Walrasian model, since the income effect implies undesirable forms of the supply and demand curves (Weintraub and Mirowski, 1994: 267). Of course, the whole method depends crucially on the preliminary axioms, which are supposed to represent 'pretty secure empirical knowledge', as Hahn put it (Hahn, 1984: 7).

Even so, the concept of time, since it introduces uncertainty, could not be integrated into the model except in a paradoxical way — and the same Hahn criticized the Arrow-Debreu model for collapsing the future into the present (ibid.: 81). Time, uncertainty and evolution will be discussed in this part.

Chapter 10 synthesizes these findings and presents criteria for a new classification of theories. Chapter 11 proceeds then to a survey of some business cycles theories, indicating some implications from early growth theories. This is not an extensive survey, but a rather precisely motivated and narrow description of the methodological procedures and the theoretical problems that are here considered relevant for an evolutionary account of the movement of the economies. Chapter 11, in conclusion, argues for a complexity and nonlinearity approach integrating historical interactions, non-quantifiable and hierarchical explanations of real life economics.

Part One treated the epistemological conditions for an evolutionist analysis and economic theory. Part Two will take now the theoretical conditions for the viability of that program.

8. The Rocking Horse: The Propagation-Impulse Metaphor for Business Cycle Theories

The problem of fluctuations cum irregularities in economic development is one of the important enigmas for contemporaneous economic science. For the early classics the explanation was evident, and oscillations were even considered as a natural consequence of the permanent conflict about the distribution of income in the society. In that sense, equilibrium was either a mere approximation or else the dramatic outcome of the degradation of the system as in the Ricardian stationary state. The abandonment and replacement of such conceptions was a decisive shift in the evolution of economic theories. That story is the theme of this chapter.

8.1. The evolution of the debate and classifications of cycles

Four major phases can be identified in the debate about growth and fluctuations, indicating the evolution of the normal-science type of consensus and classifications: a first phase, from the origin of modern economics until half of the nineteenth century, with a debate organized around Say's Law; a second one, until the end of that century, with the introduction of the first coherent heresies and the marginalist revolution; a third one, until the Second World War, witnessed major challenges to the equilibrium model, with the devastating critiques by Mitchell and Keynes; and, finally, the post-war synthesis marks the triumph of the general equilibrium approach and neoclassical economics, the orthodox view which is coming now to an end. Some main characteristics of this four phases will be indicated.

8.1.A. Say's Law under scrutiny

William Petty, one of the forerunners of early Newtonian empiricism and positivism in economics, is credited to be the first modern economist to detect the existence of cycles: 'the medium of seven years, or rather of so many years as makes up the Cycle, within which Dearths and Plenties make their revolution, doth give the ordinary Rent of the Land in Corn' (Petty, 1662, quoted in Mirowski, 1985: 13). By the time, the concept of cycle was naturally associated with agriculture yields. One century later, Adam Smith, drawing from his Physiocratic inspiration, enlarged the concept of yield and stock to the whole economy, providing the new born science with some capacity of measurement and identification of functional relations. His concept of cycle was determined by the circulation process of commodities, corresponding to the development of the new generalized market relations: 'When the profits of trade happen to be greater than ordinary, overtrading becomes a general error both among great and small dealers ... . Sober men, whose prospects have been disproportioned to their capitals, are likely to have neither wherewithal to buy money nor credit to borrow it, as prodigals whose expense has been disproportioned to their revenue' (Smith, 1776: 406). The disproportion of capital requirements and projects and the role of profit were indicated in a very simple manner, stressing the characteristics of the new times, the emergence of industrial capitalism; but they were nevertheless seen as an accident, and the overall optimism remained unchanged for a period.[1]

The following authors suggested two alternative visions of that society and its economic relations. Ricardo took a much more abstract approach, and deduced conditions for long run proportions of distributive shares among the social classes in the development of the capitalist economy. Lags were introduced for the sake of realism, and some disproportions were possible in this universe of diminishing returns, but the general trend was defined as some sort of entropic evolution to stationary degradation. Say, at the same time, defined his alternative optimistic vision of a market economy which would not suffer either from major contradictions or from crises, since 'It is worthwhile to remark that the product is no sooner created than it, from that instant, affords a market for other products to the full extent of its own value' (Say, 1803-I: 167).

Two critiques were opposed to this type of explanation. Malthus attacked those notions of equilibrium, suggesting a new theory of development and doom, based on his theory on population and resources.[2] Alternatively, Sismondi, who presented the first extended version of the principle of effective demand (Sismondi, 1819: 131), criticized Say and Ricardo in a very effective way (ibid.: 344, 372), rejecting their 'metaphysical abstractions' (ibid.: 351), their laissez-faire conception and namely the notion that individual self-interest leads to social welfare (ibid.: 50) and their mechanistic concept of causality (ibid.: 125).

By the half of the century, the transition from the old regime was under way, capitalism was dominant in Europe and in the world markets, and Say's Law was accepted in economic science as the expression of such achievements. One of the main consequence of this triumph was the abandonment of a significative part of the agenda of the classics, namely their concern about distribution — that is, the social differentiation in the concrete society — as the main theme for economics. Even if the fluctuations in growth were not coherently addressed by the classics, this modification in the agenda meant that equilibrium metaphysics would thereafter consider the very concept of fluctuations as essentially irrelevant or unexplainable by science.

8.1.B. The marginalist revolution and the first global heresies

The second half of the century was marked by the marginalist revolution, extending the mainstream consensus about Say's Law. The marginalist revolution was fuelled by the analogical incorporation into economics of the energetic physics of Joule and others, created by the 1840s, introducing the maximization principle and the notion of conservation of utility and force, and this made possible a new formalization capacity and an impressive intellectual development. As a consequence, this neoclassical paradigm unified a very peculiar behavioural approach, the 'legal atomistic' identification of every agent under the same type of objective function, with the correspondent mathematical formalism. In conclusion, the equilibrium paradigm was reinforced by the idea, proved under severe restrictions in the models, that every factor is paid according to its own contribution to the product, and therefore the problem of distribution no longer exists. As far as the cycles are concerned, orthodoxy assumed therefore that the existing fluctuations could only be minor accidents, irrespective of the general equilibrium of the economy. This Walras synthesized with the contention that equilibrium was the only relevant framework of economics and that statics was its method.

Not surprisingly, the heresies after Sismondi and Malthus challenged the whole corpus of this doctrine. Marx attacked Say's Law and equilibrium economics, suggesting a schema of reproduction based on the classical problem of social distribution and accumulation and leading under realist conditions to the natural state of disequilibrium. Moreover, he developed a theory of exploitation and alienation which implied that, in normal situations of business, social inequality was the privileged form of growth. Investment was considered as the prime mover of some seven to ten years cycles in the replacement of the systems of machines in the carrier industries — like cotton — and Marx even considered some other hypotheses about longer fluctuations of the economy.

Juglar published, after and independently from Marx's first sketches, a new theory of fluctuations, adding a broader perspective to Smith's 'overtrading', eventually influenced by his own medical practice: he identified a cycle of 5-7 years of 'expansion' and 1-2 years of 'liquidation' in the commercial statistics of the most developed European countries, and concluded that the social body suffered from successive states of prosperity and depression, in a sequential causal chain (Juglar, 1862: 202; 1893: 23), concurring with Marx that fluctuations were part of the normal evolution of capitalism. Similarly to Quesnay's *Tableau Economique*, Juglar analysed the economic evolution in terms of the circulation of value defined in analogy to organic processes (circulation of agricultural product for Quesnay and commodities and prices for Juglar). But both heresies were ignored by mainstream economics, which for a long time did not include the inquiry on cycles, since they were considered irrelevant or simply impossible. By the end of the century, the organic analogies were generally displaced by mechanical determinations.

8.1.C. The third wave of heresies

The third period, from the beginning of the century until the end of the Second World War, was dominated by the dispute between general equilibrium theories and some major new arguments developed within the academic discipline: the 'classics' versus Keynes debate.

Keynes, strongly opposed to Ricardian economics and to the general equilibrium paradigm, classified the theories according to a very simple criteria: they could be either Ricardian or heretic, accepting or rejecting the metaphysics of a self-adjusting mechanism in the evolution of societies. And this antinomy called for an aggressive strategy: 'thus if the heretics on the other side of the gulf are to demolish the forces of the nineteenth century orthodoxy ... they must attack them in their citadel' (Keynes in a 1934 radio broadcast, quoted in Freeman and Soete, 1994: 28). This was of course what Keynes tried to do with considerable success, but without defining a comprehensive theory of the cycle, since the ontological and epistemic status of the marginal efficiency of capital was not very clear.

Anyway, Keynes departed from orthodoxy in several crucial themes. First, this central explanatory variable[3] encapsulated a combination of endogenous and exogenous factors in a (non-declared) nonlinear model, therefore abandoning the positivist criteria. Second, since demand is psychologically determined and supply is technologically determined, maladjustment can arise within a group of factors (the 'animal spirits') or between groups of factors and, naturally, instability is the normal state of the model. Keynes's refusal to present the model in a formal way was not innocent: it was in fact impossible, since the evolution itself was determined by these combinations of economic and extra-economic factors, in an unpredictable mix comprising several non-quantifiable factors and complex interactions. Keynes's was indeed the first major argument for a complex approach in economics after Marx's.

Another alternative school was American Institutionalism, particularly relevant for this survey on business cycles, since it did produce the most persistent research program on the matter for some decades. Mitchell worked on a detailed methodology and chronology of cycles, and in fact classified theories according to a criterion similar to the one Keynes would later on imply in his *General Theory*. After his first major study on cycles (1913), Mitchell suggested in 1927 the following classification of theories: (i) physical theories (Jevons's theory of agricultural fluctuations from sunspot, Sombart); (ii) emotional theories based on psychological errors (Pigou); and (iii) institutional theories, including the expectations of fluctuation in profit (Veblen), monetary fluctuations (Kuznets), increase in production of equipment goods simultaneously with a decrease in the marginal demand price for consumption goods (Aftalion), maladjustment of demand for investment and current savings (Tugan-Baranowsky), and overproduction of industrial equipment (Spiethoff) (Mitchell, 1927: 49 f.).

Two features of this classification are relevant for this study. First, it deals with the theories as holistic explanations. In this context, the distinction between endogenous and exogenous variables is not definite and even less fixed once for all. Only with a major break with this tradition, under the influence of Frisch, Hicks and Hansen among others and the incorporation of the prescriptions of the Vienna Circle in the first research program of the Cowles Commission, did such an antinomy come to be considered as a central characteristic of the methodology for the next stage of the debate about cycles.

Second, Mitchell had a clear synthetic intention and his work was designed to provide a comprehensive account considering all sorts of factors in historical detailed descriptions. His procedure for the classification of theories was implicitly followed by Haberler, in an important book published in 1943, which was considered by other authors as the most important theory of cycles of the time. Haberler and Hansen were then two of the most well known theorists of cycles, and later on both embraced Keynesianism. According to Haberler's classification, six types of theories should be considered: (i) monetary theories based on the fluctuations of supply and price of gold (Hawtrey); (ii) overaccumulation theories, a rather eclectic classification including Wicksell, Cassel, Hayek, Mises, Robbins, Spiethoff, Hansen, Schumpeter, Harrod, Pigou and others; (iii) theories on the variation of the production costs (Mitchell, Beveridge, Fisher); (iv) under-consumption theories (Malthus, Sismondi); (v) psychological theories (Keynes, Pigou, Mitchell); and (vi) agricultural fluctuations theories (Jevons) (Haberler, 1943: 13 f.).

This classification was obviously more complete than Mitchell's, but it was also more dubious. It included very large and heterogeneous groups and different time scopes which do not help to identify the specific problematic of the authors and, furthermore, some of the classifications were ambiguous (Mitchell as a defender of 'psychological theories', for instance). This classification departed from Mitchell's as it considered the more concrete impulse-propagation mechanism which was being argued as the correct approach to the inquiry. The full adoption of this procedure would change the whole analysis of cycles.

8.1.D. The great synthesis: mechanism is back

The debate between Keynes and the neoclassicals was superseded after the end of the Second World War by a synthesis, which in fact began with the IS-LM version of the *General Theory* by Hicks and defined orthodoxy for at least two generations. This synthesis reinvigorated the general equilibrium approach and allowed for the incorporation of the econometric methods in the research program of neoclassicals.

As far as the theory of cycles was concerned, two short and influential papers provided the arguments for the transition from holistic systems of explanation to strictly organized mechanical models discriminating endogenous and exogenous variables and their different causal capacity: Slutsky's 1927 article, and Frisch's 1933 celebrated contribution to the *Festschrift* of Auguste Cassel.

The Slutsky-Frisch approach considered two types of mechanisms: an impulse generation, from exogenously given factors, and a propagation system, endogenously damping the oscillations. Econometrically, the impulse could be delt with as a stochastic variable, whereas the propagation system of simultaneous linear equations would represent the desired property of natural convergence towards equilibrium. According to this view, theories should be classified following the type of impulse suggested and the internal mechanism described in the model. In 1951, Hansen endorsed this vision of the cycle as the response of the economic system to random, 'erratic' external shocks (Hansen, 1951a: 6), and divided the causal explanations in seven categories: (i) aggregate demand theories (Malthus, Hobson); (ii) credit and confidence external shocks (Stuart Mill, John Mills, Marshall); (iii) investment irregularities (Schumpeter, Tugan-Baranowsky, Cassel, Robertson, but also Jevons, Juglar, Sismondi, Rodbertus); (iv) investment demand schedule (Wicksell, Fisher, Keynes); (v) impulse theories (Pigou, Aftalion, Clark); (vi) monetary disequilibrium (Hawtrey, Hayek); (vii) lags, leads, sequences creating oscillations (Mitchell) (ibid.: 270 f.). Hansen intended to combine all those elements in a modern theory of the business cycle (ibid.: 491). In the same vein, Burns classified the theories according to their stated propagation mechanism (Burns, 1969: 10-3). Many other economists, even in spite of their distances to the mainstream, adopted this kind of taxonomy based on the distinction between endogenous and exogenous variables (Sherman, 1991; Dore, 1993). This became the traditional typology (for example, Guesnerie and Woodford, 1992: 289).

As a consequence of the adoption of this Slutsky-Frisch analytical framework, the overall theoretical orientation and purpose of business cycle theory was changed: rather than studying the reasons for structural change and evolution, the theory separated the problem of growth from that of the oscillation, attributed growth to the chapter of comparative statics or moving equilibrium and restricted the cycle to the analysis of the stochastic error term and to the properties of the mechanism of equilibration. That double decomposition became the hallmark of orthodox theories of economic turbulence.

For some time in the 1940s, the direction of the research program was still defined by the linear acceleration models of endogenously produced cycles,[4] but this extreme unrealist approach was abandoned and rapidly replaced by the concept of oscillations driven by purely exogenous shocks. By the 1970s, virtually all the available Equilibrium Business Cycles (EBC) theoretical models were developed in this framework. In the eighties, the development of the new Real Business Cycle (RBC) models confirmed once again this choice, although considering another factor as responsible for the exogenous effect. In the late eighties, the development of the new growth theories opened the way for new conceptualizations in economics, including increasing returns and positive feedbacks, but that conception has not been developed as a new theory of cycles.

The metaphorical origins of this new paradigm will be now presented, before its formal discussion in the next section.

8.1.D.1. The mechanical metaphor of equilibrium

Frisch, the leader of the early econometric program who shared with Tinbergen the first Nobel prize for economics (1969), was fully aware of the limitations of the equilibrium concept and of its mechanical roots. Yet, he thought that the concept was essential in order to produce operational statements and to apply the econometric methods, decisive for the progress of the discipline. In his seminal 1933 paper on impulse and propagation systems,[5] Frisch used the mechanical analogy in two different senses.

The first one was explicitly acknowledged: in order to explain Schumpeterian innovations as an example for the exogenous and more or less continuous shocks, Frisch used his 1931 metaphor of a pendulum[6] suspended from a receptacle permanent filled with water. A pipe was prolonged until the end of the pendulum and opened so that a regular movement was created, preventing the cessation of the oscillation. These movements represented Schumpeterian innovations, 'operating in a more continuous fashion and being more intimately connected with the permanent evolution in human societies' (Frisch, 1933: 203). Frisch added that the analogy had been decisive for him to understand the theory: 'Personally, I have found this illustration very useful. Indeed it is only after I had constructed this analogy that I really succeeded in understanding Schumpeter's idea' (ibid.). Indeed, Frisch consulted Schumpeter, who accepted this imaginary model as a convenient representation of his own ideas. The metaphor is very important, and Schumpeter referred to it several times and always approvingly;[7] it is nevertheless paradoxical, since innovations were presented, against the spirit and the letter of Schumpeter's *Business Cycles*, as an exogenous and quite regular factor.[8]

The second mechanical analogy was more important and pervasive: the concept of equilibrium itself, defined later on as a stationary 'mechanical motion' (Frisch, 1935-6: 101). In the same paper, Frisch indicated that equilibrium could be defined as a mechanical concept or as a 'social concept'; of course, the second concept was not used, since it was assumed that economies were always equilibrated.

Ignoring this social interpretation, Frisch's impulse-propagation scheme incorporated the physical and mechanical notions of movement and rest, and made possible the use of the more sophisticated tools of statistical analysis: the theory provided a synthesis between determinism (the propagation system) and the stochastic view (the impulse system) and defined a clear causal relation. This constituted a new departure for mainstream economics, and Samuelson correctly interpreted this change of paradigm as comparable to the transition from classical mechanics to quantum physics (Samuelson, 1947: 243).

Samuelson's 1947 doctoral thesis — *The Foundations of Economic Analysis* — constituted another essential building-block for this paradigm. Once again, he did not ignore the role of metaphor, and the very first phrase of the book vindicated the constructive role of analogies between different bodies of knowledge, claiming for a general unification of theories as regards those mechanicist central characteristics (ibid.: 9). Although by that time exclusively concerned with the deterministic part of the Frischian paradigm, Samuelson's specific contribution added two core concepts to the paradigm. One was the notion of causality: as the (propagation) system was representable by a system of simultaneous equations, no variable could be said to distinctively influence the others, given that they were all simultaneously determined since the equilibrium conditions were set and the equilibrium state was stable and defined once for all. Therefore, causality was only attributable to exogenous variables or changes in the parameters:

> Within the framework of any system, the relationship between our variables are strictly those of mutual interdependence. It is sterile and misleading to speak of one variable as causing or determining another. Once the conditions of equilibrium are imposed, all variables are simultaneously determined. Indeed, from the standpoint of comparative statics, equilibrium is not something which is attained; it is something which if attained has certain properties. The only sense in which the concept of causation is admissible is in relation to changes in external data or parameters. (Samuelson, 1947: 14-5)

Of course, this is the pure mechanical conception: the mechanism defines its own equilibrium as the articulation of all its internal movements and mutual influences and, unless there is an externally motivated change, the obtained result is permanent. The solution of the equations is the formal representation of the equilibrating mechanics. This is also the common synthesis between positivist determination and the assumption of stability: since equilibrium rules the system, disturbances can only be imposed from outside, as stated by Hayek (1933: 42-3).

The second important concept is, as in Frisch, the definition of equilibrium as a stationary situation to be described by a static theory and namely as a particular case of dynamics, as that case in which no change will happen (Samuelson, ibid.: 223, 243, 270-1, 282). Samuelson argued that two different meanings of statics were possible: statics as a degenerate case of dynamics or else as two completely separate classes, dynamics as all those cases which are non-static. Not surprisingly, he added that what was involved was 'only a verbal problem of definition' (ibid.: 269). It is not: a crucial difference exists between both concepts, but Samuelson simply followed orthodoxy in that matter. Samuelson's own example is the most telling: he argued that a bullet of a canon is in equilibrium, be it at rest or be it in the 'medium course' of a shot (ibid.: 283) — but the ontological difference between both cases requires no argument, and of course there is no meaningful sense for the 'equilibrium' of a shot (it follows approximately one trajectory, but never at the same velocity, and will be destroyed at the end).

Indeed, the concept of equilibrium as a state of rest is the only meaningful proposition compatible with the mechanical metaphor, because the nature of the object is conserved, and in that sense it was fully endorsed by Samuelson. The reason for doing so was invoked at the very beginning of the book by another metaphor: all unstable equilibria are transient, and stability is the natural situation to be studied and verified: 'How many times did the reader see a standing egg? From the formal point of view, it is frequently worth considering the stability of non-stationary movements', wrote Samuelson (ibid.: 11, my translation). Walras and Columbus were finally vindicated by Samuelson's egg.

The consequence was to reduce economics and the growth or cycle theories to very limited set of phenomena, excluding realist assumptions about uncertainty: Samuelson acknowledged Knight's and Keynes's work on uncertainty but concluded that they were inconsequential for the determination of the equilibrium values of the variables (ibid.: 114), and specifically rejected Knight's critique of the physical metaphor,[9] dismissing it as a confusing paper (ibid.: 267).

Since his 1939 *Value and Capital*, Hicks followed the same path, restricting his inquiry to comparative statics, assuming that the system was always in equilibrium and even ignoring the processes of convergence towards it; as for Samuelson, the system of equations represented the 'mechanical periodicities' (Hicks, 1939: 131). Of course, since this was an intellectual construction, the author accepted that 'in another wider sense' the economies were always out of equilibrium or in an 'imperfect equilibrium' (ibid.: 131, 134), and that equilibrium would be indeterminate if realistic assumptions were introduced about the existence of monopolies or State intervention (ibid.: 83-4). Put in other terms, the equilibrium paradigm's Titanic shipwrecks against the iceberg of reality.

Curiously, in spite of the important role of his book in the progress of the neoclassical program, Hicks expressed later on important reservations about the equilibrium paradigm. The first argument was that equilibrium models were limited to the stationary case, useless for practical purposes:

> With every step that we have taken to define this equilibrium model more strictly, the closer has become its resemblance to the old static or even

> stationary equilibrium model; its bearing upon reality must have come to seen even more remote. It has been fertile in the generation of class room exercises; but so far as we can yet see, they are exercises, not real problems. ... They are shadows of real problems, dressed up in such a way that by pure logic we can find solutions for them. (Hicks, 1976: 183)

The representation of time and change is evidently the main difficulty, since when time is plotted as a spatial coordinate, something is ignored: the difference between past and future and therefore uncertainty (Hicks, 1976: 135). As a consequence, equilibrium methods restricted the field of the inquiry to timeless and changeless systems (ibid.: 140; 1979: 45, 82). The second important reservation by Hicks was about the heuristic capacity of the equilibrium paradigm, which encouraged economists to waste time with abstract constructions ignoring history, and being 'practically futile and indeed misleading' (Hicks, 1976: 143).

Another example of major contributors to the definition of the program who later on developed a sceptical view of the project is that of Hahn, who was responsible with Arrow for a major development in the program and who acknowledged that the notion of equilibrium originated in mechanics. These authors considered the Paretian equilibrium as 'the most important intellectual contribution' of economics to social sciences (Arrow and Hahn, 1971: 1). But, after some years, Hahn argued that the program could not develop an acceptable account for the convergence and stability conditions since:

> the whole subject how to attain equilibrium has a distressing ad hoc aspect. There is at present no satisfactory axiomatic foundation on which to build a theory of learning, of adjusting to errors and of delay times in each of these. It may be that in some intrinsic sense such a theory is impossible. But without it, this branch of the subject can aspire to no more than the study of a series of suggestive examples. (Hahn, 1982: 747)

Arrow made the same point, and so did Hutchison, criticizing the 'fantasy' and 'blackboard concepts' developed by the equilibrium paradigm (Hutchison, 1992: 86). This rather impressive late reactions against the mechanical metaphor, namely by some of its more distinguished creators, is a brilliant evidence of the current crisis of the neoclassical program. But, then, how was it possible that such a metaphor dominated the research for generations?

8.1.D.2. Mechanics and its lever

The first building block of the new approach was the 1927 article by Slutsky,[10] translated ten years later in the influential *Econometrica*, the voice of the new synthesis between probabilistic (quantum) and deterministic (energetic) economics. According to Slutsky, two possible origins could generate the cyclical behaviour: the existence of deterministic and structurally created cycles, or the impact of random shocks on a stable structure. The second cause was the object of his study. Two years before the publication of this article, Irving Fisher, arguing that there was no cycle but only a 'dance of the dollar' above and below trend, had concluded that the oscillations were nothing more than 'Monte Carlo Casino's cycles' (Fisher, 1925: 191-2): Slutsky provided some mathematical foundations for such an argument.

Slutsky considered pure random shocks and even suggested to 'give up the hypothesis of the superposition of regular waves complicated only by purely random components' (Slutsky, 1937: 107). In his paper, the irregular fluctuations were created by a single process of random disturbances. Using several series from the Russian lottery and other sources, Slutsky was able to prove that the simple summation of random processes could create cyclical patterns with approximate regularity (Slutsky, 1937: 105 f.). This was just the opposite to Kondratiev's belief that the shocks could hide but never create an illusion of the wave (Kondratiev, 1992: 267); in the context of the Slutsky's series, the random causes created a fictitious cycle, and the author suggested that moving average methods could as a consequence show cycles where none existed. But Slutsky also showed that the summation of this random causes could imitate for some cycles the harmonic series of a small number of sinusoidal curves, but that after a limited period a new and radically different regime would be established (ibid.: 123). These so sharp changes are not, of course, the observed phenomena for the majority of the historical periods.

The Slutsky scheme may be interpreted in several distinct ways. In the original sense, it is a model generating artificial cycles, which exist only in the statistical representation by imposition of the linear filtering (moving averages of other) procedures. Otherwise, if the economy is considered to be like a dampening and stabilizing system, as in the Frischian model, then the random shocks can be conceived of as being averaged and dissipated by the very functioning of the economy, and the cycle is supposed to happen in real terms. This is the interpretation followed by more recent models, such as Lucas's: the linear filtering of random shocks (of money supply) creates autocorrelated fluctuations, and therefore involve all other variables in a cyclical process.

This example proves that the argument was compatible with the idea of a damping internal mechanism leading to a normal state of equilibrium, to which some external shocks were impinged creating oscillations: that is why it was incorporated in the neoclassical tradition. Frisch, aware of Slutsky's article when he developed his 1933 model, argued in the context of the dominant metaphor: the economy could be modelled as a mechanism (the markets) with a stable rest state and a force towards equilibration (the market clearing processes). If the economy was conceived as gravitating around equilibrium and cycles were viewed as the outcome of shocks moving the economy away from the centre of gravitation, then the articulation of some degree of realism and a general equilibrium approach was possible. Moreover, as Frisch openly proclaimed, 'If fully worked out, I believe that this idea will give an interesting synthesis between the stochastic point of view and the point of view of rigidly determined dynamical laws' (Frisch, 1933: 197-8). Thus, the theory of cycles became a central point of the articulation of neoclassical economics after the probabilistic quantum revolution, and introduced the parallel econometric revolution.

Frisch considered the intrinsic structure of the market economy to have a damping tendency determining the length of the cycles; but the cause of the oscillation, nevertheless, were the external shocks, the 'source of energy in maintaining oscillations, ... a stream of erratic shocks that constantly upsets the continuous evolution, and by so doing introduces into the system the energy necessary to maintain the swings' (ibid.: 197-8). The problem was then to explain how those Slutsky shocks were accumulated and transformed by the weight system provided by the internal mechanism (ibid.: 202-3). The analogy illuminating this process was the movement of a rocking-horse, as firstly argued by Wicksell and Ackerman and then repeated and developed by Frisch (ibid.: 198): the deterministic part of the economy is represented by the damping propagating mechanism (the wooden horse), while irregular cycles are created by the impulse system of stochastic and external shocks (the force applied to the horse).

The mechanical analogy inspired the Frischian synthesis of determinism and randomness: external causal forces create the impulses, while the propagation system is the mechanism which accounts for the stabilizing properties and the convergence towards equilibrium. The requirements of orthodox epistemology were met: causality was clearly defined and attributed and, although the primary cause was considered unknowable since exogenous, stability and equilibrium are guaranteed by a controllable specification of the model, which was the domain of a practicable econometric study. The inquiry was therefore restricted to the definition of the hypotheses about the behaviour of the shocks, to estimate the deterministic part of the model, to generate a series from it and to compare such a series with reality in order to obtain a confirmation of the theory. The model is formally presented and discussed in the next sections.

8.2. How does the horse rock?

Three conflicting approaches inspired from the thirties and the forties the discussion about the cyclical behaviour of real capitalist economies: first, the models derived from the multiplier-acceleration principle, second, the Frischian type of model, both assuming a linear specification, and finally some nonlinear models of endogenously generated growth and cycle. This section presents the relation between the first two types of models, and discusses the reason for the final predominance of the Frischian approach.

8.2.A. A general linear model of cyclical potentialities

Consider a very simple general structural linear model of an abstract economy (adapted from Kenkel, 1974: 323 f.), where y_t is the dependent aggregate variable, and x_t the independent aggregate variable. Lags are considered in both cases: r for y_t and s for x_t, and a stochastic term is included, e_t, representing the traditional specification either for the shocks or for omitted variables or measurement errors. The following equation gives the state of the economy at t:

$$b_0 y_t + b_1 y_{t-1} + ... + b_r y_{t-r} + a_0 x_t + ... + a_s x_{t-s} + e_t = 0 \qquad (8.1)$$

or

$$y_t = -(b_1/b_0)\, y_{t-1} - ... - (a_0/b_0) x_t - ... - (a_s/b_0)\, x_{t-s} - (1/b_0)\, e_t \qquad (8.2)$$

and considering now the system describing the evolution of the economy with $r+1$ equations, simplified and organized in a matrix form,

$$Y(t)=C\,Y(t-1)+D\,X(t)+V(t) \qquad (8.3)$$

where obviously $Y(t)$ and $V(t)$ are $((r+1)\times 1)$ matrices, $Y(t-1)$ is $((r+1)\times 1)$, $X(t)$ is $((s+1)\times 1)$, C is $((r+1)\times(r+1))$ and D is $((r+1)\times(s+1))$. The coefficients of C and D are, of course, $c_{1i}=-b_i/b_0$, and $d_{1i}=-a_{i-1}/b_0$, and $v_{i1}=-e_i/b_0$. It is assumed that lim $C^t=0$ for $t\rightarrow\infty$.

Now, (8.3) can be easily solved for $Y(t)$. Since

$$Y(t) = (I-C\,L)^{-1}\,[DX(t)+V(t)] \qquad (8.4)$$

where L is the lag operator. The general solution is:

$$Y(t) = C^tY(0) + DX(t)+CDX(t-1)+C^2DX(t-2) + ... +C^{t-1}DX(1)+V(t)+CV(t-1)+ C^2V(t-2)+ ... +C^{t-1}V(1) \qquad (8.5)$$

The equation (8.5) is the composition of three distinct parts:

(i) $C^t\,Y(0)$, which are the initial conditions of the model. In this case, $Y(0) = [y(0)\ y(-1) ... y(-r)]'$.
(ii) $\Sigma_i C^i\,D\,X(t-i)$, with $i=0,...,t-1$, which are the exogenous contributions to the evolution of the system.
(iii) $\Sigma_i C^i\,V(t-i)$, with $i=0,...,t-1$, which are the weighted stochastic influences on

the evolution of the system.

The cyclical behaviour is determined by a conjunction of causes from these components:

(i) by the factor including the initial conditions, since it can exhibit an oscillatory pattern under some circumstances. $C^tY(0)$ can be expressed as a linear function of the roots of C: if any of these roots is complex, C^t will provoke oscillations.

(ii) the exogenous influence on the system is a linear combination of the previous values of the independent variable. Two different situations can be considered. If the roots of C are complex, then it will provoke an oscillatory behaviour. But still if the roots are real, a possible oscillatory movement depends on the behaviour of the exogenous variable, with the cycle being externally created (or damped). This is the case dealt with by the Political Business Cycle theories.

(iii) finally, the effect of the stochastic variable. Under certain assumptions about the distribution of the vectors of $V(t)$, for example, that v_t are IID, $E(v_t)=0$ and v_j and v_i are uncorrelated for any $j \neq i$, W being the covariance matrix of $V(t)$, and considering $Z(t)=\Sigma_i C^i V(t-i)$, then:

$$E\,[Z(t)\,Z(t)'] = E\,[\Sigma_i C^i\,V(t-i)]\,[\Sigma_i C^i\,V(t-i)]' \quad \text{with } i=0,...,t-1$$
$$= W + C\,W\,C' + ... + C^{t-1}\,W\,C'^{t-1} \tag{8.6}$$

and generally,

$$E\,[Z(t)\,Z(t-j)'] = C^j\,E[Z(t)\,Z(t)'] \quad \text{for any } j \neq t \tag{8.7}$$

the vectors $Z(t)$ are correlated and so there is a time dependence between any $Z(t)$ and $Z(t-j)$. If the roots of C are complex, $C^jE[.]$ oscillates. But even if the roots of C are real, the cycle is still possible due to the correlation of $Z(t)$ with any $Z(t-j)$. This is the Slutsky effect: a linear combination of stochastic independent variables, can produce an oscillatory behaviour.

In summary, if all or any of the roots of C are complex, the three components will create internally generated fluctuations; if the roots are real, the stochastic element and the exogenous variable can still determine externally generated cycles. Both Samuelson's and Frisch's interpretations of the cycle constitute particular cases of this general linear model.

8.2.B. The acceleration model

In 1939, Samuelson presented a multiplier-accelerator model in a very simple form (Samuelson, 1939: 265 f.). The system includes three equations; Y is income, C consumption, I induced private investment and G is governmental expenditure; a, the marginal propensity to consume, and b, the acceleration coefficient or the 'relation', are constant:

$$Y_t = G_t + C_t + I_t \tag{8.8}$$

$$C_t = a\,Y_{t-1} \tag{8.9}$$

$$I_t = b\,(C_t - C_{t-1}) \tag{8.10}$$

In other words, Samuelson's system is a particular version of (8.3), where $r=1$, $c_1=a(1+b)$, $c_0=-ab$, $V(t)=0$, $D=I$, $X(t)=G_t$. As previously demonstrated, the only possible origins for cycle are complex roots,[11] since G_t is considered non-oscillatory, and for simplification equated to one by Samuelson. Samuelson rejects, by this specification of the model, the effect of the stochastic variable as a possible cause for oscillations, and rejects also the effect of the exogenous variable as a possible cause. The only source of fluctuations is the effect of initial conditions and the endogenous variable.

As shown by Samuelson, there are four different regimes, according to the values of the coefficients. In two regimes there is non-oscillatory convergence or divergence (with either distinct or repeated real roots), in two others there is oscillatory convergence or divergence (with complex roots lying inside or outside the unit circle). And only under the restrictive assumption of a certain value for the parameters we can have permanent cycles. As this outcome corresponds to the observed situation better than explosive or damped cycles, it is obvious that this model only explains fluctuations if there is some celestial entity setting the desired and precise constant coefficients, so that the oscillation be perpetually repeated and stable.

It is very difficult to accept either the assumption or the conclusion. Furthermore, in spite of the fact that the model has the attractive peculiarity of explaining the behaviour of the system by internal relations between the variables, it cannot explain irregular and nonperiodic timing in fluctuations. In other words, it is irrelevant for a real world analysis. This is the reason for the general rejection of the model: it does not explain cycles (Baumol and Benhabib, 1989: 78; Zarnowitz, 1985: 544). As a consequence, theories based on the accelerator were forced to develop complementary hypotheses: for instance, Haberler's theory stressed the role of the expectations-driven accelerator creating a disproportion in production of capital goods and moving from the boom into a crisis, but added the role of innovations to explain the lower turning point (Haberler, 1943: 345-6).

The acceleration principle introduced a peculiar version of the physical metaphor, as recognized and praised by Hicks: 'The acceleration principle established an analogy between economic fluctuations and the 'waves' which are so elaborately studied in physics; thus a vast amount of knowledge and technique acquired for that purpose by applied mathematicians and physicists suddenly become relevant to the economic problem' (Hicks, 1950: 4). To the mechanical pendulum was now added the concept of friction and viscosity. Still, a more complete model was necessary in order to integrate the various motivations for the cycles. Frisch provided such an explanation in the neoclassical equilibrium framework.

8.2.C. Rocking again

What is here called Frisch's model is, in fact, the 1933 version by Frisch of the impulse and propagation systems, using the important insight by Slutsky about the possible active role of exogenous shocks in the creation of cycles under trivial conditions of randomness. It is also a particular case of the general linear model previously presented.

In its standard formulation, the model considered the propagation system to be defined by $CY(t\text{-}1)+DX(t)$, and the impulse to be generated by $V(t)$, following the specification of the general linear model. Oscillations created by previous reasons *i* (initial conditions) and *ii* (the known exogenous variable) were excluded, and only causes *iii* (the stochastic random effect) were considered. Unlike Samuelson's multiplier-accelerator model, in this case the linear stochastic difference equations exhibit some more realism and shape irregular cycles, although this is still more restrictive than Slutsky's original approach, which also considered but did not discuss a possible cause (*i*).

This type of model very rapidly became the canon in orthodox economics: it was easily tractable from the mathematical point of view, it rejected the purely endogenous explanations which could not explain the observed irregularity of the cycles under a linear specification — that was the case of the multiplier-accelerator model —, it combined endogenous and exogenous determination and rejected the 'inadequate behavioural foundations of the early [nonlinear endogenous] models' (Boldrin and Woodford, 1990: 190-1), since these did not assume optimization. A large body of research could therefore be based on this assumptions and techniques, and its success established the pattern of normal science for long time.[12]

Furthermore, the propagation-impulse schema provided a clear cut distinction between endogenous and exogenous variables, which defined the following classifications of theories and organized the separation between the growth theories and the theories of cycles. In short, the price for the crucial role the theory of cycles played in the reorganization of the neoclassical paradigm was its own downgrading to an inferior epistemological status, since there was not much causal explanation to be defined from an external, unknowable and essentially irrelevant and uninformative stream of events.[13]

This was why, in spite of the character of their analysis of long term evolution of capitalist societies, these models did not consider structural evolution, since technical and social change was excluded: in the acceleration context, structural change was ignored by the initial truism declaring that, under the *ceteris paribus* of no modification of technology, then an increase in national income requires an increase of capital, given the stable capital-output ratio, and in the Frischian framework it was ignored by the trendless simulation and by the structure of propagation.

Anyway, the equilibrium concept was saved, and cycles, under the names of 'oscillations', 'fluctuations', 'shocks' or others, were incorporated into the main metaphor. Several decades later, this schema was still supposed to legislate about the disputes, as in the Keynesian-Monetarist controversy,[14] and presided over the development of the mainstream models of cycles.

8.3. Classification of classifications: problems and alternatives

The acceptance of the Frischian approach implies a double decomposition (trend/cycle and impulse/propagation) and a distinct articulation of modes of causality, since two different mechanisms were supposed to be simultaneously at work: the propagation (which must be explained) and the impulse system (which cannot be explained, but about which some crucial hypotheses must be established). Accordingly, the theories are classified in function of their statements about each one of this mechanisms, and could be separated in two main groups. The first group includes the research programs on the propagation mechanism, leading to the growth theory taking the trend as the relevant object to be studied and restricting dynamics to comparative statics, or leading to theories of convergence to equilibrium, whereas the second one includes the research programs on the impulse system, leading to the specific theory of cycles, based on static assumptions.

Of course, this distinction is highly artificial and any theory claiming for some realism is forced to make statements about both domains. So, the current practice has been to classify theories according to their central governing mechanism, be it internal or external, and to provide a rather eclectic and descriptive list of theories, each one being a specific instance of the class (this is for example the procedure of Burns, 1969).

As a consequence of this procedure, two of the essential features of the paradigm were undermined. The first is the very distinction between endogenous and exogenous variables, since the direction of the determination is opposed in the two research programs, and in fact a variable defined as endogenous in one model can be considered as exogenous by another. This imposes the instability of the classifications, since they depend on the particular theory and not on the paradigm itself.

The second consequence is even graver, since the epistemic separation of the study of cycles and growth eventually prevents any of them from reaching relevant results. The growth theories concentrated on the trend, viewed as the result of technical progress embodied in autonomous investment, or the process of accumulation and its impact on productivity, plus some structural variations (for example, population), the slope representing the equilibrium value of growth; in other words, the causal structure was considered partially or wholly exogenous. Simultaneously, the cycle theories concentrated on other exogenous impulses on the trend, and the distinction between both types of shocks is necessarily formal and constitutionally ambiguous. As a consequence, each theory explained that it could explain nothing, and the paradigm was reduced to a petition of principle: there are fluctuations because the macro-variables representing the economy do fluctuate.

The classification of Zarnowitz takes this point of departure (Zarnowitz, 1985: 532 f.), suggesting two main groups of theories: (i) theories of self-sustaining cycles, including (a) theories of vertical (production of capital and consumption goods) or horizontal maladjustments (periods of development of new capital goods), (b) Keynesian theory of uncertainty and the schedule of marginal efficiency of capital, (c) theories on wage and price dynamics, and (d) disequilibrium models with capital accumulation (including the early nonlinear models); and the second group being (ii) theories of cyclical responses to monetary or real exogenous disturbances (EBC and RBC). The first group includes the theories dominated by the propagation mechanism (or the first and eventually the second causes of the general linear model), and the other type of theory is concentrated on the impulse system (or the third cause).

The general methodological limitations and consequences of this sort of classification were criticized above. There is, nevertheless, another important difficulties: this type of classification fits well if the hypothesis to be classified is by itself Frischian, that is, if it accepts and works on the presumption of the distinction between the propagation and impulse systems. In this case, the classification is always accurate. But there are some important exceptions. Two examples of this kind of difficulty are illuminating.

The first case is Schumpeter's theory of cycles. As it will be argued in Part Three, Schumpeter's own views supported the neoclassical tradition on equilibrium; consequently, he supported the orthodox view on the trend and on the nature of the economic mechanism. But when defining innovation as the impulse system *and* the propagation mechanism, Schumpeter implicitly departed from those traditional definitions, since he included a variable he insisted to classify as an endogenous factor and which still could behave like an exogenous variable. As a consequence, the classification of his theory is not clear for a Frischian criterion.

The second case is far more conclusive. Keynes's theories are either considered 'psychological' (a strictly 'endogenous' factor for every 'agent', which must be considered exogenous by economic theory) and an impulse theory as in Haberler's classification, or are considered as theories of the propagation mechanism of the investment schedule, as in Hansen's and Burns's classification. In fact, both are correct, since the marginal efficiency of capital is one *and* the other: this is why Keynes's approach cannot be captured by a Frischian taxonomy.

The reason for the difficulties in this classification is, as indicated, that non-Frischian theories represent an attempt to cope with reality in a complex way and no deterministic model can fully represent their analyses. This is the substantial reason for arguing for a general mode of classification which takes into account not only the description of the conclusions of the theories, but also their own process of conceptualization and the nature of their premises. These insights will be reconsidered later on when a new set of classification criteria will be presented.

9. Premises and Implications of the Double Decomposition

The dominance of the Frischian approach to business cycles is based on the essential role it played in the formalization and progress of the neoclassical paradigm. This chapter develops this argument in the domain of the two main methodological questions previously addressed, equilibrium and endogeneity, and namely of their implications for the analysis of the relations between trend and cycle.

The resilience of the equilibrium approach was explained by the importance, availability and efficacy of the energetic metaphor in the constitution of the neoclassical research program. New instances will be presented now evidencing the importance of this mechanistic vision in economics: imposing its strict analogies, this physical metaphor requires a strong determinism as the only legitimate form of causality, therefore distinguishing between endogenous (explained and possibly explanatory in a very weak sense) and exogenous (explanatory but not explained) variables. Furthermore, the Newtonian characteristic of the system suggests a stable causal relation and a reversible and timeless system of 'legal atoms', the desirable properties of the mainstream's ideological vision.

This system will be now discussed.

9.1. The Dreamland of equilibrium

Equilibrium has been defined in economics in two major senses: what may be called the *position* concept implying the static state of rest of some system,[15] and the *movement* concept relative to the evolution between two successive states of rest, such as the effect of a price variation on the space of the supply-demand relation. Both concepts are non-trivial, since they suppose a strong hypothesis of rationality and a vision of the mechanistic functioning of the economy. Moreover, in the framework of a non-dynamical universe, the equilibrium point has the properties of uniqueness and stability, restricting that mechanism to a very specific structure.

As Arrow and Debreu pointed out for a Walrasian framework, once the price equilibrium is attained — by the solution of a system of simultaneous equations — there is no change in the system while the expectations of the agents are not changed (Arrow and Debreu, 1954: 265 f.). But this solution supposes that there is no time: all decisions are concentrated in a single moment, all the agents wait in order to buy or to sell until the market is cleared so that they maximize their utility — all trade occurs in the eternal present. As a consequence, the auctioneer is the only agent without an utility function: in the fable of the bees, he or she is the only unselfish being.

This very elegant and powerful result requires, nevertheless, a large set of strong assumptions, embodied in the following axioms:

1. The economy is static or defined by a stationary process.
2. The organization forms, the set of available products and the set and ordering of preferences are constant.
3. Producers or consumers are exclusively defined as such.
4. All operations are simultaneous.
5. The production set is convex.
6. The behavioural rule is profit and utility maximization.
7. The price is the only relevant information flow, and it is complete in order to make possible decision making.
8. There is no uncertainty.

The next sections will present the critiques of these axioms by Mitchell, Keynes and Kaldor. Then, rationality, uncertainty and dynamics will be discussed in turn.

9.1.A. Indomitable critiques: Mitchell, Keynes, Kaldor

Many decades before the crisis of mainstream economics and namely before the admission by some prominent neoclassicals of the incompatibility between the persistence of business cycles and the concept of the return to equilibrium,[16] the theoretical discussion about fluctuations provided some relevant evidence and critical arguments. These critiques highlight some of the central problems of the paradigm of equilibrium.

9.1.A.1. Mitchell: the dispensable hypothesis of equilibrium

Mitchell developed the first consistent research program on cycles, as distinct from the neoclassical orthodoxy — in fact, he did so previously to the formulation of the Slutsky-Frisch approach — and was naturally confronted to the concept of equilibrium, which he rejected. His main arguments against the 'dreamland of equilibrium'[17] were the two following ones.

On one hand, Mitchell and Burns argued that the state of equilibrium does not exist: it cannot be observed in the economic series, and therefore the assumption about some mechanism of deviations and return to the equilibrium path is a mere unsubstantiated intellectual construct. Furthermore, the definition of the model discriminating between endogenous and exogenous factors is also ambiguous and arbitrary:

> To say that business cycles are departures from and returns toward a normal state of trade or a position of equilibrium, or that they are movements resulting from discrepancies between market and natural rates of interest, will not help, because we cannot observe normal states of trade, equilibrium positions, or natural interest rates. Nor, when we start observing, can we tell whether cyclical movements are due to factors originating within the economic system or outside of it. (Burns, Mitchell, 1946: 5)

The empirical analyses of business cycles in multiple series, a massive effort by Mitchell's NBER, was based on the assumption that not only a general equilibrium pattern was irrelevant, but also that each concrete cycle should be analysed as a historical individuality (ibid.: 43). The conclusion is thus severe: neither equilibrium normal states have any ontological significance, nor does the distinction between endogenous and exogenous factors has any epistemological relevance. As a consequence, Mitchell rejected the general equilibrium theory.[18]

On the other hand, Mitchell warned against the mechanical analogy, responsible for the pervasiveness of the equilibrium account, and suggested an alternative view:

> Doubtless it was a mechanical analogy which gave its vogue to the notion of economic equilibria. Everyone admits that analogies, though often most suggestive in scientific inquiries, are dangerous guides. The usefulness of the analogy in question was greatest and its dangers least when the economists were treating what they called 'static' problems. Such problems can be given a quasi-mechanical character .... But the problems of business cycles are the opposite of 'static'. ... Yet there is a different conception of equilibrium which may help us — the equilibrium of a balance sheet, or better, of an income and expenditure statement. Such a statement has nothing to do with mechanical forces, and that is a safeguard against false analogies. It deals with pecuniary quantities, and they are genuine elements in our problem. (Mitchell, 1927: 186)

This alternative solution is, of course, a very limited one. It states simply that every economic system can be a posteriori equilibrated in an accounting manner, but that technique can be adapted to any level of disequilibrium by the defined technicalities of the balance sheet. In that case, the final equilibrium situation is artificial since both columns add to the same value whatever the reality of the economy may be. This alternative is merely argumentative, and did not take any relevant place in Mitchell's study of the concrete cycles: the difference between the columns indicates the amount needed to produce equilibrium, and is therefore part of the ongoing action, while of course the concept of equilibrium implies a situation in which no action is necessary.

Schumpeter's critique of Mitchell's position concentrated on this second aspect and on his descriptive and allegedly 'non-theoretical' methods, but he also mentioned the first topic (Schumpeter, 1952: 337-8 n.). The core of Schumpeter's argument was that since Mitchell accepted that all agents look for pecuniary gains, then they were supposed to be rational, and the system they conform should also be rational, that is equilibrating.[19] The rationality assumption was for Schumpeter the bodyguard of the equilibrium approach: of course, if one falls, the other follows.

Koopmans criticized Mitchell on the same grounds. Since hypotheses are part of measurement, Mitchell's stance was denounced as an artificial 'measurement without theory' (Koopmans, 1947: 162-3; Hansen, 1952: 305-6). But this was a rhetorical and inexact argument based on a polemic misperception, since Mitchell's careful empirical analysis was not atheoretical, as the just quoted critiques on the equilibrium paradigm have shown.

9.1.A.2. The methodological insights of Keynes

Keynes went even further than Mitchell, in the sense that he explicitly challenged the pillar of the rationality postulate. In a 1937 article defending his writings from some neoclassical opponents, Keynes argued that two main differences identified his position: the first was the rejection of an all comprehensive rationality, the 'Benthamite calculus' which underlies the fiction of certainty, and the second difference was his own methodological contribution, since 'This psychological law was of the utmost importance in the development of my own thought, and it is, I think, absolutely fundamental to the theory of effective demand as set forth in my book'[20] (Keynes, 1937: 220; also 222-3).

Even if this method was not extensively discussed in his work, it can be summarized as the rejection of the traditional model of reasoning, and as the appeal to facts of observation (that is why he stipulated for instance a nominal wage), and as the rejection of the pervasive method of *ceteris paribus* clauses. Only such an empirical analysis could highlight the evolution of expectations and behaviours, since on an a priori basis there could be made no meaningful statement about expectations. According to Keynes, the rationality postulate should be replaced by some realistic assumption about the instability of human behaviour.[21]

Equilibrium is discussed in the Keynesian context in two different meanings. The first is the equilibrium value of some exogenous variable, corresponding to the maximum probability of the distribution of its values (Vercelli, 1990: 234). But since that distribution is not determined in the model or otherwise known, no assertion can be made about this equilibrium. The second meaning is referred to a multiplicity of paired values of endogenous and/or exogenous variables. But these equilibria are deprived of uniqueness and attractiveness: for instance, each value of the rate of interest is equilibrated in that sense with a corresponding level of employment. This is parallel, in a sense, to Mitchell's definition translated to a global theoretical framework: every situation of a system can be photographed and the state of the corresponding variables can be postulated as an equilibrium. As a consequence, the notion of equilibrium loses all its semantic pertinence and far-reaching implications.

It is well known that the more relevant consequences of this approach were, according to Keynes, the affirmation of the indeterminacy of the result of human action and of consequent intrinsic uncertainty of the social systems. In order to describe that conception, Keynes used a new metaphor which indicated the extreme variety of solutions and possible scenarios that could occur in a closed universe: the kaleidoscope which, when moved, always produces surprising combinations. Later on, Shackle generalized that metaphor in order to define Keynes's methods as kaleidoscopic. Keynes formulated the example in the following terms:

> Nevertheless we must not argue that an expansion of the currency influences relative prices in the same way as the translation of the earth through space affects the relative position of the objects in its surface. The effect of moving a kaleidoscope on the coloured pieces of glass within it is always almost a better metaphor for the influence of momentary changes in price. (TM: 81)

In spite of Keynes's own considerations about the achievements of his method, the result was not so conclusive and this very metaphor indicates the reason. A considerable confusion between the dynamical properties — implying disequilibrium — and the comparative static account, frequently used by the author, allowed for many different and contradictory interpretations; even worse, some of them were not clearly unauthorized by Keynes himself, such as the influential IS-LM equilibrating mechanism. Based on this, part of the Neo-Keynesian school reintroduced equilibrium in a matter of years.

The IS-LM scheme was presented by Hicks to a meeting of the Econometric Society at Oxford in September 1936, some months after the publication of the *General Theory*, at the same time of other papers on the topic by Harrod and Meade. Hicks wrote many years later that the diagram was only designed for expository purposes and that 'I am sure that if I had not done it, and done it in that way, someone else would have done it very soon after' (Hicks, 1979: 73 n.). Furthermore, Keynes 'did not wholly disapprove of what I had made of him', argued Hicks on the basis of a letter by Keynes sent some months afterwards (Hicks, 1976: 141). Nevertheless, the friendly but imprecise letter did not implicate a full endorsement of the interpretation,[22] which was clearly contradictory with the whole theory. In fact, the 1937 paper by Hicks did not hide the purpose of introducing Keynes in the neoclassical universe: 'Income and the rate of interest are now determined together at P, the point of intersection of the curves LL and IS. They are determined together; just as price and output are determined together in the modern theory of demand and supply. Indeed, Mr. Keynes's innovation is closely parallel, in that respect, to the innovation of the marginalists' (Hicks, 1937: 135). This was indeed very far from Keynes's own thought: in a previous letter (December 1934) to the same Hicks, he had stated that 'Walras's theories and all others along those lines are little better than nonsense' (quoted in Skidelsky, 1992: 615). The transformation of the *General Theory* in another version of the general equilibrium paradigm was in fact a major shift from its original purpose.[23]

In that sense, Hicks himself pointed out the two major modifications that the IS-LM scheme implied for the original Keynesian theory: first it excluded time and uncertainty, and therefore reduced the GT to a particular case of general equilibrium,[24] and second it introduced an alternative vision of causality, which is sequential and non-deterministic for Keynes and simultaneous and deterministic for the Hicksian scheme (Hicks, 1979: 74).[25] Hicks accepted later on the Keynesian claim that general equilibrium is not realist and therefore is not general:

> I doubt if there is any concept of equilibrium usable in economics which is truly 'general', in the sense that there are no choices which might conceivably be open to the actors but which have not been, for the purpose of the model, deliberately closed for example, collusion between agents. The Walrasian equilibrium itself, which is commonly regarded as a pattern of general equilibrium, is not general in this unrestricted sense. (Hicks, 1979: 79 n.)

Another critique on Keynes's *General Theory* was developed by Schumpeter, this time on the opposite grounds of his previous critique of Mitchell's position. Schumpeter argued that, since Keynes restricted himself to the short term analysis, he could neither understand nor explain the most important phenomenon, economic mutation: 'All the phenomena that affect the creation and change of this apparatus [the industry], for example, the phenomena more influent over the capitalist process, remain in consequence out of consideration' (Schumpeter, TGE: 381). In fact, this limitation severely restricted the methodological impact and clarity of the General Theory. But Schumpeter's own position was itself contradictory and, although accepting the importance of Keynes's analyses,[26] he could not recognize all the consequences of Keynes's vision about the nature of the dynamical forces of the capitalist society. Schumpeter could not follow Keynes in his important insights against the general equilibrium approach, although he had more than substantial reasons for doing so.

9.1.A.3. Kaldor: moving away from equilibrium

Kaldor represented a third challenge against the mechanical metaphor. Departing from the traditional definition of Paretian equilibrium, Kaldor stressed the methodological consequences of the assumption of equilibrium and then argued that the introduction of realism and of time implied the conception of an evolving economy, moving away from equilibrium. Both points are very conclusive: as he insisted, 'the "equilibrium story" is one in which empirical work, ideas of facts and falsification played no role at all' (Kaldor, 1985: 11), and those axioms constitute a 'blind alley' (ibid.: 57).

Specifically, as the economy is a continuously evolving system, like any ecological system, its time path cannot be determined and the equilibrium concept plays no part in its analysis (ibid.: 12). And, since the neoclassical story depends on the equilibrium concept, it must impose an exogenous explanation to the fluctuations:

> The very notion of equilibrium, particularly of long run equilibrium, amounts to a denial of this [endogenously created changes in the economy] — for this notion asserts that the operation of economic forces is constrained by a set of exogenous variables which are given from the outside, so to speak, and which remain stable over time. ... Indeed, as is often emphasized, the exogenous variables that determine the nature of equilibrium are independent of history in their most important characteristics. Any given constellation of such exogenous variables, whatever the initial situation, will inevitably lead, perhaps through a succession of steps (succession of 'temporary equilibria') to a unique point of final equilibrium. ... Continuous growth can only be thought of within this intellectual framework as a steady state, where every thing grows in exact proportions ... . (ibid.: 61-2)

This conclusion explains the actual predominance of comparative statics as the instrument for the analysis of dynamically changing real economies: cycles can only be conceived of as externally driven in order to save the philosophy of equilibrium, so that evolution can be captured in different moments as the fluctuation around the equilibrium path. Finally, the equilibrium paradigm excludes any complete explanation of the very movement towards equilibrium, since the first impulse is exogenously driven.

Kaldor did not try to repair the damaged concept of equilibrium and suggested a wholly different approach to economics, based on the evolutionary metaphor and on the concept of increasing returns, thus conceiving an economy moving away from equilibrium (ibid.: 63-4).

9.1.B. Rationality and equilibrium

The three main arguments produced in these previous critiques of the equilibrium metaphor can be synthesized as follows: (i) Mitchell's argument about the unrealistic assumption of a 'natural' unobservable state of equilibrium stated the irrelevancy of such an hypothesis for the concrete study of cycles; the argument was supported by Kaldor and is implicitly accepted by Arrow; (ii) Keynes's argument against the rationality postulate, which is still the most controversial, rejected that general hypothesis about economic behaviour; and finally (iii) Kaldor's argument stated that in the equilibrium approach all the fluctuations (away from the equilibrium path) are externally motivated, and furthermore that the existence of a 'natural' equilibrium in a dynamical evolution must be conceived of as a steady state growth (represented by the trend). These arguments synthesize the case against the orthodox program of identification or of modelling the accumulation process.

The rationality postulate, which is considered in this section, is the jewel of the Crown of neoclassical theory: since the marginalist revolution, and resisting through the tempests of the econometric revolution, it has been a consistent justification for the maximization calculus, for the aggregation procedure and for the assumed equilibrating properties of the system. In fact, Walrasian rationality implied not only the optimizing behaviour of all agents, but also a very restricted form of that behaviour, since no single agent can benefit from specific informational opportunities (information is universal, evident and unique, and the agents must wait for the final decision of the auctioneer about the settlement of equilibrium), a part of the system does not generate and maximize profit from its intervention (the auctioneer himself) and no agent defines his set of objectives based on any other type of instrumental rationality (for example, institutionally or strategically defined). In this scheme, no government or oligopolies or any form of association of interests can exist, and the universe is restricted to atomistic agents submitted to uniform rationality and behaving homogeneously. Of course, if these fictions are dropped the result becomes indeterminate, multiple equilibria are possible, there are transactions costs or trade outside the optimum and, in general, there are non-convexities. On the other hand, the Pareto conditions on equilibrium also suppose that not all individual rational choices are fulfilled, since singular improvements would still be possible if the constraint on the other's welfare ceases to be imposed. In other words, equilibrium is neither a meaningful solution nor a logical implication of the aggregation of rational behaviours.

A most decisive precondition for the rationality axiom is that the universe of agents be homogeneous in behaviour and that all those agents be completely autonomous and self-sufficient. In that case, their relations may be ignored, and the economy is considered as the working of a single representative agent in each economic activity: the total independence of the agents is the logical requirement for the assumption of perfect rationality. But such a construction is obviously flawed: methodological individualism is denied by the idea of the homo economicus, since in this account no agent is a recognizable and distinct individual; and the aggregation of economic actions, which is supposed to settle the equilibrium, implies that humans do not know each other and yet trade with each other — in spite of completely ignoring the other's intentions or strategies. If otherwise individuals are considered and endowed of the human powers of purposeful action and of choice, the system becomes indeterminate and equilibrium a chimera. The 'legal atomic' independence of the agents is the experimentum crucis of the general equilibrium paradigm.[27]

Moreover, any other behavioural account beyond the heroic hypotheses of neoclassical rationality may imply a situation of non-equilibrium, since simultaneous optimization can no longer be fictionalized. Finally, this concept of rationality[28] — the world reduced to the same behaviour of the same type of agent, all gifted with negligible market power, the same tastes and resources — implies that there is no meaningful sense for the concept of 'choice'. Even worse: in such an hypothetical case, societies would get trapped in a Ricardian-type of stationary state, since there are no forces for change and evolution — in fact, different expectations, different endowments, different information capacities and availability of inputs are more realistic and efficient for creating innovation opportunities than the atomistic rational behaviour presupposed in the equilibrium paradigm. Furthermore, as Keynes pointed out, the general framework of decision making cannot be restricted to the case of alternative and measurable consequences, and qualitative elements and bounded rationality are always present.[29]

Therefore, the Schumpeterian story, be it proclaimed by the author as aimed at improving the neoclassical theory, was in fact deeply contradictory with the rationality postulate. Schumpeter recognized once this to be the case (Schumpeter, TGE: 192), even if he did not draw any general conclusion from it. The consideration of real time and real agents on the one hand, and on the other hand of rationality — restrictively defined as the maximization behaviour — are quite contradictory: time is change and choice and cannot be represented by the synchronicity of the decisions of maximization automata.

One recent and brave attempt to rescue the concepts of rationality and equilibrium is Sargent's book on bounded rationality. In spite of the misleading title, the author sustains once again the Rational Expectations hypothesis, although recognizing that it cannot explain the selection of an equilibrium when several possible equilibria exist, the regime changes or new sources of dynamics or disturbances such as major crises (Sargent, 1993: 24 f.). In spite of this poor record, Sargent still argues for the general equilibrium based on rational expectations: 'Why have economists embraced the hypothesis of General Equilibrium? One reason is that, if perceptions about the behaviour of other people are left unrestricted, then models in which people's behaviour depends on their perceptions can produce so many possible outcomes that they are useless as instruments for generating predictions' (ibid.: 6). In other words, the adherence to General Equilibrium is an imposition of the computability of the models; therefore the nature of its propositions about the world is subordinated to the limitations of the models supposed to explain it.

9.1.C. Certainty and equilibrium

In the previous discussion about the rationality-equilibrium connection, one of its main consequences was left aside: the notion of certainty in economic behaviour. This section deals with this concept, developing Kaldor's argument and discussing one of the actual neoclassical answers to this critique.

The debate about certainty and uncertainty informed the 'years of high theory', that period from 1926 until 1939 when Knight, Myrdal, Lindhal, and mainly Keynes developed detailed critiques of the social and individual behavioural assumptions of the mechanical metaphor.[30] These authors did not accept the axiom of certainty, namely that price information flows determine all available and useful information, and that preferences, organization and production sets are constant.

There are in fact two possible alternative concepts of certainty and uncertainty: for the first, uncertainty is bounded by some strict hypotheses about expectations — in other words, uncertainty is reducible to a certainty condition, for instance under the form of a known distribution of probabilities; for the second, uncertainty is epistemologically autonomous of certainty. The opposition is rather sharp, since the first version implies the reduction of the uncertainty concept to that of mathematical risk, as some precise distribution is assumed: that is, for instance, the strategy of the Rational Expectations models. Keynes argued extensively against this procedure (Keynes, 1937: 212-3), and so did Knight, defining uncertainty as a non-measurable value (Knight, 1921: 19-20) and Shackle, arguing that the irreversible quality of time prevented any meaningful knowledge of the probability distribution of economic actions (Shackle, 1990: xii). Instead, for the second approach, uncertainty is not an eventually transitory state of limited knowledge, but the normal situation of knowledge of complex systems.[31] Keynes emphatically indicated the nature of uncertainty:

> By 'uncertain' knowledge, let me explain, I do not mean merely to distinguish what is known for certain from what is only probable. The game of the roulette is not subject, in this sense, to uncertainty ... .The sense in which I am using the term is that in which the prospect of a European war is uncertain ... or the obsolescence of a new invention, or the position of private wealth owners in the social system in 1970. About these matters there is no scientific basis on which to form any calculable probability whatever. We simply do not know. ... I accuse the classical economic theory of being itself one of these pretty, polite techniques which tries to deal with the present by abstracting from the fact that we know very little about the future. (Keynes, 1937: 214-5)

Furthermore, the definition of a probability distribution requires the knowledge of a complete list of all possible outcomes and that the probabilities of all instances sum one: this is unrealistic for a large number of cases, since uncertainty is generally originated by subjective valuations or even by the very existence of simultaneous and multiple processes of choice in the economies. Certainty implies for instance the implausible assumption that money cannot be a reservation of value, which is necessarily influenced by subjective decisions. That celestial mechanics is history without evolution; certainty is only possible without time and without people.

The strategies of defence of the equilibrium and certainty paradigm have been based in two alternative arguments. The first one, as indicated, was the revision of the equilibrium program defining uncertainty as a probability distribution, which maintains the deterministic pattern of the theory but does not immunize it from the previous critiques. Hicks suggested in 1939 a quantitative procedure in order to reorganize the certainty axiom, with the introduction of an 'allowance for risk' to the agent in order to accommodate uncertainty (Hicks, 1939: 125-126). This defines a 'certainty equivalent', comparing the anticipations with a theoretical state of certainty (Baumol, 1970: 60 f.; 89). Arrow developed later on a generalized procedure in order to maintain the Walrasian system: the commodities in all future periods or possible environments should be considered as different commodities traded in the present, so that all contracts are made at one single point in time (Arrow, andHahn, 1971: 245), although the scheme is accepted as unrealist:

> We can imagine, for the moment, that markets exist today for deliveries of commodities at any date in the future. Then there will be an equilibrium for the system in which supply and demand are equated for all periods of time. ... Thus, on the hypothesis that markets exist for all commodities at all times, the general competitive model implies a time path of equilibrium dynamics. (Arrow, 1988: 276)

This is the point where Laplace's demon meets the Big Bang: for his infinite intelligence, all the future states of nature are collapsed into one moment, the present. But this means that all agents must have a complete list of all future states of the environment and be able to held the correct — and therefore identical — beliefs about those potential states, and still that all agents be able to compute the probabilities of each of these states. Moreover, it implies that once the contracts established for the present and for the future, no change is possible for the entire history. In other words, time flows without change and the future may be anticipated under the condition of being fully known to the agents: uncertainty is completely eradicated from the system.

This strategy was discussed by Arrow, as he recognized the difficulties in the equilibrium dreamland: in his theory of information, he changed the assumption about commonly shared information of all Walrasian agents, and allowed for uncertainty as the outcome of the system. Uncertainty, 'a condition which is all too obviously true in economics' (Arrow, 1984: 44), was modelled as either (i) random terms, or (ii) ignorance about the parameters, or (iii) ignorance about the structure of the equations. Uncertainty was therefore considered as a state of knowledge, and information defined as a 'negative measure' of uncertainty (ibid.: 138), in what was described as a 'considerable revision of General Equilibrium Theory' (ibid.: 139-40).

Radner reconsidered the general equilibrium model with the introduction of uncertainty and concluded that the computational capacity should therefore be virtually infinite in order to maintain the Arrow-Debreu equilibrium (Radner, 1968: 31; 1982: 925; 1989: 310). If otherwise institutions are considered, or the limits of computational capacity are identified, the general equilibrium paradigm falls. Silvestre added that even small deviations from the unrealistic standard assumptions of general equilibrium could lead to movements away from equilibrium, which will no longer converge (Silvestre, 1993: 105 f.). All these remarks by Arrow, Radner or Silvestre share the same concern for the revision of the model under more realist assumptions, but the conclusions are very pessimistic about the final result of the project.[32]

Other possible strategies of defence of the core hypotheses of the program followed an alternative path, simply denying any relevance to realism and evading the whole problem: this was in fact Friedman's solution in the 1953 essay arguing for instrumentalism. This was very frankly supported by Solow: 'The fundamental difficulty of uncertainty cannot really be dodged; and since it cannot be faced, it must simply be ignored' (Solow, 1963: 15).

This radical Cartesian shift in the neoclassical program moved it away from any possibility of empirical testing, rejected realism of theories and realisticness of assumptions and exiled it to pure tautological proof. Better than anyone, Lucas described this epistemology under the following considerations:

> One exhibits understanding of business cycles by constructing a model in the most literal sense: a fully articulated, artificial economy which behaves through time so as to imitate closely the time series behaviour of actual economies. (Lucas, 1977: 11).

Or else:

> On this general view of the nature of economic theory then, a 'theory' is not a collection of assertions about the behaviour of the actual economy but rather an explicit set of instructions for building a parallel or analogue system — a mechanical imitation economy. A 'good model' from this point of view, will not be exactly more 'real' than a poor one, but will provide better imitations. (Lucas, 1980: 697)

Abandoning the ambition of a real science, general equilibrium has been consequently reduced under the New Classicals to the department of mechanical models and demonstration by simulation. But, for those studying cycles, which are suspected to be real and have observable instances, this is of course no adequate alternative: uncertainty, which is a qualitative phenomenon, cannot be reduced to a measurable quantity, just as subjectivity cannot be reduced to objectivity through any correspondence rule.

9.1.4. Dynamics and equilibrium

For the majority of the models in the framework of the neoclassical program, the assumption about the existence of a unique general equilibrium rest point holds, such that supply equals demand and all markets are cleared. Static or comparative static analyses have been conceived of as the pertinent tools for describing the properties of those systems. The fortress of the program has been protected by the impressive walls of the certainty, the rationality and the equilibrium postulates, and minor changes could be explored if faithful to the core hypotheses, such as assuming uncertainty as an expression of mathematical risk under a known probability distribution, as a certainty equivalent. Nevertheless, the analytical properties of the derived models are still too limited.

In order to overcome those limitations, a major modification was introduced by the development of dynamical equilibrium analysis, namely by the inquiry into the stability properties of dynamic systems. Dynamics was rather loosely defined: 'We may say that a system is dynamical if its behaviour over time is determined by functional relations in which "variables at different points of time" are involved in an "essential" way' (Samuelson, 1943: 59). But this definition is ambiguous, given the fact that its mathematical formalism subjugates the substantial meaning, since it makes possible two different and opposed conceptions of dynamics and statics: (i) statics can be defined as a special degenerate case of dynamics, or (ii) dynamical systems can be conceived of as all those not static. In the first case, with the usual notations,

$$dx_i/dt = f_i(x_1, \ldots, x_n) \quad \text{where } i = 1, \ldots, n \tag{9.1}$$

the system is static when f_i=0 for any i. In this case, the solution involves annihilating the time dimension, and equilibrium is defined as the limit of the dynamical behaviour:

$$lim\, x_i\,(t; t_0) = k \quad \text{as} \quad x \rightarrow \infty \tag{9.2}$$

The convergence towards this point of equilibrium is the crucial property of the system. This is no more than a 'precise analogy with mechanics' as Harrod claimed: dynamics is defined as the rate of increase at one point of a particle under the action of determined forces, and statics is that particular case in which no force is acting (Harrod, 1960: 279).

Of course, if dynamics is defined by the action of forces any variation may create damaging instability, but Samuelson argues that unstable positions are only transient, and that reality is essentially stable: just as the egg cannot stand by itself in an unstable position, all disequilibrium is transient (Samuelson, 1947: 5, 11, 182). As a consequence, structurally unstable models should be rejected.[33] But this is contradictory with the traditional multiplier-accelerator models, which can only be conservative (that is, maintaining the harmonic oscillation in order to represent the persistence of the cycle) under very specific and unchangeable values for the parameters: those models are structurally unstable, since the smallest variation of the parameters may imply a qualitatively different topological behaviour of the system.

In the second case, we have for the dynamical case the variables $x_i = x_i(t;\, t_0)$ and the general unspecified system $h_i(x_i, \ldots x_n)=0$, while for the static alternative systems no time variable is involved. Equilibrium is defined for this dynamical case as an attractor (be it a mechanical rest point or a strange attractor) satisfying a set of static conditions (market clearing), but there are no forces or mechanism establishing automatically this situation, that is, there is no stability or convergence property.

The two concepts are substantially opposed, as the previous example of Samuelson's bullet showed. Of course, appearances may fool someone: an apple is an apple either if it is being eaten by Mr. Newton under the tree or if it is catastrophically falling towards his head; but still there is a crucial difference that Mr. Newton is supposedly going to detect very soon, and the same applies a fortiori to the example of the bullet.

The steady state trajectories are conceived of on the basis of the first of these analogies with dynamics: as a sequence of equilibria — the ideal representation of the trend (Schumpeter, BC: 156; 553) — the trajectory represents constant ratios and exponential growth of the essential variables, under the assumption that rational expectations are fulfilled and a Paretian equilibrium is sustained.[34] As previously argued, this approach is easily reduced to the comparative static methodology.

The incorporation of the alternative concept is however much more difficult, since there is no assumed equilibrating mechanism. Yet, it describes a realistic representation of the different states in nature, and therefore assumes the relevancy of a diversity of models dominated by structural change and non-equilibrating dynamics. This is true for the society as a whole, and it is also true for the economic systems, which are a product of morphogenetic evolution. The assumption of fixed input-output coefficients, or of a stabilizing property of the general linear model (9.4), may not even hold for the short term: steady growth is an unrealistic extension of the unrealistic metaphor of equilibrium.

In other words, the generalization of the concept of equilibrium to trajectories and to time is only possible under additional severe restrictions which require a certain behavioural assumption and rules out structural change by definition: of course, this is not a suitable approximation for the study of real dynamical processes. Once again, here it is the obliquity error: similarity does not entail causality. In fact, in physics, the secondary subject, the concept of equilibrium was originally related to deterministic experimental phenomena, in which uncertainty was supposed to be fixed in each case (for example, the measure of the orbit of the Earth around the Sun), but such an assumption is neither possible nor relevant in economics.

The generalization of the equilibrium concept from the rest point to the steady growth trajectory, and finally to any form of attractor, is not a solution either, since the generalization modifies and undermines the concept to the point that it becomes inapplicable. Poincaré generalized the concept of equilibrium from a state of rest to a closed curve, and later on Lorenz generalized it to from the closed curve to a closed region, but there we have already an unpredictable variety of aperiodic motions which are indistinguishable from genuine randomness. In this context, the concept of equilibrium loses all its analogical properties so useful to mainstream economics, since the only equilibrium states to be modelled are now the limit sets receiving these aperiodic and apparently erratic trajectories. And since from the attractors we cannot trace back the equations, the systems are unknowable: consequently, eventual equilibrium is not even recognizable.

In this nonlinear world, where Samuelson's second concept of dynamics is identifiable, the only parallel concept to that of equilibrium is the attractor, as Baumol and Benhabib argue: 'An attractor is what most of us might describe as the equilibrium or limit time path of a stable dynamic system, whether or not that system is chaotic' (Baumol and Benhabib, 1989: 91). But the attractor concept cannot reorganize and indeed it destroys the equilibrium metaphor, since it is conceived of as a metastable state in the context of the evolution of a system, which is quite unpredictable and in which uncertainty is an increasing function of the variations of the parameters and the unknown initial conditions.[35] Therefore, in these nonlinear processes, the traditional concept of equilibrium is lost (Granger and Terasvirta, 1993: 14). In short, the attractor is only compatible with the equilibrium metaphor under the constraint of a very low dimensional system — a fixed rest point or a limit cycle, asymptotically stable. In superior dimensions, the notion of attractor can be no longer equated to equilibrium, since in general there is no structural stability in those conditions, as proved by Smale.

This particular question has been at the centre of some strong debates between physicists and economists eager to appropriate the new gadgets of attractor theory. The first meeting of the new-born Santa Fé Institute, in 1987, was precisely dedicated to that matter and quickly witnessed the sharp difference between both families of scientists. As one of the participants in the meeting made very clear, the assumptions of equilibrium shared in the neoclassical program are incompatible with the notion of strange attractors:

> Let me state things somewhat more brutally. Textbooks of economics are largely concerned with equilibrium situations between economic agents with perfect

> foresight. ... The examples of chaos in physics teach us, however, that certain dynamical situations do not produce equilibrium but rather a chaotic, unpredictable time evolution. (Ruelle, 1991: 113)

Another participant at the same meeting, the physicist and Nobel Prize winner Philip Anderson, put his arguments in an elegant way:

> From William Brock's summary and José Scheinkman's and Thomas Sargent's discussion of the concept of the Arrow-Debreu theory, we learned that even theories which appeal to the concept of 'equilibrium' do not necessarily avoid the apparently random fluctuations in the course of time which characterize of driven dynamical systems in physics. In physics these are called 'non-equilibrium' systems; a liberal education on the various meanings of the word 'equilibrium' was a bracing experience for all. (Anderson, 1988: 265)

As Anderson indicated, the time symmetry and the probability of equilibrium, current in some physical systems, are paralleled by time asymmetry and probability of disequilibrium in economics (ibid.: 267). Furthermore, the participants at that meeting could hardly accept the general equilibrium postulates of rationality and perfect foresight.[36]

*

Static equilibrium is the foundation of neoclassical economics. Its epistemological requirements — certainty, behavioural rationality, atomism and aggregation properties — are intimately connected and mutually dependent. The critiques reviewed in this section challenged this theory, stressing the irrealism of these hypotheses (Mitchell), their implications and shortcomings (Keynes) and the error of their causality concepts (Kaldor), and can be applied to describe accurately the limits of the rocking-horse methodology for the analysis of cycles. Two strategies of defence were discussed, the first making possible the transformation of the certainty postulate in order to include fixed and known 'uncertainty', the second one rejecting any ambition of realism in economics. None of them solves the contradictions of the 'dreamland' approach.

The reason for the Frischian ontological and epistemological distinction between propagation and impulse systems can now be perceived in all its dimensions: it is the very condition for the instrumental definition of positive causality and for the computation of the attributed model, however arbitrary the attribution may be. Otherwise, if both were conceived of as part of the organic whole, the impulse system could be modelled as being able to create meta-regime transitions and to interfere with the structure of the propagation system, and this one should be modelled as being able to generate some of the impulses. In order to prevent such a disruption of the paradigm, morphogenesis and complexity are excluded by the positivist reaffirmation and separation of the context of motivation from the context of equilibration.

9.2. Trend and cycle: the first decomposition

The distinction between trend and cycle is the analytical prerequisite of the Frischian approach to fluctuations: in the first decomposition of a real time series, trend is considered as the equilibrium path and the cycle as the residual. This operation corresponds to good common-sense Cartesianism allied to traditional positivism: subdivide the difficulties into parts in order to study each of them and to create the entities suitable for empirical work submitted to positivist criteria. The whole construction was previously criticized, since the behavioural foundations (optimizing rationality) and the contextual assumptions of equilibrium are neither adequate nor realistic. Now it is time to reconsider the problem in a distinct framework: the analytical procedures for operating the distinction between cycle and trend are reviewed, in order to conclude that there exists neither a satisfactory method nor a coherent economic theory justifying the common mathematical procedures.

The previous sections developed an external critique of the rocking-horse paradigm, opposing the global theoretical basis of the method. The two following sections recapitulate those critiques and show their implications to the study of trend and cycle and some alternative methods, but the third will present an internal and powerful although implicit critique, developed within the orthodox theory itself.

9.2.A. Mitchell and the NBER method

The acceptance of the distinction between trend and cycle implied a considerable epistemological revolution in economics, since the evidence of life suggests that the trend is the abstract entity, the intellectual construct, and that the cycle represents both growth and change. In effect, the cycle is observable in time series and economic behaviour; it is the cycle which is discussed in policy-making. The trend is discernible or conceivable as the representation of the evolution of some long term tendencies, such as the evolution of accumulation, population or knowledge and technological competence, but anyway growth expresses itself under the concrete form of irregular cyclical movements. Thus, the problem of decomposition of the time series into trend and cycle is a mere consequence of the early conjectures of equilibrium economics, and it is not by itself an analytical imposition of the realism of models or of theories. It is because equilibrium is so powerful a world-vision that the core of economic explanation must be rescued from perturbations and encapsulated in a deterministic trend as the stable locus of equilibrium: philosophy supersedes economics and ideology dominates the technique.

As a consequence, the first decomposition was addressed by very rough statistical devices as a technical question with no major economic justification. The review of the first methods used in order to deal with the problem confirm this conclusion: (i) Warren Persons, by 1919, eliminated seasonality from time series, then fitted a straight line using the least square method and computed the values of deviations expressed as percentage values of the original series; (ii) Frickey limited himself to eliminate seasonality and to express the series as link-relatives, each value as the percentage of the previous one, but in some circumstances this may produce the vanishing of the trend; (iii) Kuznets computed decade averages to define the trend; and (iv) Kondratiev used a more sophisticated procedure for his long waves evidence, calculating a nine years moving average on detrended deviations — the trends were described as linear or nonlinear curves without any economic justification for each choice — then a new five years moving average was computed and the resulting data was finally multiplied by weighted coefficients expressing the relation between initial deviations and averaged deviations. But these methods were unsatisfactory, both by their assumptions and by their assumptions and by their spurious consequences.

Mitchell, the founder of the American NBER (National Bureau of Economic Research), which was for a long time the main centre for the research on cycles, developed a more sophisticated statistical method for the inspection of trend and cycle. As previously indicated, he did not accept the equilibrium assumption, and was therefore entitled to conclude that:

> Secular trends of time series have been computed mainly by men who were concerned to get rid of them. Just as economic theories have paid slight attention to the 'other things' in their problems which they suppose to 'remain the same', so the economic statisticians have paid slight attention to their trends beyond converting them into horizontal lines. Hence little is yet known about the trends themselves, their characteristics, similarities, and differences. Even their relations to cyclical fluctuations have been little considered. (Mitchell, 1927: 212-3)

The obvious conclusion is that the statistical fitting of a trend curve is not a trivial choice. A log-linear trend assumes growth at a constant rate, a straight line with positive slope assumes permanent growth, the choice of the hyperbolic curve assumes a diminishing percentage rate of growth. Every choice is theoretically significant, and in fact most determinant to the following conclusions: a mechanical fit (like the standard least squares procedure) can hide cycles, or affect the turning points (Burns and Mitchell, 1946: 38, 270, 309), and not always does it have clear a economic interpretation. The importance and indeed the operational significance of the paradigmatic assumption of equilibrium can now be grasped: only such a type of hypothesis can avoid further questioning about the theoretical nature of the trend which was fitted to the time series, as this is hierarchically ruled by the preliminary epistemological assumption of the decomposition procedure — equilibrium is the Inquisition Court of statistical decomposition.

For Mitchell, the consequence was that a theoretical and historical basis was needed for any analysis of cycles and that the trend should be fitted to those assumptions:

> If an investigator fits a trend line in a mechanical manner, without specifying in advance his conception of the secular trend or of cyclical fluctuations, he may get 'cycles' of almost any duration. But an informed investigator who is seriously studying cycles of a given order of duration will use whatever guidance he can get from history and statistics ..., seek to mark off in advance the cycles or traces of cycles that correspond to his basic conception, then choose a trend line that cuts through and exposes the cycle in which his interest centres. (Burns and Mitchell, 1946: 37-8)

Mitchell went on arguing that the criteria for fitting a curve should not be based on any form of statistical minimization of the deviations, but rather on previously defined conceptual hypotheses, under the form of some historical explanation.[37] The existence of the trend should then be reassessed from factors which simply act in longer schedules than the business cycle: population variations, changes in education, technological knowledge and equipment, or the quantity or quality of resources. The trend is thus causally related to the cycle, and as a consequence they are indistinguishable. But this is not the end of the story.

Burns differentiated between the 'primary trend', the statistical construct of an exponential curve fitted to the data, and the 'trend-cycles', the computed decade rates of growth, accounting for some long-wave type of oscillation (Burns, 1934, 174 f.): the concept of equilibrium was alien to this methodology, since the trend itself was considered to have 'undulatory contours' (ibid.: 32, 44). Mitchell developed a some more sophisticated approach, later called the NBER Method: specific cycles for each series and reference cycles for the economy were computed considering periods established from turning points defined by visual inspection of the series and from historical evidence; the cycles were computed from trough to trough, and the values of the variable were defined as percentages of the average value of each sub-sample. As a consequence, the 'intra-cycle' trend — the global growth during the sub-sample period — was considered, but the 'inter-cycle' trend — the evolution distinguishing one cycle from the other — was not (Burns and Mitchell, 1946: 26; 56; 131). This was a minor inconvenience, according to Mitchell, since the first year of one period is always the last year of the previous one.[38]

By the decade of 1960, the NBER Method was transformed and the econometric approach dominated: the NBER researchers adopted a 'growth-cycle' approach, defined as a 'fluctuation around the long run growth trend of a nation's economy, that is, a trend-adjusted business cycle' (Moore and Zarnowitz, 1986: 772). A nonlinear specification for the trend is used, based on an interpolation between segments of series determined by a 75-months moving average (Zarnowitz, 1992: 213).

9.2.B. Goodwin: nonlinearity and the decomposition problem

In 1953, in one of his early papers, Richard Goodwin suggested that the economy could not be considered in any steady growth process, but should be analysed by the variation of rates of variation (Goodwin, 1982: 112). In what constitutes the major development since Mitchell, Goodwin defied the traditional separability of trend and cycle: 'If we decompose the behaviour of such a model into trend and cycle, the result must be regarded as purely descriptive and with no identifiable counter-part in the model' (ibid.: 113; also 122). There is no theoretical justification for the trend adjustment, every decomposition is arbitrary and depends on the linearity assumption of the superposition of causally independent[39] impulse and propagation mechanisms.

Alternatively, Goodwin suggested the Schumpeterian notion of a unified theory of growth and cycle. 'Specifically, it is the vigorous boom which does generate the trend, and it is this leap forward into new levels of output which governs the subsequent slump' (Goodwin, 1982: 117), and the analysis should 'start from the hypothesis that growth generates cycles and that cycles interrupt growth' (Goodwin and Punzo, 1987: 106). The precise notion of causal interdependence remains a model specification.

The influence of Keynes and Harrod is evident in this conception of the cycle as the concrete form of growth;[40] the Schumpeterian inspiration of this critique is also obvious in the concept of a single entity including both phenomena. Developing this vision and rejecting the equilibrium approach to the trend, Goodwin provided a considerable basis for a new wave of modelling and theoretical analysis of fluctuations.

But the major improvement was nevertheless his critique of the linearity assumption, which is so essential to the credibility of the decomposition procedure: if the economy is conceived of as a set of nonlinear and complex relations, then one must recognize that the traditional decomposition is irrelevant, arbitrary and misleading. This is a rather definitive critique and opens a new way for the interpretation of cycles as organic phenomena, more akin to biology than to physics and requiring the tools of bifurcation analysis, chaos and complexity in order to understand the morphogenetic features (Goodwin, 1985: 7; Goodwin, Kruger and Vercelli, 1984: x).

9.2.C. Nelson and Plosser: RBC and the stochastic trend

One of the most important features of the debate between the New Classical Equilibrium Business Cycles type of model (EBC, Lucas, Barro, etc.), based on an impulse system of stochastic monetary shocks, and the Real Business Cycles approach (RBC, King, Nelson, Plosser, etc.), is relevant for the discussion of the decomposition procedure. In an important paper published in 1982, Nelson and Plosser argued that macroeconomic time series are not stationary fluctuations around a deterministic trend, but instead non-stationary stochastic processes with no tendency to return to a deterministic path (Nelson and Plosser, 1982: 139).

This alternative conceptualization does not reject the decomposition method, but emphasizes some of its essential problems. The macroeconomic variable is still described as

$$y_t = {}^p y_t + {}^c y_t \tag{9.3}$$

a sum of the permanent and the cyclical components. If ${}^p y_t$ is considered as a trend and ${}^c y_t$ as a stationary process, the traditional and EBC distinction between the two elements means that: 'One is tempted to interpret the permanent component as the natural rate of output, and the temporary component as the deviation of output from the natural rate' (Campbell and Mankiw, 1987: 115).

Nelson and Plosser argued that this interpretation is not accurate if it implies a deterministic permanent component of the series. Criticizing the traditional method of detrending the time series against time (or a polynomial on time) and of considering the residuals as the cycle, these authors denounce the misspecifications derived from the application of a deterministic model to an integrated stochastic process, which can generate pseudo-periodic behaviour and whose auto-correlation function of the residuals depends on the sample size (ibid.: 140). Furthermore, the specification of a deterministic trend implies the doubtful exclusion of long run uncertainty.

The stochastic trend approach was developed by Stock and Watson (1988a,b), who hypothesize that variations in trends are responsible for an important part of the fluctuation of output in U.S. series: as Beveridge and Nelson (1981) and Nelson and Plosser (1982), they state that business cycles can be defined as adjustments to changes in the growth paths. If this is the case, a regression analysis and the first decomposition mislead the researcher (Stock and Watson, 1988b: 149).

The reason is obvious if we consider the two classes of processes studied by Nelson and Plosser: (i) the TS process (trend-stationary), where ${}^p y_t = \alpha + \beta t$, and ${}^c y_t = c_t$, α and β being parameters and c_t a stationary process with zero mean, of which the EBC models are consistent examples and (ii) the DS process, a non-stationary process whose first or higher order differences are a stationary process. In the first case the variance of the forecast error is finite, the uncertainty being therefore bounded; in the second case, the forecast error is unbounded (Nelson and Plosser, 1982: 122 f.). In the first case we have a deterministic system superimposed by a stochastic disturbance, and decomposition is redeemed; in the second, the deterministic decomposition is dramatically wrong.

For the series Nelson and Plosser considered, the null hypothesis of being DS processes could not be rejected. Therefore, they concluded that there was enough evidence to argue that non-stationarity arises from 'accumulation over time of stationary and invertible first-differences' (ibid.: 160) and that even if a decomposition is assumed, ${}^{p}y_t$ cannot be considered to be a deterministic trend but should be viewed as a non-stationary stochastic process. Friedman's hypothesis of permanent income provides an example for this type of DS process, where

$$y_t = {}^{p}y_t + {}^{c}y_t \quad \text{where} \quad {}^{p}y_t = y_{t-1} + u_t \text{ and } {}^{c}y_t = v_t \tag{9.4}$$

The permanent component is a random walk with innovations u_t (with $\text{cov}(u_t,u_s)=0$ for any s≠t) which contains no transitory movements; it is a stochastic trend, while the cyclical component is a stochastic independent process, v_t, which is serially uncorrelated; ${}^{c}y_t$ has no trend. The additive relation between both components is assumed (Beveridge and Nelson, 1981: 156-7) and, of course, after differentiation the trend is still present (u_t). But further hypotheses are needed to make possible a general decomposition and the estimation of the parameters, and two different assumptions are suggested. If the two families of innovations are perfectly uncorrelated, then estimation is possible on the basis of the autocovariances and variances of the first differences (Nelson and Plosser, 1982: 154-5); the RBC models assume this possibility of decomposition, that is, are restricted to those assumptions. If alternatively one assumes the restrictive assumption that the innovations are perfectly correlated, arising from the same source as one single process, estimation is trivial (Beveridge and Nelson, 1981: 151 f.). Nevertheless, as Stock and Watson proved, the two strategies produce largely different estimates of the effect of an innovation on the long term trend, the difference being two to one (Stock and Watson, 1988b: 156). But if these assumptions are relaxed, then the parameters cannot be identified (ibid.: 155; Zarnowitz, 1992: 184).

The whole conceptual construction of stochastic trends undermines the idea of deterministic processes, and does not lead to a general method: the computation of the trend and the estimation of the parameters imply severe restrictions on the definition of the nature of the random behaviour. Moreover, the results of the procedure depend on the explanation of those random shocks: if the larger part of the real fluctuations is explained by the trend variations, then the cycles will be considered as empty categories of residual white noise; alternatively, the assumption about a single source for the stochastic shock — the monetary and technological disturbances — implies the dominance of the impact of the cyclical component. The arbitrariness of the choice is additionally emphasized by the narrowness of the epistemological foundation of the random term concept.

9.3. Endogenous, exogenous and semi-autonomous variables: the second decomposition

The Frischian methodology developed a classification of variables according to their belonging to the impulse (exogenous) or the propagation (endogenous) systems, and provided an economic rationale for such a distinction. This section will examine some arguments against that claim. The second decomposition implies an epistemological contradiction since it leads to the primacy of endogenous theories, given that equilibrium and immutability define the space of intelligibility, whereas the causality for the maintenance and shape of oscillations is simultaneously attributed to external sources (random shocks). Moreover, even the cognitive status of this single random variable is not clear and two main interpretations were presented. Frisch interpreted it as a measure of error, as an error-in-variables: 'Following Frisch, each of the variables may be conceived as the sum of two components, a "systematic component" or "true value" and an "erratic component" or "disturbance" or "accidental error". The systematic components are assumed to satisfy the regression equation exactly. ... The erratic component is taken as an error in the literal sense of the word' (Koopmans, 1937: 5-6), and every regression may be interpreted as the unveiling of these two components. Alternatively, Koopmans and Tinbergen interpreted the random error in a much more general way, as error-in-equations: 'The disturbance in any one equation is here looked upon as the aggregate effect of many individually unimportant or random variables not explicitly recognized in setting up the behavioural equation in question' (Koopmans, 1957: 200). As always more prudent, Tinbergen admitted: 'The error term is introduced as a catchall for less important independent variables and for measuring errors of both the dependent variable and the independent variables. ... Essentially, the introduction of an error term is a second best setup and in a way a *testemonium paupertatis*' (Tinbergen, 1990: 201).

Of course, the implications of both interpretations are devastating to the impulse-propagation model, since it is not possible to discern the explicatory capacity of errors of measurement — a feature of the measuring apparatus and not of reality — or of the alleged incapacity of the model (or of the modeller himself) to incorporate the relevant variables. The remaining theory is therefore doomed to be a testemonium paupertatis of the statistical knowledge.

Some attempts to combine explicative influences of exogenous and endogenous variables for the sake of realism are more literary than methodological,[41] and had no significant impact on the orthodox dichotomy. Orthodoxy is itself the problem.

The next argument proceeds as following. First, a very common case of confusion induced by such a classification is presented: the analysis of innovation. Second, two theories, by Schumpeter and Keynes, will be briefly described from the point of view of their considerations about endogeneity. Third, a general category of semi-autonomous variables is presented, in order to deal with the classificatory incoherences just mentioned. Finally, the conclusion indicates new criteria for classification.

9.3.A. Where does innovation innovate?

The traditional distinction between impulse and propagation mechanisms is supposed to clarify the role of each variable and to add powerful logical rules for the formulation of every theory. The reality shows that such an expectation cannot be fulfilled, and the example of the treatment of the concept of innovation is conclusive.

The definition of innovation in Schumpeter's theory of cycles is the first example evidencing the inappropriateness of the dichotomy. It was considered by the author to be endogenous, and thus belonging to the explained propagation mechanism; so do Guesnerie and Woodford (1992: 289). In the same sense, Grandmont and Malgrange classify a large group of models as internally generating fluctuations, as those of Schumpeter, Haberler, Hayek, Wicksell, or the Keynesian tradition (Grandmont and Malgrange, 1986: 3). Hicks distinguished between external causes (wars and revolutions) and causes explainable by economic reasons, such as innovation (Hicks, 1950: 63). But Goodwin classifies innovations as external factors, just like wars (Goodwin, 1982b: 121); Haberler, although stressing the conventional rather than logical character of the distinction, classified innovation as exogenous — indeed, he added that the ideal of purely endogenous explanations for the cycle was inherited from Laplace (Haberler, 1943: 543). Hansen considered the progress of technology, like the growth of population, wars, autonomous monetary interventions, as 'outside impulses' (Hansen, 1952: 303), even if he carefully discriminated in a previous text between the exogenous condition of the changes in technology and the endogenous condition of innovative entrepreneurship (Hansen, 1951a: 308). But the general methodological problem survives these attempts of definition. Moreover, some authors state that the condition for the coherence of an explanation is endogeneity in the sense that technical innovation and change should be completely explained by the model (for example, Rosenberg and Frischtak, 1986: 10, 22).

Apparently, this is a dialogue of the deaf: the very authors indicated as producing endogenous explanations for innovation classify it as an exogenous factor, or at least accept the separation between impulse shocks and internal mechanisms of innovation itself (Hansen, Wicksell). To add some more confusion to the picture, Hansen called the internal mechanism of entrepreneurship the 'impulse', to be distinguished from external technological shocks. And Schumpeter considered innovation as an endogenous feature of the economy, indeed its most important feature, but that did not prevent his acceptance of Frisch's pendulum analogy which defined innovation as exogenous, in spite of his own reticences.

Some recent developments in the literature, inspired by Schumpeter's work, distinguish between exogenously generated inventions and endogenously processed innovations. This distinction is no doubt analytically useful, but it cannot discriminate clearly all the involved processes, since the majority of the inventions is created within the economic institutions of capitalism. New growth theories even suggest that all growth is created by investment or by specific types of investment, thus accounting for the innovation process. For the sake of precision and realism, the researcher cannot avoid treating innovation as a complex and inter-linked endogenous *and* exogenous variable; moreover, as the frontier is defined by the model itself, a prudent attitude should be followed in relation to the presumed implications of such a distinction on the explanation of reality.

Obviously, some of the difficulties and contradictions in the different authors' considerations about the nature of Schumpeterian innovation were partly due to Schumpeter himself, mainly to his lack of clarity about the scope of the model to which the variables are referred to. This general element of confusion is no doubt a consequence of the amateurish metaphorization and of the modelling procedures unable to avoid the error of oblique transfer, since what some authors consider to be an endogenous variable in the economic sphere in society is for others an exogenous attribute of the specific mathematical form of the model, which can only treat a substantially reduced part of that sphere.[42] The trade-off between formal rigour or tractability of models and the consideration of a glimpse of reality is a first choice to be made by the theorist, and such a clarification makes possible the comparison of different types of models.

In this context, it must be noted that the theory and the specific model developed by analogy to the theory may — and normally will — held incongruent hierarchies of causality and, therefore, different concepts and frontiers of endogeneity. The very common example of governmental expenditure or bank activity and regulations illuminates this point: their exclusion from simple models (for example, Lucas's models), thereby implying their formulation in the framework of the random error, do not necessarily mean that they are unexplainable and lie outside the scope of economic theory, but merely that those are the specific attributes of a specific model. The same applies for all the relations of power: for instance, for Sherman the concept of institutions of capitalism forms the endogenous system and, as a consequence, business cycles are completely explainable by internal causes since all causes are simply by definition internal to the system (Sherman, 1991: 5); naturally, the author classifies the theories of Marx, Mitchell, Keynes and Kalecki as endogenous. This choice is of course theory-dependent, and not a feature of some dubious objective nature of the models themselves.

The endogeneity-exogeneity antinomy depends therefore on three conditions: on the theory-model relation and specification defined as the space of the model, on the evolution of the disciplinary boundaries as a whole defined in the space of the theory,[43] and on the space and time dimensions under consideration.[44] The reason for the unending discussion about the topic may then be easily grasped: while the neoclassicals define their choices in the exclusive space of the models, the distinctive features of the Marxian and Keynesian visions are defined in the space of the theories and Schumpeter's in the time dimension: they are paradigmatically incompatible.

9.3.B. Evolutionary accounts of endogenous systems

The antinomy just addressed has been a theme of economics since the very early classics. For Marx, the last of the classical authors, the space of the theory was privileged, and his analysis of cycles provides a clear illustration: after the generalization of the market and in the epoch of 'industrial mechanics', the rotation of capital under the form of the renovation of equipment and circulation of value was supposed to create expansions and contractions (Marx, 1885: 94). But the capital was defined as a social relation, a condensation of the forms of domination over wage labour, thus incorporating the historical and political conditions in the evolutionary account of the cycles, departing from the Newtonian analogy of the orbits of the planets: the effects become causes.

If this approach is accepted, then the status of the distinction between exogenous and endogenous variables is considerably changed. Other evolutionary theorists dealt explicitly with this argument.

9.3.B.1. Schumpeter: economic mutation

In the 1951 preface to Schumpeter's *Ten Great Economists*, his wife Elizabeth wrote that he shared with Marx one central vision of the economic process as created by internal factors,[45] and this point had been made several times by Schumpeter himself. This suggests, if the analogy is true, a very broad view of the economic process a dialectic of contradictions.

In effect, Marx considered the most important economic variables — rate of profit, rate of exploitation, capital — to have what will be called here a status of *semi-autonomy* is his model: they were internally explained, but were also referred to crucial historical factors and conflicts which were impossible to summarize in strict economic terms or even in some cases to quantify. This was not an exceptional feature in the theory but rather the rule: commodity itself and value were considered to be representations of social relations. A limited economic analysis and quantified measure, understood in a positivist way, was considered to be insufficient and indeed incompetent to scrutinize reality, and namely those social relations. Yet, these concepts indicate meaningful relations and are building-blocks of the theory.

Schumpeter was recognizably close to this vision in his evolutionary approach. In his 1941 preface to the Spanish edition of *Theory of Economic Development*, he stated that:

> first, one must deal with capitalism as a process of evolution, and that all the fundamental problems came from the fact that it is an evolutionary process; and, second, that this evolution does not consist neither in the effects of external factors (including political ones) on the capitalist process, nor in the effects of a slow growth of capital, population, etc., but in that sort of economic mutation, I dare to use a biological term, which I have called innovation. (Schumpeter, 1941, in 1911: 14-5)

As a general rule, Schumpeter described innovation as an endogenous factor. But, since he did not provide any theory for the creation of innovation or for the bunching of the new combinations, some scientists felt entitled to make the distinction, as Hansen, between the original conception — the invention, an exogenous, non-explained process — and the economic diffusion of the innovation, under the impulse of the opportunity-seeking entrepreneurs. Then, innovation is a falsely endogenized variable and yet it is not a purely exogenous one.

The same concept appears in other texts by Schumpeter. For instance, he distinguished between exogenously (chance) motivated discovery of new territories and endogenously determined processes of exploitation of those areas (Schumpeter, BC: 499), just like the general process of innovation that he explicitly used as an analogy for territorial expansion. And in the last page of his *Business Cycles* (1939), he explained the 'disappointing features of the present Juglar' by external factors, and argued that such an event was to be generally expected in 'transitional stages'. In a precious footnote, he then added that those factors were considered external 'in the narrow sense adopted for the purposes of this book', since 'In a wider sense those factors and the mentality or moral code behind them are not external to the process of economic evolution but part of it' (Schumpeter, BC: 1050 and n.).

Two conclusions are relevant. First, Schumpeter clearly separated the context of the model and the general framework of his theory: for the first, economic mentality is undoubtedly an external factor, for the second it is considered as a contextual internal variable or a function. Second, this clarifies the procedure of model-making but it does not define the precise status of the central variables. If they are defined as *semi-autonomous*, their discriminatory role is lessened and as a consequence they would assume the apparently more imprecise status of a central explanatory variable which, as in the case of innovation, cannot be totally determined either by the model or by the economic theory itself. In that case, some sort of prudent combination should be put forward, such as Haberler's: one serious theory must include exogenous and endogenous variables, being unsatisfactory if only internal factors are comprised, and irrelevant if only external factors are considered (Haberler, 1943: 10).

In any of the cases, the Schumpeterian concept of innovation is the example of a major failure of the Frischian-type of classification, confirming the previous point: Frischian classification is only adequate to classify Frischian models. Outside those boundaries, such a classification fails and suggests confuse and ambiguous conclusions.

9.3.B.2. Keynes and the animal spirits

Keynes's theory presented three classes of variables which cannot be simply classified as endogenous (or exogenous): (i) the three psychological schedules (propensity to consume, desire for liquidity determining the rate of interest, expectation of future yields of capital assets determining the marginal efficiency of capital); (ii) the wage unit, including the social conflict about the determination of the wage; and (iii) the quantity of money. The aggregate impact of those variables is the generation of uncertainty, non-measurable and non-quantifiable uncertainty, or the influence of 'animal spirits'.[46]

Furthermore, as the model is not a conservative one, as in the case of the multiplier-accelerator model under certain fixed parameters and in the general case of the system obeying to the thermodynamics of the first law, uncertainty is both internal and external. This defines the nature of the variables and of the concepts of causality in the theory, providing the most clear examples for the current argument about the definition of the epistemic status of the variables. As Shackle argued:

> But a variable which is in all connections and throughout the argument to be treated as independent implies by this status a whole philosophy of explanation .... Keynes has, in fact, ... implicitly rejected the closed dynamic model of the type invented, or borrowed from physics, by Ragnar Frisch and developed with fertile zest by Harrod, Domar, Kalecki, Samuelson, Kaldor and Hicks. For in those models each variable has, in effect, its own determining equation, each in turn is exhibited as dependent on some of the others, and we have an insulated, closed and complete set of a very few variables mutually determining (once some 'initial' values are given) the skein of time-paths they shall follow. For Keynes, by contrast, there are economic wind and weather, in the form of politics, invention, fashion and the incalculable movement of expectation, great forces quite outside and unshaped by the economic ship whose course we seek to understand and control. These are ultimate and truly 'independent' variables, focused and canalised in their effects as the marginal efficiency of capital, the rate of interest and the propensity to consume. (Shackle, 1967: 158-9)

Economic 'winds', indeed, since uncertainty and unpredictability is intrinsically generated by the system and the rational behaviour of non-maximizing agents; but those variables cannot be simply modelled as 'truly independent', since they are effects as well as causes — the 'animal spirits' are a metaphor for complexity.

9.3.C. The class of semi-autonomous variables

The concept of semi-autonomous variables was formulated in one of the last papers by Kalecki (1968), when dealing with the incorporation of the phenomena of trend and cycle.[47] Technically, the two semi-autonomous variables he presented are exogenous to the model: 'It may be called a semi-autonomous variable because we shall not try to relate it to any other variables entering our argument' (Kalecki, 1968: 167-8). The two cases are a variable representing a part of capitalists' consumption, 'a certain slowly changing magnitude dependent on past economic and social developments' (ibid.: 168) and another representing 'additional stimulus to investment which is a direct outcome of innovations' (ibid.: 173-4), that is, the effect of positive feedbacks. Kalecki argued that the trend in the rate of growth was driven by these two variables rather than by the endogenous system itself (ibid.: 178, 183). The conclusion of the paper states that semi-autonomous variables and changing parameters should be central points in future agendas for research.

It is quite obvious that these variables cannot be reduced to the status of random shocks, which are residuals of ignorance or otherwise errors of measurement, since for this case they represent identifiable processes, although they cannot be currently modelled. In particular, it is their historical character, and namely the fact that they imply a connection between the economic space which was modelled and larger indeterminate social factors and unique events, that forces their representation as exogenous variables. Yet, they are not exogenous in the sense that they are well known social processes, which can be represented as sequential causal process influenced by the economic development, as causes *and* as effects. The modelling limitation of the model and of the available techniques must not determine the theory.

Semi-autonomous variables are therefore a *testemonium paupertatis* of the Laplacean dreams of perfectly comprehensive modelling, and the live testemonia of the necessity of historical analysis of economic series. In particular, this class of variables should be considered in the formulation of meaningful hypotheses about:

a. Structurally unstable models, in which some crucial changes may affect the topological behaviour of the whole phase portrait;
b. Evolutionary accounts of the structural change, that is, of historical and irreversible processes;
c. Non-quantitative variables which are pervasive in economics.

These cases are evident from the previous arguments. A fourth case defines the role of those semi-autonomous variables in the relation between economics and other sciences, as far as the explanatory content of theories is concerned:

d. Semi-autonomous variables are as those relating several levels of abstraction and/or of determination in the models.

In particular, if a theory is defined in the space of a broader social science, semi-autonomous variables are those that make possible the incorporation of historical and social features which should not be modelled as the pure outcome of endogenous economic systems, if the economic explanation is to be meaningful. For instance, the initial process of innovation (and not the whole process of diffusion) is exogenous to the strictly considered domain of economics, but endogenous from the point of view of a broader historical — economic and social — explanation of structural change under capitalism. In other words, the difficulty for the incorporation of those variables in the operative models is not only caused by limitations in the modelling techniques, but mainly by the real impossibility of representing the whole process of innovation under a single variable: something must be missed so that modelling becomes possible, so that the process may be simply represented either by an exogenous or by an endogenous variable.

As opposed to the instrumentalist epistemology, this class of variables is part of an evolutionary epistemology which recognizes several legitimate forms of determination and therefore seeks to increase the empirical and cognitive content of the theories and models accepting the demonstrative role of arguments defined in the intersection of economics and other social sciences. The implication is a reconceptualization of the endogeneity-exogeneity dichotomy.

Finally, if the relation between the variables is not represented by the linear specification in which all the influences (exogenous and endogenous) are simply added (or logarithmically added, or in some way combined under the assumption that their relation is knowable and fully specified), and if instead an indeterminate nonlinear relation is considered, then all the explanatory variables should be operationally considered as semi-autonomous. And since nonlinearity is pervasive in economics — since there are so many agents interacting, taking subjective and rationally bounded decisions — this class of variables is the crucial one in order to develop some realist models.

As far as Marx, Keynes or Schumpeter are concerned, semi-autonomous variables are the centre of their general theories. This is not the consequence of the *ad-hocness* of their work, but instead the demonstration of the deepness and productivity of their hypotheses.

9.4. The endogeneity-exogeneity antinomy reconsidered

The strict distinction between exogenous and endogenous variables is not a trivial imposition of the logical treatment of models. Its specific role is closely determined by the mechanical metaphor, and particularly in cycle analysis by the rocking-horse dichotomy between impulse and propagation systems, and by its underlying concept of equilibrium. In this Newtonian positivist world, where the intelligible is the immutable, a closed system is indeed considered as the ideal form of endogeneity. One typical model of this type — Samuelson's multiplier-accelerator — was previously discussed.

But this approach is theoretically flawed, since the source of the change lies outside the scope of the model and of the theory, and furthermore the description is not realist. It imposes a strict definition of endogenous and exogenous variables, which cannot be applied in some relevant instances, as in the cases of Schumpeter's innovation, Keynes's concept of marginal efficiency of capital, marginal propensity to consume or liquidity preference, or Marx's rate of exploitation, profit or capital. It excludes structural change and, in fact, dynamical evolution and time. Kuznets, dealing with long series, recognized this problem: for the long time perspective, demographic, technological and social variables become relevant, and therefore 'it is inevitable that we venture into fields beyond ... economics proper' (Kuznets, 1955: 28).

But what is hereby suggested is a much broader implication than Kuznets'. In fact, the abandonment of the equilibrium paradigm for the analysis of cycles implies a revision of the role and distinction between endogenous and exogenous variables. This position was argued in a brilliant insight by Veblen:

> Social advance, specially as seen from the point of views of economic theory, consists in a continued progressive approach to an approximately 'adjustment of inner relations to outer relations'; but this adjustment is never definitively established, since the 'outer relations' are subject to constant change as a consequence of the progressive change going on in the 'inner relations'. (Veblen, 1899: 133-4)

Evolutionary theory implies a morphogenetic approach to societies, and as a consequence tends to combine endogenous and exogenous variables, whose distinction depends more on specific models than on the theory itself. The dichotomy must be weakened and challenged, considering the semi-autonomous variables which embody a complex combination of social relations for the explanation of historical events. For evolutionary theories, the strict exogeneity condition is meaningless, while for mechanistic theories it is the condition for production of propositions about causality, which is deterministic but indeterminable. In the same sense, the condition of strict endogeneity is empty for evolutionary theories, while for mechanistic theories it is the very condition for intelligibility.

And, finally, the Frischian approach and the general equilibrium model are supported by the linear specifications of models: once in a nonlinear world, the stability of the endogenous-exogenous antinomy collapses, since interactions are possible and all variables can create fluctuations. Prigogine argued precisely this point about economics:

> Clearly, a social system is by definition a nonlinear one, as interactions between the members of the society have a catalytic effect. ... The fact that each actor influences the behaviour of each other leads to complex, nonlinear processes involving different populations (white collars, blue collars, consumers, etc.) and different economic functions (services, industry, etc.). This in turn gives the system access to divergent evolutionary paths leading to structures and organisations. (Prigogine, 1986: 503)

Non-equilibrium systems create order and movement, and the evolution of the world is explained as a destructive and creative process and is not recognizable in the museum assumed by the classical mechanical metaphor of positivism. There is no certainty in that world, rationality is bounded and behavioural assumptions are predominantly descriptive, explanations are limited, deterministic laws collapse and equilibrium is no longer the typical outcome of the systems. Nothing is lost but the illusions: recognizing reality is by itself an escape from dreamland, even if the price for change is abandoning the comfortable fictions and to rediscover the difficult task of science.

*

This chapter discussed the strategy of the double decomposition, the cornerstones of the rocking-horse metaphor: the distinction between trend and cycle, separating the dimensions of growth and of fluctuation, and the distinction between exogenous and endogenous variables, separating legitimate causality and intelligibility. This strategy leads fatally to defeat: arbitrary choices and methods replace the inquiry about reality.

Real history, which is irreversible, which includes human and social choices and conflicts about domination and coordination, which is sequential and yet knows brusque mutations, must be treated with historical methods: that is the place and function for the concept of semi-autonomous variables. Indeed, no *perpetuum mobile* mechanism can represent humanity. Therefore, decomposition, be it between trend and cycle or between impulse and propagation systems, must be firmly rejected.

The survival of the decomposition procedure is a clear indication of the limitations of the current theories of the cycle: on the one hand the deficiency of the theoretical criteria to discriminate between the trend and the cycle, or their arbitrariness, and on the other hand their restriction to the linear representation, or their irrelevancy, mean that positivist models renounce to explain the real world — but that was supposed to be the alma mater of science.

10. New Criteria for the Classification of Theories: An Intermediate Balance Sheet

So far, the discussion of some central methodological issues in the analysis of cycles and growth has been related to the available classifications and depended on their criteria. But several instances were found to challenge these standard definitions, such as the definition of innovation or consequently the classification of theories based on cycles generated by innovation. Indeed, this case demonstrated the intrinsic incompatibility between the coherent Frischian equilibrium paradigm and the analysis of structural change.

The equilibrium approach, under its extreme forms of certainty and optimizing rationality or even under relaxed assumptions, cannot interpret cycles and growth as interconnected parts of a global phenomenon. Consequently, the divorce between theories of fluctuations and theories of growth exiled the cycle to the domain of the unexplainable exogenous variables — and that was a substantial reason for the success of the rocking-horse metaphor, since it reintroduced equilibrium in the domain of the cycle through a new partition of the problem. The strategy of double decomposition safeguarded the axiom of harmony of the economic forces.

The most important of those exogenous variables is the one that appears in the equation as a stochastic shock element, interpretable as a measurement error or as the aggregate influence of other non-specified variables; as a consequence of this essential ambiguity, no meaningful proposition can be put forward about its economic significance. This is the fundamental reason for the current status of the statistical work on cycles or of the empirical confirmation of theories on cycles, and for the development of what Frisch in his last years dubbed 'playometrics' (Frisch, 1970: 163).

A coherent theory of the cycle and growth, of structural evolution and change, must be rooted in alternative methodological foundations and requires a new procedure of classification of models and hypotheses, which is the subject of this chapter. Since the critique of the rocking-horse approach was already developed in the last two chapters, the next sections present an intermediate balance sheet and a set of criteria for the comparison of theories.

10.1. The conceptualization of the system of hypotheses

The description of the characteristics of a theory is strictly dependent on the paradigm producing the classification and discriminating the relevant features. So, a Frischian theory can be adequately classified by a convenient description of its impulse and propagation mechanisms, but the same method will produce spurious conclusions if applied to different types of theories. Thus, a more general approach is needed.

The first building block for such a general approach is the definition of three criteria identifying the premises of the theory or of the model. For each criterion, two stages are identified for clarification.

10.1.A. First criterion: the metaphorical innovation of the model

One general category of models has been very commonly found in the reviewed literature: the models inspired by the mechanical analogy, the pattern so pervasive in the positivist paradigm of neoclassical economics. The analogy proceeds in a very limited way since, in order to establish similarities and differences and to rule about the relation between the model and the economy it is supposed to describe, it is forced to restrict the economic behaviour to quantifiable, mechanic and harmonious relations. Since the rationality postulate is embodied in this model, every social behaviour is considered to be economically determined by the maximization rule, and the coherence between the model, the theoretical economy and the real society is presupposed. The perfect determinateness of the model is considered as the mirror image of the economy which is being modelled, since both are instances of the perfect world of equilibrium. *Ceteris paribus* clauses and the identification of aggregative variables are therefore the sufficient foundations for the study of each system. Such excessive propositions and their inspiring epistemology were previously criticized: the mechanical metaphor demands no more and no less than the similitude of structure and the immutability of the analogues. As a consequence, the logic of the analogy is restricted to the oblique transfer and the consequences are devastating.

If instead a realist epistemology is adopted, a larger and more creative metaphor is required: in the case of evolutionary economics, that procedure is not only common but also necessary, since it is assumed that the set of hypotheses of the model is only able to deal with some of the social relations and cannot therefore exhaustively explain reality. Moreover, as the theory is holistic and the models are limited, the metaphor is important in order to transfer and to develop new hypotheses and, in spite of its non-demonstrative character, it contributes to the process of knowledge. In particular, the evolutionary metaphor stresses the multiplicity of behaviours, of causal determinations and of contradictions.

Type II theories, according to this criterion, are those inspired by the evolutionary metaphors, while Type I theories are those based on the mechanical metaphor.

10.1.B. The second criterion: the structure of causation

As a general conclusion of the previous part of the text, a hierarchical view of the determination relations was presented, making possible several modes of causation: strict determinism, irreversible and complex processes, functional relations, intentional causation. It is now time to argue for this case, since this is crucial to discriminate between Type I (correlationist) and Type II (causal) theories according to this second criterion.

The question of causality is in fact a constitutive part of the discussion about positivism and modern science. The difficulty in obtaining a satisfactory solution and definition of causality led some early logical positivists, as Bertrand Russell, to declare that the quest for causality was useless:

> All philosophers, of every school, imagine that causation is one of the fundamental axioms or postulates of science, yet, oddly enough, in advanced science such as gravitational astronomy, the word 'cause' never occurs ... . To me it seems that philosophy ought not to assume such legislative functions, and that the reason why physics has ceased to look for causes is that, in fact, there are no such things. The law of causality, I believe, like much that passes muster among philosophers, is a relic of a bygone age, surviving like the monarchy only because it is erroneously supposed to do no harm. (Russell, 1913: 1)

This radical agnosticism about causality led to the search for alternatives, such as defining causality as a functional relation in the framework of the mathematical formalism. Russell's indignation was nevertheless understandable given the pervasiveness of the mystified view that causality is restricted to positivist determinism, that causality equals one single and fully identified necessary condition for an event, an opinion that dominated science.

John Stuart Mill was one of the first economists to anticipate that opinion; from his writings on, the notions of invariance and contiguity of the cause-effect succession were an essential reference in economics. Applied to the domain of economic fluctuations, those definitions imply several criteria for the acceptability of theories. One of these influential criteria, that of Herbert Simon, defined the concept of causality as a logic of asymmetric relation between sentences: if A→B but B does not→ A, and it can be stated that A causes B (Simon, 1953: 50-51). Causality is therefore a determination:

> We will say that a set of conditions (and a corresponding set of laws) determines an atomic sentence a_j, if (i) we have an empirical law for each condition in the set, and (ii) in the set of the states for which all those conditions are T [true], a_j is either always T or always F [false]. We define a complete set of laws as one that determines a unique conditional state-description — that is, determines all the a's. Given a complete set of laws, we will say that a_j has causal precedence over a_i if the minimal

> set of laws determining the former is a proper subset of the minimal set of laws determining the latter. (Simon, 1952: 524)

This type of causally ordered structure is called by Simon 'self-contained structure', and a typical example of such a determination is the system of equations in the structural form whose coefficient matrix is triangular or nearly triangular (Simon, 1953: 63, 50). This is not by any means a general case, but the deductive concept of causality was largely adopted in economics for decades, being slightly transformed in order to include a deterministic account of the distribution of probabilities of a variable.[48]

The neoclassical school, and essentially the New Classicals, challenged this strong deterministic condition. For the purpose of the Frischian paradigm and its empirical work, another and weaker concept was necessary to account for the relation between exogenous and endogenous variables, and to define a new type of causal ordering. Granger provided such a theory in a path breaking paper in 1969. Unlike Simon's, his concept of causality is only relevant for stochastic processes, and is defined as a function of the predictability of the time series under observation: 'Causality. If $\sigma^2(X/U)<\sigma^2(X/{\sim}U{-}{\sim}Y)$, we say that Y_t is causing X_t, denoted by $Y_t \rightarrow X_t$. We say that Y_t is causing X_t if we are better able to predict X_t using all available information than if the information apart from Y_t had been used' (Granger, 1969: 428; X_t, U_t and Y_t are stochastic processes, $\sim Y$ denotes now the past values of such series, σ^2 is the variance of the predicted error, and U_t is all information accumulated in the universe since t-1). Linearity is assumed.

In this case, correlation is identified with causality, and spurious correlation can arise if the relevant data are not used in the regression (ibid.: 429; Vercelli, 1990: 241). Furthermore, causality is again exclusively driven by exogenous variables (Granger, 1980: 350; Zellner, 1979: 10), which was the standard definition of the Cowles program (Darnell and Evans, 1990: 116). This definition provided a powerful tool for the development of econometric confirmationism, but may lead to absurdities. Hendry gave the example of the explanation of the British inflation by the rates of dysentery in Scotland, since this variable has a better fit than the evolution of the supply of money (Hendry, 1980: 390), or by the cumulative rainfall for the same reason (ibid.: 395). Sheehan and Grieves argued that some American macroeconomic time series can be proved to Granger-cause the sunspots, not the other way round as suggested once by Jevons (Sheehan and Grieves, 1982: 776-7). In that case, spurious correlation would be treated as true causality. This type of model of causality corresponds to Type I theories, according to this second criterion.

In his *Treatise on Probability*, Keynes suggested another vision of causality, separating the *causa essendi* (ontological and deterministic cause) and the *causa cognoscendi* (epistemic and probabilistic cause), the first being a limiting case for the second, which describes the natural state of a meaningful proposition on causality. The *causa essendi* can be indicated, following Vercelli (1990: 234; the variables are as previously indicated) as Y_t being a sufficient cause for X_t with information U_t iff $P(Y_t \wedge U_t)>0$ and $t<t'$, $P(X_{t'}/Y_t \wedge U_t)=1$ and $P(X_{t'}/U_t)\neq 1$. And the *causa cognoscendi* of X_t relative to U_t exists iff $P(Y_t \wedge U_t)>0$, $P(X_{t'}/Y_t \wedge U_t)\neq P(X_{t'}/U_t)$. In this case, a rather general and non-deterministic concept of cause is involved.

The obvious difference with Granger causality was that for Keynes the causal relation should be established by the hypotheses, and not by the a posteriori verification of empirical correlations: the power of discrimination against spurious correlations lies in the theory and not in supposedly independent statistical procedures. Models proceeding in this way will be characterized as Type II theories.

10.1.C. The third criterion: the separability in the model

The third criterion encapsulates the debate on the decomposability of trend and cycle. Type I theories will be considered following the extreme assumption about a total distinction between growth and fluctuations. Type II theories are those considering the decomposition as arbitrary and irrelevant, and which study the cycles as the conditions and forms of growth. These theories get inspiration from Marx, Mitchell, Keynes, Harrod, Schumpeter, Goodwin or others.

10.2. The properties of the system of hypotheses

The three criteria above indicated describe the conceptualization of the model and provide a discrimination between types of theories and their general assumptions. The next three turn to the properties of the model as such.

10.2.A. The fourth criterion: the nature of the variables

One of the more relevant reasons for the adoption of the Frischian paradigm, just when the econometric revolution was being developed, was that it provided a theoretical application and justification for the antinomy between exogenous and endogenous variables. This distinction is the basis for Type I theories, according to the present criterion of the nature of variables.

Several attempts were made in order to answer and to eliminate the practical difficulties imposed by this dichotomy. Christopher Sims suggested a statistical procedure, using Granger causality and VAR models, initially treating all variables as endogenous: the results of the estimation, and not the preconception of the hypotheses, should indicate the variables to be designated as exogenous. According to the author, the conclusion does not depend on the theory and the results are supposed to have useful descriptive capacities (Sims, 1980: 15, 30, 32). Yet, they depend crucially on Granger's measure of correlation and cannot establish the strong causal relation normally implied by the differentiation between endogenous and exogenous variables. The neoclassical theory is based on this contradiction between a strong positivist requirement of a completely deterministic causality and the weaker and incomparable operational correlationist procedure used to confirm theories.

Another solution is suggested in this text, based on the development of the concept of semi-autonomous variables. The semi-autonomy indicates the nature of the variable: it represents social relations and movements which are not purely economic in a narrow sense and which are not strictly quantifiable, but which are meaningful if not decisive in the development of the economies. These variables are therefore essential to the desired linkage with historical analyses and other information outlining an evolutionary discussion of structural change. Type II theories are centred on this historical account.

10.2.2. The fifth criterion: the structure of the model

As previously indicated, equilibrium models depend on the postulates of rationality, certainty — or uncertainty reducible to some form of certainty — and stability, which summarize the whole method. This is Type I structuralist models. Type II are those models representing morphogenesis, including the evolution of parameters and changes in the structure of the system.

10.2.3. The sixth criterion: the functional specification of the relations

The sixth criterion distinguishes among linear (Type I) and nonlinear models (Type II). Linearity is the normal specification for the mechanistic assumption of aggregative, perfectly determined and equilibrium models. As previously demonstrated, this is required by the distinction between growth and cycle, and undermines the epistemological status of both theories, preventing them from producing meaningful statements — since the explanation is centred on the unexplainable and limited to explain why is it prevented from explaining. To re-establish the full theoretical status of cycle theory, thus considering cycles not as residuals but as the forms of growth, a more realist departure is needed. There is nevertheless a price for this nonlinearity assumption, which many scientists are unwilling to pay. At the present state of knowledge, a multisectoral system with many dynamic equations can only be solved for linear specifications, and the hypothesis of stability of the system is generally rejected in dimensions larger than two: consequently, there is a trade-off between computability and realism, between functional descriptions and semantic value. Evolution means complexity and complexity requires evolutionary and non-positivistic methods of inquiry.

The following table summarizes the six indicated criteria:

Table 10.1 *Compared characteristics of the two paradigms*

	Type I theories	Type II theories
1	mechanical analogy	evolutionary metaphor
2	correlationist causality	probabilistic (quantifiable and non-quantifiable) causality; plurality of modes of determination
3	decomposition of trade and cycle	cycles as the form of growth
4	endogenous-exogenous distinction	semi-autonomous variables
5	structuralist equilibrium	morphogenetic disequilibrium
6	linear specification and aggregation properties	complex system, holistic framework

These criteria will be applied now to some theories about growth and the cycle.

10.3. Theories and problems

The six criteria indicated in the previous chapter will be used to classify and discuss models of growth and/or cycles, in order to reconsider the methodological problems of economic structural change. Once again, the survey is not exhaustive and intends to exemplify and discuss some of the main arguments of this Part Two.

Five main families of models are considered. The first is that of endogenous models with a deterministic structure aimed at a description of the actual paths of the economies. The multiplier-accelerator models are the most relevant example of that class. The second group is constituted by the early Institutionalist research program. The third group is that of the political business cycles (PBC), a literature originated in 1943 by Kalecki and representing for a long period an exception in what concerns the formal hypothesizing of relations between political movements and economic variables. The fourth is the larger group of contemporary monocausal models, reinvigorating the Walrasian agenda and maintaining the implausible assumptions about full information and rational expectations. They are largely based on exogenous and stochastic disturbances (EBC and RBC); these models are still dominant, but their realism has been questioned since it is rather difficult to accept the idea that small unidentifiable random shocks can explain the real historical evolution and directional structural change. The last group is the family of nonlinear models, which represent an alternative strategy since growth and cycle are supposed to be dimensions of an unseparable process.

Several models do not fit exactly in the opposite Type I or Type II categories, and will show intermediate features which are most relevant to the present study. The order of presentation is not chronological, but thematic.

10.3.A. Endogenous growth theories

In the virtuous circle of the classical growth theory, profit determines the behaviour of investment, which causes growth. In order to overcome this type of tautological explanation and to understand the contradictions of the growth process, Harrod and Domar developed new models in the 1930s and 1940s which remain as landmarks for the growth theory. Harrod's system is a very simple multiplier-accelerator model in which savings was a constant percentage of the national income and equals investment, which was a part of the first difference of income (Harrod, 1939: 14 f.). The warranted rate of growth of output, G_w, was measured by the ratio of the average savings rate and the capital-output ratio, and indicated the 'moving equilibrium of advance'[49] (ibid.: 15). Since G_w could be different from the actual G, the verified rate of growth of the economy, if the value of capital-output ratio was different from the desired one, instability could occur: there is in the model an equilibrium situation, but unlikely to be attained, and unstable if attained. In particular, a cyclical behaviour is possible, since if $G_w > G$ there is a recession, whereas if $G_w < G$ an expansion is generated (ibid.: 26). Moreover, the agents are induced to behave in order to produce cycles: 'a condition of general overproduction is the consequence of producers in sum producing too little' (ibid.: 24). In gen-eral, the fluctuation of investment was supposed to be inferior to the value required by the acceleration principle.

Domar developed a similar system, but the production function took the form $Y=sK$ or, differencing in relation to time, $I = s^{-1}\, dY/dt$, where s stands for the realized investment productivity, or the reverse of the capital-output ratio (Domar, 1946: 137 f.). If the productivity of investment does not attain its potential optimum value, there is a process of junking and destruction of capital (ibid.: 144). As for Harrod's model, Domar's can generate four different regimes, resulting from combinations of oscillatory and non-oscillatory fluctuations with dampened and explosive fluctuations. Further assumptions about the value of the constants s, c or s, are needed in order to define each case, but the economic theory does not indicate any assumption about those values, besides some very general limits (the rate of savings cannot be negative). The system is unstable, and that is all that can be said. It is therefore a strict Type I theory from the point of view of the fourth criterion (endogenous and exogenous variables), but not so from the angle of the fifth (structural equilibrium is possible, although not probable).

Hansen argued for a model explaining cycles by the fluctuations in real investment, mainly autonomous and inventory investment in expansions (Hansen, 1941: 225; 317), and by the behaviour of consumption and investment in the contraction, in a cumulative process. Samuelson's multiplier-accelerator corresponds to Hansen's concepts, and it is a Type I model.

The most influential model developed in this framework was Kalecki's. Kalecki reviewed Keynes's *General Theory* as early as 1936, and criticized the very short term analysis which ignored the effect of investment decisions on expectations (Kalecki, 1936: 247 f.). If this was considered, no equilibrium would be possible since agents take autonomous ex-ante decisions, and their non-predictable effects act on real time and are irreversible. In spite of this point, Kalecki was not able to formalize expectations in this wider sense, and in his models considered the future to be anticipated as the image of the recent past.

Furthermore, Kalecki also criticized the inadequacy of the accelerator, since it did not consider other determinations of investment, namely that represented by one of his semi-autonomous variables (Kalecki, 1966: 78 f.). His dynamic version was developed simultaneously and was indeed more complete than Keynes's: Kalecki suggested instead a more sophisticated model incorporating a differential production and demand for capital goods, creating an overshooting situation and fluctuations in the degree of utilization of capacity; on the other hand, lags in the demand of investment goods could accentuate the cycle (Kalecki, 1970: 24 f.). The model included a notion of imperfect competition, and the effective demand principle led to two important conclusions: (i) it rejects the equilibrium paradigm; and (ii) it introduces a theory of distribution (Possas, 1987: 50, 92-93).

In particular, Kalecki was aware of the difficulty in the application of the model in order to analyse real series, namely of the problem of decomposition trend-cycle. The long-term trend was described as a succession of short-term situations, that is, as having 'no independent reality', so that 'the two basic relations ... should be formulated in such a way as to yield the trend cum business cycle phenomenon (...), the only key to the realistic analysis of the dynamics of a capitalist economy' (Kalecki, 1971: 165). Kalecki went as far as recognizing that the decomposition procedure he used in previous works was responsible for the error of missing the impact of technological change (ibid.: 166) — in fact, the productivity-investment relation, crucial for technical change, was represented by a semi-autonomous variable by the author. In this sense, the cycle is a property of the economic structure, while the trend represents the dynamics of its change and the variation of the parameters: causality is unified.

In spite of that recognition, Kalecki could not avoid a tautological version of the decomposition, necessary for his formal models, where for instance investment was represented by the summation of a trend and deviations from the trend (Kalecki, 1971: 177):

$$I_t = Y_t + (I_t - Y_t) \tag{10.1}$$

Possas defends this strategy, arguing the trend can be assumed to be null so that the cycle be separately inspected: the 'relative independence' of both components, and the 'basically distinct causality' permits this type of procedure (Possas, ibid.: 157, 162 n., 229; this is contradictory with the same author in p.

17). Yet, this is also contradictory with Kalecki's previous statements or with the concept of semi-autonomous variables relating trend and cycle. The problem remains and, in spite of not solving the decomposition problem, Kalecki criticized the equilibrium paradigm in the vein of Keynes and tried to model the intrinsic instability of the capitalist system, correcting or identifying some of the shortcomings of the Keynesian theory.

Pasinetti developed new models on the basis of the principle of effective demand, and argued that the theory should account for 'a complex dynamic process of growth with periodic irregularities' (Pasinetti, 1974: 54), that is, rejected decomposition. The inner relation between trend and cycle is described as a feature of an organic system:

> With this integration, an economic system as a whole emerges not so much with the character of a rigid mechanism, but rather with that of a living organism, which continuously learns from past experience and continuously faces new problems to be solved. (ibid.: 71)

Pasinetti's endogenous model of the cycle is a difference equations system which does not include random shocks: the economy may attain whatever path of exponential growth that is desired if the entrepreneurs adapt their actions, but there is no endogenous way to force them to make such an adaptation. Like in Harrod's model, instability prevails.

The neoclassical authors opposed this vision. As is well known, Solow argued that the knife-edge stability of growth of the previous models depended on the assumption of a fixed combination of labour and capital, since the production function expresses the productive capacity as a constant proportion of the capital stock (supposed to represent the factors) and 'When the results of a theory seem to flow specifically from a special crucial assumption, then if the assumption is dubious, the results are suspect' (Solow, 1956: 65). Supposing a production function with possible substitution between the capital and labour factors under a Cobb-Douglas specification, Solow concluded that the demand and supply adjustment determines the equilibrium path of the economy. The capital-output ratio is then uniquely determined as the ratio between the propensity to save and the rate of growth of the labour force, which is the equilibrium value: stability prevails.

In a later extension, Solow introduced neutral shifts in the production function and considered different vintages of capital stock (Solow, 1959: 89 f.). The technical change factor is considered to grow exponentially through time at a constant rate, and to be independent from the other factors. Under these conditions, the rate of growth of output is exogenously given by the growth of the technical progress variable and the elasticities of output respect to capital and to labour (Phelps, 1962: 549). Finally, innovation is considered under the form of new vintages of capital goods, but every vintage is supposed to embody the last technology and to be unable to produce subsequent incremental progresses (ibid.: 552). In a word, instability is avoided since it is prevented in a model which postulates equilibrium. The same suspicion applied by Solow to Harrod's model, that of depending crucially on one doubtful hypothesis, is now pertinent for his own case. These neoclassical models are Type I theories for all the indicated criteria.

The trend is therefore determined by exogenous conditions such as the technological externality and factor endowments — which constitute an unscrutinizable black box. In the heuristic abstraction of the aggregate production function, under a Cobb-Douglas specification and supposing that e^{rt} is the state of knowledge, there is

$$Y_t = A_t K_t^a L_t^{1-a} e^{rt} \tag{10.2}$$

or in log terms and differencing in relation to time:

$$y_t - l_t = \hat{A} + a(k_t - l_t) + r \tag{10.3}$$

where $\hat{A}$ is of course d(ln*A*)/d*t*. If *A* — a residual, not including the influence of knowledge — is considered to be independent of time, then we simply have that the per capita growth of output is the summation of $a(k_t - l_t)$, indicating the time shifts in the function, and *r*, indicating the time shifts of the function. The model is a Type I theory: Solow accepted that these notions depend heavily on the initial assumption of constant returns to scale and are another testemonium paupertatis of the theory:

> it is true that the notion of time shifts in the function is a confession of ignorance rather than a claim of knowledge: they ought to be analysed further into such elements as improvements in the skill and quality of the labour force, returns to investment in research in education, improvements in the technique within industries, and changes in the industrial composition of input and output, etc. (Solow, 1959: 90)

The author considered this comment to be the starting point for the later new growth theories; it is indeed a generic and important insight, recognizing the inability of traditional neoclassical models to analyse innovation.

Kaldor suggested a different agenda, since he claimed that the growth of knowledge is simply not quantifiable and that the production function is neither linear nor homogeneous. As a consequence, equilibrium cannot be assumed: income will not be distributed according to the marginal product of factors, increasing returns are pervasive and the study of distribution is again the essential problem of growth theory. According to the sixth criterion, Kaldor was suggesting a complex nonlinear Type II approach.

10.3.B. Burns and Mitchell: the Institutionalist evolutionary analysis of cycles

Mitchell's definition of business cycles was for a long period the departing reference for every study on the matter and it is a representative of the second family of theories:

> Business cycles are a type of fluctuation found in the aggregate economic activity of nations that organize their work mainly in business enterprises; a cycle consists of expansions occurring at about the same time in many economic activities, followed by similarly general recessions, contractions and revivals which merge into the expansion phase of the next cycle; this sequence of changes is recurrent but not periodic; in duration business cycles vary from more than one year to ten or twelve years; they are not divisible into shorter cycles of similar character with amplitudes approximating their own. (Mitchell, 1927: 468)

Mitchell was aware of the heuristic function of his concept, and argued this to be an advantage.[50] The trend was included, since there are economic and social processes longer than the cycle,[51] but it was referred only to those processes considered to be non influential on the cycle (Burns and Mitchell, 1946: 412-3). The careful treatment of statistical evidence was essential, since aggregation obscures the evidence, for instance in expansions, when there are several movements at a time and not every industry is growing (Burns, 1969: 64). This conclusion echoes Schumpeter.

As for the causal structure, Mitchell and Burns suggested a combination of largely independent factors (Mitchell, 1927: 419-20, 461), the most important being the profit schedule — the 'clue to business fluctuations' (ibid.: 105; also 183-184) —, the related movements of investment (ibid.: 182; Burns, 1969: 34; 38) and the Schumpeterian 'new combinations'. These authors acknowledged and praised the virtues of the combination of historical explanation with statistical evidence and mathematics: 'Of course, there is no logical opposition between the theoretical and the historical viewpoints, any more than there is opposition between causation and analytical description. On the contrary, history and theory supplement each other' (Mitchell, 1927: 57).

From this general point of view, these authors defined an evolutionary approach to the cycle, and said so: the cycle is the consequence of a culture (Burns, 1952: 25), the product of the institutions of capitalism (Sherman, 1991: 5), which can not be studied by the mechanical analogy. Processes are the central concept, and not equilibrium, stressing the 'notion of a sequence, the concept of cumulative, consecutive growth, as opposed to the Newtonian concept of equilibrium' (Mills, 1952: 119; also Homan, 1952: 179); causal relations are treated as interactions, since causes become effects (Clark, 1952: 203). Schumpeter considered their analysis as the first step for a realist version of economics (Schumpeter, 1952: 327), even if he wrongly criticized Mitchell for ignoring the equilibrium concept. In a memorial article for Mitchell, the former

NBER researcher and future Nobel prize winner Milton Friedman acknowledged this conception of cycles as the outcome of self generating institutional processes under uncertainty (Friedman, 1952: 274).

In other words, this is a Type II theory, producing explicit statements under criteria one (evolutionary metaphor), two (essential indecomposability of trend and cycle), and five (morphogenesis), and implicit statements under criterion three (nature of the variables).

10.3.C. Political Business Cycles theories

The first thorough argument about political business cycles (PBC) was presented by Kalecki in 1943, when he studied the reasons for the opposition of big business against a policy of full employment, even if profits could be higher in such a situation. The reasons were found to be political and determined by the social influence of business on the government and society as a whole: governmental interventionism could enlarge the sphere of action of the public powers, and furthermore the direction of the intervention could eventually damage private business interests in the future and that is why it should be avoided; finally, a working class under full employment was considered to be less disciplined and fearful (Kalecki, 1943: 325).

Kalecki described the current policies of deficit spending in slumps, giving way to orthodox policies once the expansion took over, and opposed this sort of counter-cyclic policies, since they did not assure full employment and were oriented to limit spending and not to develop new consumption by rising pensions or reducing indirect taxation (ibid.: 330). Thus, PBC are a clear example of the second motivation for cycles — or counter-cyclic interventions — in the general linear model previously stated, the effect of exogenous variables. Finally, these models introduced social classes into the analytical framework: the social instinct of the capitalists was what prevented their acceptance of the higher profits situation under full employment (ibid.: 326).

Three decades afterwards, in another slump period, Nordhaus redefined the agenda for the PBC research. Assuming the postulate of rationality[52] — far away from Kalecki's discussion about social conflictive behaviours — Nordhaus suggested the possibility of an intertemporal trade-off and of a public choice between inflation and unemployment (Nordhaus, 1975: 169). In this context, the current government party should direct its efforts to maximize the vote function for the next elections, subject to the inflation and unemployment restrictions. So, a cyclical pattern is imposed not because of the unrepresentativeness of the government as in Kalecki's model — the dependence of the government in relation to big business — but as a consequence of the manipulation of the cycle for electoral purposes. No social differentiation is assumed and the government is representative, but the democratic choice is myopic (Moura, 1981: 268).

Furthermore, Nordhaus assumes that the electorate has no memory about the past political choices: the successive events are independent (Nordhaus, 1975: 185). Only the policy makers know the system, unlike the voters, and act accordingly; yet, the result is sub-optimum, with lower unemployment and higher inflation than in the optimum.[53] The empirical test used to verify the model, with 1947-1972 series, confirmed the hypotheses for three countries (Germany, New Zealand, United States), rejected it in four cases (Australia, Canada, Japan, United Kingdom) and was inconclusive in two others (France, Sweden). This is far from convincing evidence.

The whole procedure is relevant as a very specific attempt to challenge the borders of criterion four, the endogenous-exogenous distinction: PBC models argue for the endogenization of the policy making, but under the constraint of a given utility function of the government and rather implausible behavioural assumptions. The result proves that Type I theories under criteria one (mechanical analogies) and five (structural equilibrium) cannot proceed to the relaxation of the distinction between endogenous and exogenous variables (criterion four) except in an incomplete and therefore incoherent way.

10.3.D. Equilibrium (EBC) and Real Business Cycles (RBC) theories

At the same time of the rebirth of a new interest in political business cycles, a new approach was suggested by the New Classicals. An influential paper by Lucas (1975) provided the first building block of this new concept in a rocking-horse framework. As in other types of models, Equilibrium Business Cycles (EBC) models describe a stationary process around a deterministic time trend indicating the 'natural rate of growth', in which serially correlated variations of the levels of the economy motivated by exogenous random shocks create the cyclical properties. This model of stochastic influences on structural relations is a trend-stationary model, based on the misperceptions of the unanticipated price signals. The exogenous shocks on the demand side have no permanent influence on the behaviour of the system. Decomposition is possible, and the residuals are the cycles.

The later Real Business Cycles (RBC) models are represented be stationary processes around a stochastic trend, and decomposition is only possible under the restrictions previously pointed out. The shocks are thus considered as real and persistent on the supply side; long run 'innovations' of the trend affect the short run cyclical behaviour of the system.

These models are briefly considered in this section.

10.3.D.1. EBC

Lucas's model is an application of the Rational Expectations Hypothesis (REH), but considering that the agents have not and cannot have full information, since they only know the particular market where they act.[54] In fact, the cycle is created by the asymmetric information of the agents, who confuse movements of the aggregate price level with changes of the relative prices of their products, and therefore increase supply.

In this case, the price at market *z* is:

$$p_t(z) = p_t + \mu_t(z) \tag{10.4}$$

where $\mu_t(z)$ is the random shocks component deviating price from its trend in market z. Its average is supposed to be zero and the variance is constant. Therefore, $V[p_t(z)]=\sigma^2+\sigma^2_\mu$. On the other hand, $p_t=p+e_t$, where p is the unobservable price movement.

So, the agents must estimate p_t and will fit the regression:

$$p_t = \alpha+\beta\, p_t(z)+\omega_t \tag{10.5}$$

where the random component is also assumed to have zero mean and constant variance. As a result, the agents' estimation provides a guide for behaviour: if there is a constant inflation, a sudden change in the price level $p_t(z)$ will be interpreted as a change in the relative prices of market *z*; if the rate of inflation is extremely variable, there will be no response. Total supply is:

$$y(t) = {}^P y_t + b(1\text{-}B)\,(p_t\text{-}p) + \lambda(y_{t-1}\text{-}{}^P y_{t-1}) \tag{10.6}$$

that is, it is the sum of the permanent component, the price surprise weighted by the agents surprise, and the last period deviation of output from the permanent component. If (10.6) is equated to the aggregate exogenous demand, the solution for the whole economy indicates the properties of the cyclical component, which is a sum of the past random shocks deviating from the trend.

The behaviour of the system depends crucially on the serial correlation of the cyclical component of output. Barro provided an extension of this model explaining the unanticipated demand shocks as changes in the growth rate of money supply, once again depending on the assumption of limited information of the agents but, as the cyclical component is not serially correlated in this case, there are no persistent cycles (Dore, 1993: 69). This is the common economic explanation for this type of model, but was soon abandoned by its authors (Stadler, 1990: 764 n.; Dore, 1993: 4-5), even if Barro maintained his scepticism about the alternative RBC models, which allegedly could not explain the dimension of the verified fluctuations in the economy (Barro, comment in Eichenbaum and Singleton, 1986: 136).

There are several important problems for the economic interpretation of this model. First, the information barrier is considered to be a short term phenomenon — some months — but it produces a persistent pattern of cycles.

Second, if the information barrier is the sole major explanation for cycles, there is no evident reason for preventing market or public provision of information to fill the gap and to react to the market opportunity, and such a measure could easily prevent the fluctuations. Finally, in the real world there are barriers to information but there are partial markets for information as well as public provision of information: if this is so, the true reason for fluctuations may well be another one.

Here we have again the paradox of mainstream epistemology: the possibilities of empirical corroboration are limited by the inclusion of nonobservable and non-testable vectors of information. Furthermore, these models imply an extreme form of instrumentalist and relativist epistemology, which is vindicated very clearly by Lucas, restricting the construction of a theory to the analogy with an artificial mechanism (Lucas, 1981: 272). In this sense, the study of macroeconomics is limited to the presentation of its analytic models, and is based on a privileged way of theorizing by simulation and by analogy from toy-models.

10.3.D.2. RBC

By the eighties, these models were superseded by the RBC type of explanation. Instead of considering random exogenous shocks on the cyclical component ${}^{c}y_t$, the RBC models also considered shocks on ${}^{p}y_t$ and its implications in the creation of the cycle. The probability distribution of those shocks is supposedly known, so that a new certainty equivalent is introduced. Given that the methodology was previously discussed, only a short comment is needed now.

The model is developed in a New Classical and REH framework, and assumes that information is publicly and costlessly available, so that business cycles are Pareto optimal outcomes of the agents' optimizing response to real variations in technology, productivity or tastes. The endogenous propagation mechanism is considered as the centrepiece of the RBC explanation: the dynamics of adjustment costs or time-to-build capital goods determines the cyclical behaviour. The production function has the general form:

$$Y_t = h_t f(L_t, K_t) \qquad (10.7)$$

where h is the technological shock,

$$h_t = h_{t-1} e^{et} \qquad (10.8)$$

where L and K are the inputs, e_t is white noise, and the equation is homogeneous of degree one. Then, the cyclical component, which is proved to be the sum of past random shocks (Dore, 1993: 83) is created by the assumed lag structure, the autoregressive process of the technological shocks. The effect of the shocks is permanent (King et al., 1987: 4). Decomposition is still possible, even if the trend is assumed to be stochastic, if perfect correlation or perfect independence are assumed between the innovations of the trend and of the cycle. Otherwise, it is not.

King et al. provided in the same paper an extension of this model to the multivariate case, using cointegrated variables and the common trend approach previously developed by Watson and Clark. The exogenous shocks with zero mean and constant variance affect the total productivity factor;[55] the production function is specified as a Cobb-Douglas, there is a representative agent and the most important origin of the fluctuations in the system are changes in the trend, which are common for the relevant variables.

Stadler tried to synthesize EBC and RBC in a single model where endogenous technological change is allowed under the form of short term shocks, but which influence the long term behaviour. The production function of a firm *i* should be:

$$y^i_t = (L^i_t)^a Z^a_{t-1} F_t \qquad (10.7')$$

where a is a constant, L is the input, F is a positive stochastic shock with a permanent and a cyclical component and Z is accumulated technical knowledge, a 'kind of capital' (Stadler, 1990: 766), which is a function of the previous period accumulated knowledge, the rate of productivity and the level of inputs. In these circumstances, a temporary change in productivity can generate changes in the accumulated knowledge, through learning-by-doing or other processes, which imply a permanent change in output. Of course, all this specification depends on the stated properties of the 'capital'. Furthermore, these families of models do not explain innovation as a real economic process or structural change as a process of growth:

a. They assume the possibility of decomposition and, in order to obtain mathematically tractable models, they require strict hypotheses about the nature of the innovation, the exogenous stochastic shock: it must be uncorrelated with the permanent random element of the trend — if the trend is specified as stochastic — and its variance must be constant. But this refers to a non-existent world: new combinations of radical change cannot be conceived of under these assumptions.
b. All markets are always cleared, so there is a persistent state of equilibrium or a strong convergence to it, and the only reason for fluctuations is the random appearance of shocks and the misperception due to asymmetric information (monetary business cycles) or to exogenously given changes in the technology (real business cycles). If it happens, change is neither explained nor explainable.

c. Finally, these models use Granger correlationist causality as well as linear specifications, and vindicate a strong concept of the endogenous-exogenous dichotomy. Under the current classification, EBC and RBC are Type I theories.

10.3.E. Nonlinear models of cycles and growth

Early nonlinear models of cycle inspired by the Keynesian tradition provided a different sort of explanation: the cycles are endogenously created by combined propagation and impulse mechanisms. Hicks, Kaldor and Goodwin presented some of the first attempts to model cycles in this framework.

Hicks's model (1950: 3) combined Keynes (the Savings and Investment functions), Harrod (the fluctuation around an expansion trend) and J.M. Clark (the acceleration principle); the accelerator was considered to be the main cause for the fluctuation, since it worked asymmetrically in expansions and contractions (ibid.: 37, 95, 101). When the accelerator exceeded unity, the equilibrium was unstable, and the model behaved as what Samuelson called a 'billiard table nonlinearity', with a floor constituted by the non-negative values of consumption and investment and a ceiling constituted by the limit of full employment of factors. The accelerator rules the expansion, which was amplified by induced investment, and the multiplier ruled the contraction; the cycle was self-perpetuating.

This approach constitutes an alternative to Frisch's model, since the aggregate net effect of the random shocks should be very limited and unable to explain the amplitude or the persistence of the cycles. The model suggested that nonlinear relations could represent the complete mechanism for generating and maintaining the cycle.[56]

Kaldor (1954) presented a model with nonlinear investment and savings functions and thus several possible equilibria points. The economy was supposed to oscillate between two attractors, and to be adjusted cyclically by the behaviour of savings and investment. Only the market of goods was considered. Benassy (1984) introduced non-Walrasian models with quantitative constraints and generating various regimes. The cycle is created by the switch of regimes and the economic response to the growth of aggregate demand produces the expansion of output and the rise of wage costs, limiting the upper turning point; the following fall of output produces the decrease of wages and thus fuels a new expansion. The regimes are determined by the levels of employment and sales; markets of goods and labour are considered.

Nevertheless, there is a common problem with these models, which is the difficult incorporation of the trend. In the case of Hicks's model, the trend is exogenously given by the rate of growth of the ceiling (the level of full employment of resources), which is independent of the cycle. If the cycles occur around a constant level, there is no growth; if otherwise there is growth, it is not explained. Therefore, the integration of both concepts is not possible:

'Professor Hicks' work provides a convincing demonstration of the fact that in adapting Mr. Harrod's theory to explain fluctuations we lose any explanation of the trend' (Goodwin, 1982b: 115).

As an alternative, Goodwin suggested a powerful endogenously generated model accounting both for the growth and for the cycle in a nonlinear framework. In his works in the fifties, Goodwin had expressed his discontent with the previous multiplier-accelerator models, which lead to explosive or damping behaviour except in one very precise case of sustained oscillations, and argued for nonlinear models maintaining the oscillations and the asymmetry of expansions and contractions. In one word, only an evolutionary and not a stationary system is adequate to model the cycles (Goodwin, 1982: 82). And since economic progress is considered not to be steady but to come in irregular but not random bursts — occurring essentially in the boom which creates the trend — Goodwin suggested a combination of Keynes's insights on aggregate demand and Schumpeter's on revolutionary innovations (Goodwin, 1982: 123). Growth was explained by innovations and shifts of the productivity function, and the lower turning point was explained by the replacement of equipment goods, since the opportunity existed and there was much redundant capital (ibid.: 137).

In 1967, Goodwin presented a new model explaining growth and cycle. It assumes an exogenously given steady and disembodied technical progress, a steady growth of the labour force, two factors, no prices, a constant capital-output ratio and a conflict between labour and capital for the determination of the level of employment and the respective shares in value added.[57] In this model, there is a permanent and irregular growth of output, which is a function of the share of labour and the capital-output ratio; cycles are generated in analogy to the Volterra-Lotka predator-prey system. Goodwin presented the advantages of this sort of models as making possible the combination of short cycles with long run growth and, since employment grows at the same rate of the labour force, the equilibrium values can be obtained:[58] 'Though the economy is never actually in equilibrium, it generates a long run constant growth rate and thus reconciles cycle theory with steady-state theory' (Goodwin and Punzo, 1987: 108); the model is unstable. In later extensions, Goodwin suggested the adoption of chaotic or catastrophe theory concepts to model the irregular burst of innovations (ibid.: 130, 144), departing from steady-growth theory.

Reviewing this model, Solow stressed its abilities as a device to train intuition and its pedagogic utility, but rejected it is an accurate description of reality, insisting on the defence of a Frischian framework: 'Perhaps the Goodwin growth cycle is not a model of the business cycle at all, at least not a model of the short run fluctuations in economic activity that the NBER dates and the financial press talks about. Those fluctuations really do seem to be dominated, usually, by exogenous and endogenous movements in aggregate demand' (Solow, 1990:

38). US data fitted by Solow indicate an evolution of the share of labour and the employment ratio which goes in the opposite direction as anticipated by Goodwin's model.

In spite of its remarkable properties, this model is still not a satisfactory solution for the problem of growth and cycle. In its first form, it is irrelevant to look for empirical evidence or for computation of the parameters and it is a simple heuristic device. On the other hand, there is no co-movement of the main variables, and employment grows at an exogenously given rate; the model does not provide a comprehensive explanation for these economic movements. Furthermore, the incorporation of exogenous technical change is a simplification which denies the Schumpeterian innovations framework.

Nevertheless, the inclusion of social conflict and the abandonment of the behavioural simplifications of an homogeneous society are great improvements. On the other hand, the combination of growth (determined by the evolution of the share of labour) and cycle, unlike in the previous nonlinear models, opens a new field of research: the cycle creates the trend, which is verified *post factum*. The criticism by Mullineaux and Peng (1993: 41) is excessive: the model does not provide an integration of a single dynamic process as Marx, Schumpeter or Keynes could have conceived of, but it indicates one of the first tools in that way and eliminates some of the confusion resulting from decades of rocking-horse domination. Growth is the instrument of its own self-organization (Goodwin, 1991: 139) and this auto-catalytic feature is the cornerstone of an evolutionist approach.

This is a Type II model, from the point of view of the first, third and sixth criteria, although it is not emancipated from the characteristics of Type I models according to the fourth and fifth criteria.

10.3.F. Models compared

The presented models do not exhaust the main alternatives neither in the theory of growth nor in the study of the cycles. As stated in the introduction to this part, the single purpose of this limited survey is to check some attempts to deal simultaneously with cycle and growth, and thus to establish a coherent methodology to assess and to improve those efforts.

The consequences of the choice of one or another type of model are substantial. The six indicated criteria for the classification of theories discriminate among important differences in the conception not only of the role of models, but also of the economy as a whole, its institutions and the relevance of the general paradigms in social sciences. The consequences, namely for policy choice, are also relevant: if random unpredictable and uncontrollable shocks explain the cycle, government activism is futile and may even be damaging or misleading. If, on the other hand, cycles are endogenously created or if the propagation mechanism is more relevant than the impulse system, then a stabilization policy is still possible even if, of course, its results can be no longer predicted by the model if nonlinearity is assumed. In that case, the impact of political decisions can only be tentatively anticipated on the basis of previous experience.

The classification is summarized in Table 10.2, where the fact that some of the models do not fall under the strict dichotomy of Type I and II is indicated. The margins name some prototypes of these categories and the combined cases are placed inside the matrix. The mentioned six criteria for classification are grouped in some more general oppositions. The shadowed areas indicate logical impossibilities.

The localization of the models is chosen according to their main feature. Schumpeter appears in two capacities: as the supporter of equilibrium (Schumpeter A) and as an inspirer of the Type II theories (Schumpeter B); this will be discussed in some detail in Part Three. The development of Type II models defines the agenda for the evolutionary program.

Table 10.2 A classification of growth and cycle models

Models	evolutionary	trend-cycle combination and semi-autonomous variables	morphogenetic models	***Type I***
mechanistic		Kalecki, Pasinetti		Frisch
trend-cycle and endogenous-exogenous variables dichotomies	Benassy Kaldor		Hicks	EBC, RBC, Nordhaus (PBC) Slutsky
structural equilibrium		Schumpeter A		Samuelson, Solow, some chaotic models
Type II	Goodwin	Mitchell, Goodwin	Schumpeter B, complexity models	

11. Conclusion of Part Two: Moving Away from the Perpetuum Mobile

Complexity is the main characteristic of social systems.

In fact, having abandoned the fictitious short term equilibrium and introduced historical evolution, there is no way back to the mechanistic metaphor. The illustration of its internal contradictions was just given: the original Frischian type of model cannot establish a clear definition of causality as required by the positivist epistemology, since it depends simultaneously on exogenous factors which are not explained and on the propagation system which is passive — but, still, it is this internal mechanism, the analogue for the clock, that represents convergence to equilibrium, the hallmark of neoclassical economics. Furthermore, the decomposition procedure heroically supposes some original superimposition of a fluctuation around a deterministic equilibrium trend, or at least that trend and cycle are independent of each other and that the relation between both is reduced to the postulated additivity of effects. If the trend is supposed to be stochastic and/or if it influences the cycle — or, simply, if history is relevant — then the decomposition is indeterminate. Either ways, fluctuations arise from behavioural or information assumptions which are incoherent under the General Equilibrium paradigm and, if these hypotheses are followed, human agency is downgraded to the status of the action of automata without consciousness or strategy. As a consequence, events are only considered if the future is disciplined to repeat the past, without humanity and without time, which is reduced to a long term trajectory of optimization, thereby allowing for no choice, error or irregularity (Smale, 1980: 108). In one word, we have the resurrection of a Calvinist form of capitalism, with absolute predestination, but now praising the envy of profit. In that case, neoclassical equilibrium is triumphantly proved.

There are then two main critiques to address to the General Equilibrium paradigm. One is its ideological character, which was stressed long time ago by Joan Robinson:[59]

> The function of economic theory, as opposed to economic theology, is to set up hypotheses that can be tested. But if an hypothesis is framed in terms of the position of equilibrium that would be attained when all parties concerned had correct foresight,

> there is no point in testing it; we know in advance that it will not prove correct. The dominance of equilibrium was excused by the fact that it is excessively complicated to bring into a single model both movements of the whole through time and the detailed interactions of the parts. It was necessary for purely intellectual reasons to choose between a simple dynamic model and an elaborate static one. But it was no accident that the static one was chosen: the soothing harmonies of equilibrium supported laisser-faire ideology and the elaboration of the argument kept us all too busy to have any time for dangerous thoughts. (Robinson, 1962: 71-2)

And the second critique follows from this: the General Equilibrium paradigm has been one of the main obstacles for the development of an empirical based economic science. Indeed, the paradigm has been accepted for ideological reasons but also for very practical ones: it allowed for computability, quantification being considered in positivism the condition for the production of meaningful propositions. The success of the rocking horse was obviously the immediate consequence of that need to simulate in social sciences the environment of natural sciences and to adapt the techniques of the fashionable bodies of physics. The rocking-horse, discriminating between stimuli and responses, provided the justification and the operational analogy for the laboratory experiments applied to the research on cycles and yet maintaining the equilibrium axiom. The quest for the perpetuum mobile was resumed: the immutability of the equilibrating mechanism was the main course in explanation, and some loosely defined external events were supposed to contribute to the causation of the cycle.

Of course, if this decomposition corresponded to real facts, the method should be welcomed. However, its mechanical application for the sake of statistical methods designed in laboratory tests — where the available series are samples from the repeatable experiment — required the acceptance of assumptions including the axioms of equilibrium, which require the axiom of rationality, which requires the axiom of certainty or a certainty equivalent, which requires the rejection of the time dimension. The acceptance of those postulates implied the rejection of falsifiability, which was at the same time defined as the most comprehensive and promising epistemology. As a consequence, the practical application of the methods and the empirical content of the research program were reduced to simulation methods, to toy-models, to statistical bric-à-brac and to theoretical playometrics. Complexity cannot be avoided if the theory is supposed to produce propositions about reality.

Two other and more relevant strategies to deal with complexity were presented. The first is the program defined by Mitchell in order to study historical, quantitative and qualitative data on cycles. The statistical procedure thus developed is very limited — the intra-cycle trend — but openly avoided the myths of equilibrium and certainty. Therefore, it challenged the behavioural assumptions of maximization or the natural rates of growth, the Panglossian variations of neoclassical economics.

The second strategy is built in the nonlinear models, and draws from a new body of physical and organic metaphors where complexity is understood as the possibility of creation of information and order in disequilibrium, as opposed to the traditional concept of the inevitability of creation of disorder if we abandon the safe ground of the First Law. This also implies a new lecture of the Second Law of Thermodynamics: instead of an ultimate destiny of heat death, the entropic curse of Maxwell, entropy and negentropy are understood as having a constructive role in dissipative structures and self-organized systems — disorder may create order.

Complexity appears from this point of view as more general than quantum uncertainty, since quantum fluctuations are small and tend to cancel each other, while in organic and complex structures there is another source of fluctuations and change which is the impact of small events on the macroscopic level, as well as self-generated changes which dominate the dynamics of the system.

The theories of complexity are now discussed from the point of view of this strategy; the topic will be extensively reconsidered in Part Four. The self-organization capacity of these systems is stressed, since for economics the current situation to be modelled is the interaction of several interdependent factors or agents, capable of complex indeterminate behaviour. This conceptual universe includes reversible and irreversible processes, stable states and bifurcations, order and disorder, fluctuations and stability in a global program for a reconstruction of science (Prigogine, 1986: 496).

If this is an adequate framework to discuss evolution, then a reconceptualization of economics is also required. 'Chaos' implies that there is a completely internal set of variables in the system which are able to create irregular fluctuations exhibiting a random appearance and creating irreversible evolution. The law of motion has the general form:

$$y_{t+1} = f(y_t) \tag{11.1}$$

where y_t has a finite dimension and the trajectory is the map $\{f: y \rightarrow f(y_t)\}$. Furthermore, this type of motion implies extreme sensitivity to the initial conditions and non-periodicity (or the existence of cycles of every order). Equilibrium accounts of chaotic systems will be presented in section one and disequilibrium concepts will be discussed in section two. A more general category of complex models, including intrinsic and extrinsic randomness, endogenous and exogenous factors, will be also presented.

11.1. Mechanistic determinism: the last dream of Laplace

The first models in economic literature to apply chaos were closely related to the classical or neoclassical tradition. In 1983, Day suggested that chaotic fluctuations could emerge in economic models not from myopia of agents but from the endogenous interplay of technological, preference and behavioural rules — and therefore be compatible with the RE hypothesis. He presented the example of the Malthusian population theory for an agrarian economy with two attractors, defined as two different regimes (Day, 1983: 201).

Day also suggested a modification to Solow's 1956 model, in order to describe capital accumulation (k) as a function of an exogenous population rate g, variable savings and a pollution effect, with a Cobb-Douglas production function. In that case, the accumulation path will exhibit a chaotic behaviour for some range of the parameters. But Boldrin and Woodford argue that the result is not compatible with the utility maximization behaviour (Boldrin and Woodford, 1990: 191 f.). These authors suggested chaotic formulations of overlapping generations models, exhibiting perfect foresight but whose solution is indeterminate (Mullineaux, Peng, 1993: 57).

Baumol and Quandt suggested an economic interpretation to the common logistic equation for the case of a single firm:

$$y_{t+1} = \omega\, y_t\, (1-y_t) \tag{11.2}$$

The authors consider the following relations: the advertising budget of the following period is a proportion of the profit ($y_{t+1}=bp_t$) which is a function of the advertising of the current period ($p_t=ay_t(1-y_t)$). If $ab=\omega$, then we have (11.3). For values of ω below 3.5, there is a stable limit cycle, but for superior values there is a chaotic behaviour (Baumol and Quandt, 1985: 7). Within these values of the parameter, cycles become so long, irregular and superimposed that they are confused with random fluctuations.

Brock and Sayers suggested a deterministic chaotic explanation to be extended to the RBC models, if these are corrected for asymmetric costs of addition of productive capacity in expansion and contractions (Brock, Sayers, 1988: 87). But in that case there is no point to explain the cycle, since it is already implicit in the assumptions about the asymmetry of the phases.

These models constitute an attempt to overcome the neoclassical legacy of the dichotomic distinction between endogenous and exogenous variables and between trend and cycle, stating a completely endogenous system able to generate irregular and apparently random cycles.[60] But, first, the results are not robust in the sense that the behaviour of the system depends on a certain range of values of the parameters and, second, the economic interpretations finally depend also on the definition of rationality. The previous conclusion stands: all claim of complete and perfect endogenization is either trivial or false.

Moreover, as previously indicated, the generalization of the concept of attractor implies the collapse of the notion of a determinate and stable equilibrium. Even if utility maximization and rational intertemporal choice is assumed, complexity may arise from the factor substitution in models of technological structures of production (Boldrin, 1988: 69) or even from the coupling of perfectly stable sub-systems (Ruelle, 1988: 199). Coevolution generally produces complexity.

Guesnerie and Woodford reviewed two types of dynamical systems: (i) optimal growth models with permanent equilibrium, determined by the initial conditions; and (ii) indeterminate equilibrium models, where there are genuinely stochastic equilibria — sunspot equilibria, the state of the system being determined by the expectations on its future state; this is the case of chaotic specifications of overlapping-generations models (Guesnerie and Woodford, 1992: 291-2). The prophecies of the agents believing in sunspot-type of evolution are therefore self-fulfilling (Grandmont and Malgrange, 1986: 4). There are rational expectations, but also an extrinsic form of uncertainty, which appears to act just like an exogenous source of fluctuations and is even compared to the Keynesian volatile animal spirits (ibid.: 4; Guesnerie and Woodford, 1992: 292). The perverse conclusion is that even under the assumption of optimizing behaviour and rational expectations, multiple equilibria become possible in a chaotic system and unpredictability becomes the rule[61] and rationality and perfect foresight are inconsequential.[62] If this is the case, the law of motion takes the form of the models of complexity:

$$y_{t+1} = f(y_t, e_t) \qquad (11.1')$$

where e_t are exogenous shocks or some other form of extrinsic uncertainty. In the general class of models of complexity, the requirement of a perfectly deterministic system is dropped: the deterministic characteristic of the chaotic regimes and *a fortiori* the features of complex regimes are unable to save the equilibrium paradigm, since the mode of determination is not known and is not knowable.

Anyway, for the moment, the empirical results obtained in the investigation about complexity in economic series are scarce and contradictory. In 1986, Brock studied the series of real US GNP and gross private domestic investment, and found no evidence to reject a null hypothesis of the generation of the series by an AR(2) model (Brock, 1986: 168 f.). The methods employed in this study, the computation of the correlation dimension and the largest Lyapunov exponent,[63] were later on developed by the author. In 1988, Barnett and Chen accepted that the hypothesis of some monetary aggregate series being generated by a chaotic system was plausible (Barnett and Chen, 1988: 201). The results were in fact somewhat inconclusive, since eight series of Divisia monetary aggregates were studied, and clear evidence for complexity was obtained in two of them, while for three others the result indicated a possible combination of white noise and complexity, and no evidence was found in the last three.

Brock and Sayers investigated the US aggregate data on pig iron production, unemployment, industrial production and also sunspots, finding weak evidence for chaos (Brock and Sayers, 1988: 72). Scheinkman and LeBaron inquired the U.S. quarterly 1947-1985 GNP and the hypothesis of a AR(2) process could not be rejected, but some evidence for chaos was found in series of stock returns, unemployment, gold and silver returns and Divisia monetary aggregates (Scheinkman and LeBaron, 1989: 214).

11.2. Second stage indeterminism

The balance sheet of cycle theory is quite inconclusive: orthodox models built on the probabilistic revolution and the first generation of indeterminism still rule. Not because their premises proved to be adequate or their axioms resisted the devastating assaults of epistemological realism, but simply because the alternatives are insufficiently developed.The alternative first models of cycle and growth were nonlinear, and therefore only tractable in very simple forms and low dimensions. But Slutsky's argument, plus its generalization by Frisch, plus the good fit the auto-regressive models can normally generate, provided a powerful alternative of stable low-order linear stochastic difference equations accounting for the fluctuations.

The revival of nonlinear models, later on, introduced a new dimension in the theoretical debate: the traditional antinomies between trend and cycle and endogenous and exogenous variables were questioned. The attempts to incorporate chaotic systems in the framework of the neoclassical paradigm, as just reviewed, were unsuccessful since the surviving concept of equilibrium is indeterminate and represents a further dissolution of the General Equilibrium metaphor. The two essential reasons for this implication are now discussed.

11.2.A. The butterfly effect

The claim that the detection of complexity implies the possibility of control and prediction is a current basis for the deterministic interpretations of complex systems. This interpretation is nevertheless contradicted by one of the essential features of chaotic regimes: the sensitivity to the initial conditions, the 'butterfly's wings effect' — a very small change in the parameters or the initial conditions implies exponential divergence of nearby orbits, and therefore unpredictable changes in the system. In other words, if the true equations of the process were known (which is not possible), even so the ignorance of the precise initial conditions would imply unpredictability.

One economic illustration of this sort of process was the interpretation of the *New York Times* about the 1987 stock crash, which is supposed to have been triggered by a fifteen minutes discussion in the House of Representatives Ways and Means Committee about a raise of 400 million dollars in the tax on corporate take-overs, an insignificant value when compared to the loss of stock values (NYT, 31 Oct. 1987, p. 17). True or false this particular allegation may be, the butterfly effect implies that the movement of variables generated by chaotic equations is really unpredictable. The basic devices for forecasting, extrapolation and estimation of a linear structural model will fail under these circumstances (Baumol and Benhabib, 1989: 79).

And, since it is not possible to monitor the wings of all butterflies in the world, this means that the economic interactions are also unforecastable.[64] The Newtonian universe collapses and the Laplacean dream — give me the initial conditions and I will give you the past and the future — is dissolved as futile vanity. Deterministic complexity is still an anti-determinism since non-determinable. And if otherwise complexity is defined as a synonymous of sensitivity to initial conditions (11.1') as Ruelle suggests, then traditional determinism is proscribed in all this class of models.

11.2.B. The Sahel effect

The second reason is the more substantial for the present case: sensitivity to initial conditions — dependence on history — creates the random appearance of the process and this behaviour may be modelled as a succession of nonperiodic jumps between different attractors. The temperature of the Sahel desert is the example used in a paper by Peixoto (1993: 2), who suggests a concrete form for the previous (11.1'):

$$dy/dt = f(y, g) + F(t) \qquad (11.1'')$$

where f(.) represents the nonlinear dynamics, and F(.) the random process. In the case of the Sahel temperature, for small values of F the system is kept oscillating around one of the two attractors; when the barrier is broken by a larger random impact, the system will eventually stabilize around the other attractor. Thus, the random process is considered to be responsible for the abrupt switch of regimes of the endogenously generated evolution.

A general category of processes including random influences and nonlinear systems is needed for several reasons. First, it is not credible that strictly deterministic phenomena explain social complexity. This point was made by Scheinkman:

> It is extremely unlikely that macroeconomic fluctuations could ever be explained by a purely deterministic model with a manageable number of variables. There are even theoretical reasons to support this view. The same property that makes chaotic systems look as if they were random — their sensitivity to initial conditions — makes it difficult to forecast future values of the variables that the agents take as exogenous. (Scheinkman, 1990: 37)

Second, the endogenous chaotic system is not identifiable by the current mathematical techniques. Therefore, the agents behave as if some of the observable variables were in fact generated by a truly random process, which means that for practical purposes the combined system of chaotic noise is relevant. From this point of view, chaos can be seen as a deterministic and indetermined origin for randomness, that is, of intrinsic randomness.

Bak and Chen suggested the concept of 'self-organized criticality' based on the metaphor of a sandpile,[65] a composite system which can develop from one metastable critical state to another — like Sahel's attractors — by the influence of minor events starting a chain of reactions and interactions (Bak and Chen, 1991: 26). In this context, the system has not a white noise component with no memory of past events ($cov(e_t,e_s)= 0$, for $t \neq s$) but has instead a long memory noise variable. This is the condition for self-organized criticality. Scheinkman and Woodford constructed a model based on this sandpile metaphor, where macro instability is generated in an industry by small exogenous changes in demand, which do not conceal each other since the structure is nonlinear, and which propagate through an inventory dynamics to create unpredictable fluctuations (Scheinkman and Woodford, 1994: 419). Large scale fluctuations are to be expected even if there is no apparent systemic reason for it (Bak and Chen, 1991: 33).

These models do not replace the linear and general equilibrium alternative, since their hypotheses and conjectures are not easily tested against empirical evidence; yet, they explore the promising avenue of the realistic assumption of a combined system of stochastic and nonlinear complex processes. The concept allowing for an economic interpretation of that type of system is that of semi-autonomous variable in a non-mechanistic world.

11.2.C. The long quest for indeterminism

For a long time, the arguments about the presence of complexity and long memory effects in economic series were depreciated. In fact, some of the first developments of complexity theory, Mandelbrot's analysis of cotton prices and stock market series in the early 60s, were completely ignored by mainstream economics. The 'Premature Fractal Manifesto' (1964) was the result of that investigation (Mandelbrot, 1987: 117 f.).

The first stage of indeterminism is described by Mandelbrot as the overcoming of the Laplacean inheritance: Laplace described astronomy as the model for exact sciences since unconditional prediction was possible, but accepted probabilistic estimation in other sciences, as a consequence of their underdevelopment. Quantum physics generalized the new epistemology of the 'first stage indeterminism': instead of probabilities being conceived as the admission of a state of ignorance, they were viewed as a legitimate description of the universe — multiple negligible or concealing causes allowed for a certainty equivalent in a broad version of the stochastic processes, given the paired assumptions of legal atomism and limited independent variety (ibid.: 121).

Mandelbrot challenged this approach. Economic and social changes and fluctuations cannot be compared to the mildness and predictability of the oscillations in the state of a gas, he argued: a second-stage indeterminism is needed. The failure of the first stage indeterminism is inscribed in its dependency on Gaussianity (ibid.: 118): supposedly independent variables are aggregated according to the Central Limit Theorem, and randomness is thought as the result of the action of millions of arbitraging agents.

But Mandelbrot argued the Central Limit Theorem may fail, even if the variables are independent but their distribution is excessively 'long-tailed', a situation of kurtosis where there is a large probability of Y_t being large. In this case, the variance of the series becomes infinite if the random variable Y has increasingly fat tails (ibid.: 121): as a consequence, the sample mean fails to converge. On the other hand, first-stage indeterminism also fails even if the distribution of Y is short-tailed but the covariance of Y_t and Y_{t+T} decreases very slowly with time. Mandelbrot adds that this mathematical pathology was already discussed by Cauchy in 1853 (ibid.: 121). And, of course, the theorem is not valid if the variables are inter-dependent, a general feature in economics.

The price movements studied by Mandelbrot showed this type of behaviour: the variance of the series changed with the samples and increased with the sample size, rather than being constant. If this is the case, the series is either a nonstationary Gaussian process or a stationary non-Gaussian process and this is therefore compatible with the hypothesis of the series being a fractal object. Mandelbrot oriented his research in this direction, which is the hardcore of the second stage indeterminism. The method he suggested will be used later on in Part Four in order to detect evidence of nonlinearity and complexity in historical series.

So did Goodwin, with his early adoption of Thom's concept of catastrophes for the analysis of morphogenetic changes (Goodwin and Punzo, 1987: 144). Jenner suggests instead a dissipative structures approach to innovation, defined as new combinations in a dynamical evolution (Jenner, 1994: 127). The example is very appropriate: innovations, the search processes and the discontinuous shifts creating new varieties of methods or goods and transforming the organization of the economy, closely correspond to the concept of the catalytic action in dissipative structures, the anti-entropic movement which explains important features of the history of capitalism. Moreover, if firms, industries, social groups and human beings are connected by so complex feed-back relations, this approach is more adequate and promising. It defines the research described in Part Four of the book.

11.2.D. From free to forced oscillations

The critique of the rocking-horse paradigm emphasized both the inconvenience of the assumption of the equilibrating propagation system and the ambiguous status of the impulse system: this Slutsky system of free movements of random particles in the economic space is irrelevant for the explanation of organized complexity, the social interaction of economic agents. The reason for the inconsistency was already noted by Marshall, who pointed out that most economic processes are comparable to forced and irregular oscillations.

An admirer of Darwin and Spencer, Marshall noted that two different concepts of equilibrium were possible: the mechanical — which he used in partial equilibrium analyses — and the biological one, related to the history of growth and decay. In fact, Marshall did not use the biological metaphor, the Mecca of the economists, and restrained his work to the mechanical systems. Against J.B. Clark, he insisted that static processes did not accurately describe a part of reality; as Clark, he thought that a new step in the direction of genuinely dynamical systems was essential; as Clark, he never took that step. But he understood, better than any of the founders of the marginalist revolution, the intrinsic limit of the endeavour. One of the most evident proofs is that when considering the industrial cycles, Marshall confronted the mechanical metaphor with realistic assumptions, in order to conclude that the equilibrium notions were non-operational:

> But in the real life such oscillations are seldom as rhythmical as those of a stone hanging freely from a string; the comparison would be more exact if the string were supposed to hang in the troubled waters of a millrace, whose stream was at one time cut off. Nor are these complexities sufficient to illustrate all the disturbances with which the economist and the merchant alike are forced to concern themselves. If the person holding the string swings his hand with movements, partly rhythmical and partly arbitrary, the illustration will not outrun the difficulties of some very real and practical problems of value. (Marshall, 1890: 346)

Now, if these partly arbitrary and partly rhythmical movements are compared to the water deposit and pendulum of Frisch's metaphor, the conclusion is obvious: Frisch described an abstract experiment in order to highlight the regularities created by free particles impinging on the system, while Marshall argued that those regularities are irrelevant from the point of view of real systems.[66] Once again, purposefulness is the distinction between the laboratory experiments and the real life events.

*

Complex mathematics is still in its infancy, and so do its economic applications, but constitutes a resourceful program. First, it implies the whole reconceptualization of random events, no longer defined as errors or unexplainable perturbations, but as being generated by the very system as unique and unpredictable changes as well as by external influences, in such conditions that both origins are virtually indistinguishable: contingency and necessity are inseparable. Second, it assumes real agents and social forces interacting in a complex world, no longer defined as optimising automata. Third, it introduces multiple feedbacks and loops and therefore challenges the linear aggregation theoretical framework. So, this is an adequate context for the explanation of dynamical stability and morphogenetic change which define capitalism.

Some of the shortcomings of chaotic models have been presented and can be addressed in theory. Sensitivity to initial conditions has a difficult economic interpretation which can be solved only if a realist interpretation is adopted, based on the dominance of cumulative and sequential causation which generates change to be described by the formation of new conditions. The irregular orbit generated by the system may be considered as the analogical representation of the historical and irreversible path followed by the economy. But as there are changes of regimes and bifurcation points — thus making possible meaningful choice — and since the origin of these changes varies, the models of complexity are adequate to explain several concurring forms of change in a structure. The evident example is once again the process of paradigmatic change in an industrial and social structure: the techno-economic paradigm dominant in each historical period may be considered as the attractor of all nearby orbits, selecting the forms of coupling of oscillations and filtering the disturbances.

The superimposition and recursive symmetry of cycles of several scales, firstly stated by Feigenbaum, presents the same sort of difficulty. The literature on cycles indicates in fact several possible modes of fluctuation — NBER business cycles, Juglar, Kitchin, building cycles, Kondratiev, long cycles — but the epistemic status of those theories is not comparable and their existence and eventual articulation is an unsettled matter.

The nonlinear and indeterminist approach provides many interesting insights about the evolution of real life economies. But its techniques are still very limited. The detection of complexity in time series is based on methods which demand detrending the series in order to avoid monotonically increasing processes — and the current procedures of detrending series were previously charged of theoretical inconsistency and mathematical imprecision, and furthermore the very concept of the separation of trend and cycle is unacceptable. So, a new procedure is needed, following Mitchell's advice: a coherent explanation is preferable to a better statistical fit supported by an obscure or ad hoc theory.

The essential limitation of this new approach is nevertheless the fact that no reconstruction of the true law is now possible, and may never be, and there are no available direct tests to check models of noisy chaos. This limitation is obviously condemned in a positivist world, but must be accepted here as the condition for realism in an alternative epistemology: it is always preferable to have a science which knows less, but still that knows about reality, than that which knows much about a comfortable fiction. Furthermore, one of the most important and apparently most ignored consequences of the adoption of a complexity approach in economic theory is that the very role of models in science is modified: a concrete model cannot prove a theory, since it is no more than a limited metaphor in the theoretical construction and since no reasoning can be proved by simple extension of analogies, even if it can produce progresses in knowledge and in the discussion of old and new conjectures.

The subject of such a program is the unpredictable and complex economy moved by quantifiable and nonquantifiable processes of structural change. In this context, no positivist knowledge is attainable. Therefore, another form of cognition is suggested, based on description and theoretical hypotheses, using the combination of historical perspective, statistical evidence, local models and a global vision exploring creative processes in real life economics. The paradigmatic shift is enormous: the quest for certainty must be replaced by the quest for wisdom, as once put by Rorty. This is a fair definition of a research program for evolutionary economics.

11.3. Metaphors, again

The precedent argument established a large number of examples of the incorporation of the physical metaphor into economics, namely the energetic concepts of the 1840s into the marginalist revolution and the neoclassical program defined since the 1870s. This has been established as the dominant paradigm in economics for a very long period, a normal science orienting the study of any economic problem, including growth and fluctuations, defining the legitimate forms of causality and proof, confirming the central idea of order in the society and defining the behavioural rules of the *homo economicus*. Its core concept, equilibrium, is the analogue for the Newtonian version of the mechanics of the closed Solar system.

Several economists challenged this approach. Marx presented crises as deviations from equilibrium and considered the manifold of social institutions of power conditioning the reproduction of the relations of production. Since the crises may lead to the restoration of equilibrium or else to its break and therefore to social change, Marx, after Sismondi, relaunched the historicist critique of Ricardian economics and argued for a new science closer to the Darwinian metaphor of evolution and change. So did Marshall, who considered that in the long run 'organic growth' and non-quantifiable factors and changes should be taken into account, and that 'static equilibrium' was irrelevant in order to understand such processes (Marshall, 1890: 461).

The organic metaphor has three central characteristics which are decisive for this application:

a. First, *organic systems are open systems*, since they are in permanent exchange with the environment and as a consequence may prevent the entropic evolution. As a consequence for the economic analogue, events and facts matter, no single social process can explain or represent the evolution of the system: the optimization of a function under constraints is a formalism restricted to the universe of closed systems and rejected in the case of the biological metaphor. Optimality is irrelevant in biology, since the possible range of genotypic variation is unknowable. Teleology is therefore rejected.[67]
b. But *the biological metaphor does not ignore equilibrium*, although the concept is corrected. Several definitions are possible, as Lotka registered already in 1925: the stationary state concept of equilibrium, moving equilibrium, kinetic equilibrium (velocities tend to zero), dynamic equilibrium (forces are balanced), energetic equilibrium (virtual work tends to zero; Rosser, 1992: 201), Bertalanffy's equifinality[68] and ecological equilibrium should also be added (equilibrated change of energy). In particular, equilibrium in Darwinian biology describes several different processes: there is *a phenotypic equilibrium* if a constancy in the gene ratios of several generations is attained (under the assumption of the existence of a large population and no mutations or changes in the environment), but there is a simultaneous *genotypic disequilibrium* which is the condition for the creation of new forms and species. Equilibrium is not a *deus ex machina* as in the mechanical metaphor.
c. Optimality and equilibrium are sometimes stated as general features of the organic evolution, partially given the cultural influence of the positivist paradigm and partially because those *characteristics are confused with the real stabilizing properties of organisms*. In particular, natural selection eliminates the most structurally unstable life forms and tends to increase the degree of structural stability of the other forms, although the complete elimination of instability is impossible and even undesirable (Vercelli, 1982: 179). In organisms, like in all dissipative structures, the *principle of homeostasis* — the chemical balances maintained by the cells — is the stabilizing property which explains the maintenance of life. It is not derived from and it is not an analogue for mechanical equilibrium, since there is evolution, life and death; but still it is a form of stability, since the structure and its hierarchical organization are protected.

Córdon suggested three levels of energetic-material integration of live reality: the protoplasmic, the cells and the animals, each level being derived from the evolution of the previous ones (Córdon, 1982: 30, 32). General laws can be established from the internal movement and coherence of each level, and analogies can be drawn across species, since the microstructure and metabolism of cells of different organisms is similar, but new properties are revealed at each level (Gaill, 1987: 246; Waldrop, 1992: 82; Davies, 1989: 149). As a whole, this picture presents the dynamism, integration and evolutionary history of hierarchical systems.

The incorporation of the evolutionist metaphor is therefore an effective instrument in order to eliminate strict determinism, to consider mutation and history or complexity and qualitative factors. In organisms, as in all dissipative systems, disequilibrium is the condition for events; so it is in economics. Moreover, the dissipative structures approach challenges one of the cornerstones of the equilibrium paradigm defined for economics since the Fable of the Bees: equilibrium, which means total order and the Paretian satisfaction of all agents in economics, means destruction of life in organic and thermodynamic systems. Convergence towards stability, the desired evolution in orthodox economics, means the entropic failure in dissipative structures.

In this sense, the dissipative systems approach suggests a new interpretation for complexity in societies, claiming that this is the most general case, since no complex system is permanently stable. In complex systems, disequilibrium and change are the conditions for the creation of new forms of order, and economic systems are no exception.

If the concept of equilibrium is metaphorized as in biology (phenotypic equilibrium in genotypic disequilibrium) or in thermodynamics (order out of chaos), it should be better replaced in economics by those concepts of order or *coordination*, which are the analogues for the homeostasis principle. Coordination is defined as the function of institutional and social structures which organize and bound the evolution of societies: history is back.

11.4. Coordination: the creative properties of stability and instability

As referred to throughout the last chapters, the concept of equilibrium was defined by several different authors in complementary but also in contradictory ways: as a measure of the state of two inter-related variables in a system, as a method for the analysis of the properties of the evolution of a system, as the historical feature of market economies, as a value judgement. In other words, equilibrium was defined in strictly fictional terms just as it was presented as a description of real processes or as a prescription; it was defined in normative as well as in positive economics, and this pervasiveness has been a major source of confusion.

In the analytical context, the strongest and most common in instrumentalist mainstream theories, equilibrium is defined as the propriety of a model. Existence, uniqueness and stability must define such a system so that equilibrium be meaningfully interpretable. Now, these properties can never be simultaneously met in the reality of economic evolution.

Nevertheless, there are social and institutional properties accounting for some 'gravitational' effect (Stengers, 1987b: 344) or, borrowing a metaphor from Waddington, some form of chreodic canalization of development that indicates stability characteristics. Several economists noted these features. Marx argued that social equilibrium was restored after the crises. Schumpeter argued that new forms of stable equilibrium emerged in the depression phase of the cycle. And Keynes argued that capitalism was a self-adjusting process to under-employment equilibria, and that no explosive fluctuations appeared to dominate: 'it is an outstanding characteristic of the economic system in which we live that, whilst it is subject to severe fluctuations in respect to output and employment, it is not violently unstable. Indeed it seems capable of remaining in a chronic situation of subnormal activity for a considerable period without any marked tendency either towards recovery or towards complete collapse' (Keynes, GT: 249). In particular, in Keynesian theory, only dynamics can account for the tendencies and counter-tendencies for change, as Robinson[69] or Pasinetti[70] argued.

Before dealing with the problem of structural stability and instability in economics, the argument must be presented in the general framework of dissipative systems. As previously, structural stability is defined as the property such that a perturbed system is topologically isomorphic to the unperturbed one; in open economic systems, perturbations may frequently arise either from exogenous forces which are pervasive or from the internal organization of the system itself, which includes variables representing many different agents endowed with capacities of choice and decision. The plurality of interdependent agents is by itself a route for chaos; the classical case of the three-body system in mechanics proves that complexity may arise from rather simple interactions in these conditions (Xia, 1994: 289 f.). Furthermore, if humanity and choice are considered, complexity is established also at a second level, that of purposeful and strategic action: it is *organized complexity*.

In ecological systems, coevolution of several different species in the same niche implies that the system tends auto-catalytically towards complexity. Moreover, as each single adaptive or selective change in one species may affect all the other species as well as the fitness landscape of the niche, the chaotic outcome is rather general. The pertinent question is why are not the ecological systems driven to de-regulation and explosion. Kauffman answered that the reason is the *weak connectivity* between species, that is, that the system by itself, and not only each species, has self-sustaining and self-generating autocatalytic properties (Kauffman, 1988: 137). In other words, the coordination of the system is embedded in its functioning so that complexity does not destroy the niche.

In fact, one of the striking facts of many dissipative systems is that complexity is compatible with stability, and even that greater complexity is associated with greater stability:

> In short, complexity and population stability may well be associated, but no causal arrow need point from complexity to stability: to the contrary, if there is a generalization, it could be that stability permits complexity. (May, 1973: 76)

In particular, according to May, increased trophic complexity is parallel to increased community stability. Several reasons may explain this outstanding fact. One is that the natural systems are not arbitrary complex systems, but those selected by a long process, which excludes the most unstable forms (ibid.: 3-4). A second reason is that the coevolution may select those narrow regions of space where there is relative stability or the weak connectivity property pointed out by Kauffman. In other words, the complex systems may frequently be at that region of the edge of chaos or of turbulence: 'Instead, all these complex systems have somehow acquired the ability to bring order and chaos in a special kind of balance. This balance point — often called the edge of chaos — is where the components of a system never quite lock into place, and yet never quite dissolve into turbulence, either' (Waldrop, 1992: 12).

In social systems, there is still a further major reason for expecting such related complexity and stability properties of the system, which is social coordination: the institutional system is designed to minimize the effects and costs of change and yet to produce change — each innovation in semiconductors does not require an entirely new generation of computers; each change in law does not alter the process of selection of judges or the functioning of courts, and it does not necessarily question the legitimacy of the justice system. Coordination acts as a purposeful system actively maintaining the stability of the economies and societies, managing the weak connectivity between some of its components and implying restrictions on the behaviour of others.

This general feature of social complex systems was previously described as the dynamic stability (convergence) and *simultaneous* structural instability (divergence if some bounds are surpassed) properties: the systems may tend to the gravitational centre which has been called *equilibrium* in simple models, but are also able to change topologically, creating new orders out of equilibrium. The social coordination process, which is associated to a system of power, is designed to maximize the conditions for dynamic stability and to minimize the conditions for structural instability. Of course, there are bounds in the technological, economics or social possibilities open to each system at each time, so that coordination is in fact limited to driving the system across a limited number of possible bifurcations.

The regions of structural stability and instability are therefore the main features to be analysed in complex systems. History indicates that major changes are possible and indeed occurred in the past — the Industrial Revolution, the electricity revolution, the microelectronic revolution — and change is permanent even in socially coordinated systems. For a given society, structural change without coordination is dangerously damaging; coordination without change is doomed to fail.

This approach is taking momentum in economics. In the discussions of the 1987 inaugural meeting of the Santa Fé Institute, one of the working groups concluded that the 'remarkable stability properties' of capitalist economies may be more apparent than real, since every business cycle crisis had different features and led to new situations (Anderson et al., 1988: 248). In other words, history really matters. The conclusion is that the social and historical processes are more complex than complex systems in formal models: in simple mathematical models representing the interaction of several oscillations, complexity may easily arise, but the ecological or economic evolution proved that there are complementary factors — the coordination processes — that account for weak or selective connectivity so that complexity and stability are simultaneously created. On the other hand, structural change is a permanent process in real economies and societies, although the meta-structure is only altered from time to time: complex evolution is bounded by the coordinated parameter setting most of the time, but the changes may cumulatively lead to a major and revolutionary shift in the global structures and the very mode of coordination. These qualitative and cumulative changes have been inappropriately described in economics as cycle and trend, while they are intertwined modes of oscillation generated by a single historical complex phenomenon.

This suggests an alternative to the orthodox account of the equilibrium behaviour of the systems, based on an adaptive dynamics of self-organizing systems. Alternatively to the Frischian paradigm, in this evolutionary paradigm the propagation system is creative since it generates changes from its complex relations, and the impulse system is semi-autonomous and not completely exogenous. Turbulence, self-organization and coordination are one and the same process in economic systems.

The metaphor of self-organization is a powerful one and attracted the attention of many different economists, from Arrow (1988: 281) to Goodwin (1990: 45). It is potentially a new way out of the ideological bias of the normal science in economic: it is time to move away from the perpetuum mobile, and such a rupture is the theoretical condition for the study of the historical processes in economics. Two authors, Schumpeter and Keynes, discussed those new avenues.

Notes of Part Two

1 David Hume, a friend and correspondent of Smith, wrote him once about the cyclical desperation of concrete people: 'We are here in a very melancholy Situation: Continental bankruptcies, universal loss of Credit, and endless Suspicious ... . The Carron Company is reeling, which is one of the greatest Calamities of the whole; as they gave Employment to near 9.000 people. Do these events any wise effect in your Theory? Or will it occasion the

Revisal of any Chapters?' (letter dated 27 June 1772, quoted in Mirowski, 1985: 18). Apparently, the chapters were not revised.

2 The difference between Malthus and Ricardo was also a matter of methodology and of definition of the very purpose of political economy. In a letter of January 1817 to Malthus, Ricardo argued that 'It appears to me that one great cause of our difference in opinion on the subjects we have so often discussed failure in "effectual demand" is that you have always in your mind the immediate and temporary effects of particular changes, whereas I put these immediate and temporary effects quite aside, and fix my whole attention on the permanent state of things which will result from them.' The answer by Malthus, two days later, was highly indicative: 'I certainly am disposed to refer frequently to things as they are, as the only way of thinking of making one's writings practically useful to society .... Besides I really think that progress of society consists of irregular movements, and that to omit the consideration of causes which for eight or ten years will give a great stimulus to production and population or a great check to them is to omit the causes of the wealth and poverty of nations — the grand object of all enquiries in Political Economy' (quoted in Skidelsky, 1992: 419).

3 'The essential character of the Trade Cycle and, specially, the regularity of time sequence and duration which justifies us calling it a cycle, is mainly due to the way in which the marginal efficiency of capital fluctuates' (Keynes, GT: 301, also 299). Other factors considered by Keynes were states of uncertainty (influencing the preference for liquidity), the useful life of durable goods or the disproportion in the production of capital and consumer goods.

4 Some different endogenous but nonlinear models were also developed in the Keynesian tradition by Hicks, Kaldor or Goodwin. But, as they implied the rejection of the maximization and 'rationality' hypotheses, they were not incorporated in the mainstream. They will be presented and discussed later on.

5 Pigou preceded Frisch in the separation of initiating impulses and the 'mechanics of the cycle' (Schumpeter, 1927: 291). But Frisch provided the first systematic econometric and theoretical development of the impulse-propagation theory of the cycle. Tinbergen also shared this conception about the cycle: 'I think the general econometric view now is to state that the phenomenon of the rather irregular movements we are accustomed to call business cycle must be understood as the cooperation of essentially two elements, both relevant to the explanation, viz, a mechanism capable of performing characteristic cyclical movements ... and exogenous "shocks" as initial movers' (Tinbergen, 1951: 59).

6 Yule previously used the example of a pendulum, pelted with peas in order to simulate the cyclic movements: random events maintained the oscillation (Morgan, 1990: 82). Jevons referred to a vibrating ship hit by waves (ibid.: 93).

7 The authors discussed the matter at least since 1931, as their correspondence shows, and Schumpeter accepted his friend's version, although expressing some doubts which were quite revealing of his discomfort (Louçã, 1997a).

8 There an implicit explanation for Schumpeter's acceptance of Frisch's model, when he compared his own sharp remarks against Kalecki's model with Frisch's: 'If this [the 1933 Frisch's] model had been associated by its author with a claim to representing the cyclical process, objections *mutatis mutandis* to those formulated above against Kalecki would again have to be urged. But since it is not intended to be another perpetuum mobile theory of business cycle but the presentation of a piece of mechanism, we can not only enjoy its simplicity, but also use it to demonstrate the possibility of a distinct type of oscillation' (BC: 189). Of course, this is not accurate: Frisch's model was in fact similar in conception to Kalecki's, and the same comments would apply. The fact that it is a 'toy' model is certainly less important than the friendship between Schumpeter and Frisch, which prevented any criticism and recommended the agreement. Indeed, this is not the single case of Schumpeter's bias towards his friends' theories, Spiethoff being another example. In the case of Frisch, it is known that Schumpeter even envisaged to learn Norwegian to correspond directly with his colleague in his own language (10 May 1935 letter by Schumpeter, in Frisch Archives, University of Oslo Library).

9 Knight wrote in 1921 a powerful book against the epistemology of the mechanical metaphor, arguing that the concept of equilibrium implied that the causal relations be representable by a system of simultaneous equation in analogy to equations of motion. Since such analogy is inaccurate, the equilibrium concept is misleading and there is no real tendency towards

equilibrium (Knight, 1921: xxii-xxiii; or 1935: 161-2, 166). Furthermore, in mechanics the equilibrating forces are concrete, while in economics they are abstract and impossible to verify. The conclusion is obvious: 'Our general conclusion must be that in the field of economic progress the notion of a tendency towards equilibrium is definitively inapplicable to particular elements of growth and with reference to progress as a unitary process or system of interconnected changes is of such limited and partial application as to be misleading rather than helpful. ... Probably we must go further and reject entirely the use of the mechanical analogy, the categories of force, resistance, and movement, in discussing basic historical changes' (Knight, 1935: 184; also 1921: 201). Knight's main argument about the specificity of economics was the purposeful character — therefore, intrinsically non-mechanic — of human and social actions and choices. If that is the case, economics cannot accept any analogue for the First Law of Thermodynamics, and the stability properties of a system are nothing more than theoretical conventions, rejected in the framework of genetic-historical methods and in evolutionary accounts (Knight, 1956: 25). These elements of critique were later developed by Keynes, Shackle and Robinson, some years later on.

10 Slutsky was a researcher at the Conjuncture Institute, directed by Kondratiev, and a teacher at Moscow University. Unlike Kondratiev, he was not repressed and kept his post until his death in 1948 (Allen, 1950: 209; Morgan, 1990: 66-7). His 1927 paper, with a large summary in English, was quickly circulated in the West and had a large impact (Hendry and Morgan, 1995: 15).

11 The reduced form is a second order linear difference equation. The general solution is the sum of a particular solution and the general solution of the homogenous system: the first representing the equilibrium path and the second the disturbances.

12 The argument was enthroned for instance by the paper of the Adelmans, 'proving' in the context of a Klein-Goldberger model of the American economy that it was dynamically stable, and that fluctuations should be imposed from the exterior in order to mimic the cyclic behaviour (Adelman and Adelman, 1959: 596 f.). Both types of variables and of causation should therefore be included in order to explain reality, according to this view.

13 'What reassurance could be gained from knowing the economy was "stable" if the economy was being defined as a small, passive (and ultimately insignificant for the question of stability) fraction of the social existence? What is a "stable economy", if all macroeconomic fluctuation comes from outside the economy' (Mirowski, 1985: 69).

14 For the Keynesians, the impulse is created by investment, while for the Monetarists the investment is considered to be part of the propagation mechanism of autonomous monetary impulses.

15 This is not a trivial definition, since it permits the inclusion of subjective expectations and it implies general consistency of social action: 'A market is in equilibrium, statically considered, if every person is acting in such a way as to reach his most preferred position, subject to the opportunities open to him. This implies that the actions of different persons trading must be consistent' (Hicks, 1939: 58).

16 'The presence and persistence of cyclical fluctuations in the economy as a whole of irregular timing and amplitude are not consistent with a view that an economy returns to equilibrium states after any disturbance' (Arrow, 1988: 278; also 1994). Arrow suggests the return to Keynesian insights and the development of nonlinear dynamics.

17 The term was coined by Schumpeter (HEA: 1000 n.).

18 In his major 1927 book, Mitchell declared emphatically: 'Nor can the idea presented in many theories that business cycles represent alternate rupture and restoration of economic equilibrium be included in our working construction. Men who take as their point of departure the theorem that economic forces tend to establish a stable equilibrium may conceive the main problem to be how this fundamental tendency is overcome at times and how it presently reasserts itself. I have not chosen that point of departure. Hence it is no part of my task to determine how the fact of cyclical oscillations in economic activity can be reconciled with the general theory of equilibrium, or how this theory can be reconciled with facts' (Mitchell, 1927: 462).

19 Schumpeter's argument, in a book organized as a tribute to Mitchell's memory, was presented in a very vivid manner, from a concrete discussion between the two: 'He never would listen to the argument that rational schemata aim at describing the logic of certain forms of behaviour

that prevail in every economy geared to the quest of pecuniary gains — a concept he understood so well — and do not at all imply that the subjects of this rationalistic description feel or act rationally themselves. And I shall never forget his speechless surprise when I tried to show him that his great book of 1913, so far as the bare bones of its argument are concerned, was an exercise in the dynamic theory of equilibrium'; and, more precisely, in a note to the same page: 'For what else are his "recurring readjustments of prices" to which he returned again and again but imperfect movements of the economic system in the direction of the state of equilibrium?' (Schumpeter, 1952: 329; 329 n.).

20 This is why Keynesianism is not considered as an Equilibrium research program by the New-Classicals. This psychological motivation of the economic behaviour is not compatible with the equilibrium assumption and the aggregation procedure (Barro, comment in Eichenbaum and Singleton, 1986: 136). Barro is of course right.

21 Mini makes very strongly this case: 'There is no concept of equilibrium in Keynes because his framework of analysis rests on facts, and on the fact par excellence: the ignorance that pervades decision-making. If the very actors of economic life are ignorant, it ill befits the student of this aspect to be dogmatic' (Mini, 1974: 255-6). And also: 'And if the price to pay for attaining the truth is the abandonment of that "peace of soul and pleasure" that rationalism gives, Keynes leaves no doubt as to his preference. For the mechanistic characteristics of his predecessors he substitutes an economics more consonant to that "instability due to the characteristics of human nature" [from the *General Theory*] ...' (ibid.: 260).

22 Keynes wrote in his letter: 'I found it very interesting and really have next to nothing to say by way of criticism'. But he added then: 'In the case of inducement to invest, expected income for the period of investment is the relevant variable. This I have attempted to take account of in the definition of the marginal efficiency of capital. As soon as the prospective yields have been determined, account has been implicitly taken of income, actual and expected' (letter of 31 March 1937, xiv: 79). So, even when approving Hicks's summary, Keynes maintained his uncertainty approach. Some days later, he wrote another letter to Hicks: 'I do not really understand how you mean interest to be determined by saving and investment ...' (ibid.: 83). It should be also noted that Keynes, normally very attentive to his correspondence, took five months and an half to comment on Hicks's schemes, indicating at least that he did not rate the matter very highly in his scale of priorities. In that he was wrong.

Joan Robinson wrote: 'J.R. Hicks was one of the first, with his IS-LM, to try to reduce the GT to a system of equilibrium. This had a wide success and has distorted teaching for many generations of students. J.R. Hicks used to be found of quoting a letter from Keynes which, because of its friendly tone, seemed to approve of IS-LM, but it contained a clear objection to a system that leaves out expectations of the future inducement to invest' (Robinson, 1973: 79). Of course, the IS-LM interpretation is the prototype of what Robinson called 'bastard Keynesianism'.

23 Skidelsky argues that most of the Keynesians, such as Hicks, Meade, Hansen and Harrod, invented or accepted this version since it cleared the ground for the priority of policy discussions (Skidelsky, 1992: 538). This is certainly accurate, but it is still true that Keynes sometimes resisted against those attempts of transformation of his theory, namely when he warned Harrod: 'I am frightfully afraid of the tendency, of which I see signs in you, to appear to accept my constructive part and to find some accommodation between this and deeply cherished views which would in fact be only possible if my constructive part had been partially misunderstood' (letter to Harrod, xiii: 548). The enigma is then why was he frightfully afraid of those versions but did not react to Hicks'. Of course, Keynes did not consider very highly Hicks's work, and that may indicate an explanation (about his reaction to the 'emptiest platitudes' of *Value and Capital*, see Moggridge, 1992: 553).

24 According to Hicks, the IS-LM 'reduces the GT to equilibrium economics; it is not really in time. That, of course, is why it has done so well' (Hicks, 1976: 141). Here is how Mini summarizes his case: 'The LM curve resembles Keynes's treatment of the financial aspects of capitalism as Apollo resembles a satyr. For the LM curve re-enthrones, as geometry always must, naive calculation and stability where neither exists' (Mini, 1974: 252).

25 Pasinetti argued that 'The Hicks reinterpretation also helps to illustrate how the replacement of causality ordered relations with a system of simultaneous equations is not used only as a purely

found device but as a medium to reintroduce a basically different interpretative model of economic reality' (Pasinetti, 1974: 47).

26 'Even if the "catastrophe theory" of Keynes is totally different from the one of Marx, both have in common an important characteristic: in both the catastrophe is motivated by inherent causes of the functioning of the capitalist system, not by the action of external factors' (Schumpeter, TGE: 382).

27 'If one maintains the fundamentally individualistic approach to constructing economic models no amount of attention to the walls will prevent the citadel from being empty' (Kirman, 1989: 126).

28 Several programs try to address these limitations of the concept of rationality in neoclassical economics. Game theory formulates rationality as opposing strategies under well-defined rules, that is, abandoning the self-sufficient homo economicus, but still some contradictory results may emerge from those rules, such as a conflict between the rational choices of agents and the optimal situation. Experimental economics argues for the 'deepening' of the concept of rationality (Vernon Smith, 1991: 878). The results are mitigated.

29 Keynes, in a letter to Towshend: 'Generally speaking, in making a decision we have before us a large number of alternatives, none of which is demonstrably more 'rational' than the other, in the same sense that we can arrange in order of merit the sum of aggregate benefits obtainable from the complete consequences of each. To avoid being in the position of the Buridan's ass, we fall back, therefore, and necessarily do so, on motives of another kind, which are not "rational" in the sense of being concerned with the evaluation of consequences, but are decided by habit, instinct, preference, desire, will, etc.' (Keynes, xxix: 294). Decision following the habit was particularly stressed by Keynes.

30 'The fatal defect of the older conception was its assumption that men possess adequate knowledge, that they can act in the light of reason fully supplied with its necessary data. But this assumption is contrary to all experience. It is the false analogy from celestial mechanics, the unconsciously wrong and misleading interpretation of the word "equilibrium"' (Shackle, 1967: 136). Shackle named that period of controversy against the 'older' conception the 'years of high theory'.

31 Einstein, in a 1921 lecture to the Prussian Academy, stated that 'In so far as geometry is certain, it says nothing about the actual world, and in so far as it says something about our experience, it is uncertain' (Einstein, quoted in Mini, 1974: 21-2). How powerful was the positivist scientism so that this sort of realist and prudent assessment was despised for so long!

32 Hicks accepted long ago that if expectations are to be considered, then the stability condition would be severely weakened or denied (Hicks, 1939: 256), and Hahn accepted that no equilibrating forces exist in the dynamic systems: 'The main conclusion is rather pessimistic: we have no good reason to suppose that there are forces which lead the economy to an equilibrium. By that I mean that we have no good theory' (Hahn, 1984: 11).

33 Structurally stable models may be defined as Varian: 'Let $f: X \rightarrow N$ define a vector field on some state space X. Then, roughly speaking, this system is structurally stable if small perturbations in the function of f do not change the topological structure of the vector field $dx/dt=f(x)$' (Varian, 1981: 107; or Smale, 1980: 88; Gabisch and Lorenz, 1989: 162; Gandolfo, 1980: 403). Obviously, Samuelson's multiplier-accelerator model is structurally unstable.

34 'An equilibrium path is a path along which, if individuals have given expectations with respect to, for instance, relative prices, etc., in the future, and they act according to those expectations, those expectations are in fact fulfilled. A disequilibrium path is a path along which people behave according to certain expectations, but those expectations are not (in general) fulfilled' (Stiglitz and Uzawa: 1969: 6).

35 'The amount of information one needs to gather about the initial conditions increases exponentially with time', and so does uncertainty (Bak and Chen, 1991: 31).

36 'The issue is slightly confused by the fact that "equilibrium" as understood by economists does allow for time dependence in the form of anticipation of the future. Here, however, we question the standard economic assumptions and specifically the perfect "foresight" of economic agents in the presence of sensitive dependence on the initial conditions' (Ruelle, 1988: 199). In the same meeting, Palmer summarized his questions to the economists under

three topics: why do economists downgrade the psychological, political and social forces? why do they keep the rational expectations hypothesis even if it is obviously wrong? and how can they model innovation with formal models of a fixed number of variables? (Palmer, 1988: 258 f.). Brock answered that the rationality postulate should be used since it leads to equilibrium (Brock, 1991: 127), that is, the means are justified by the ends.

37 'The trends which promise the most important additions to our knowledge are those which correspond to rational hypothesis, although they may not "fit the data" so well as empirical constructions which are difficult to interpret' (Mitchell, 1927: 230).

38 Other authors are critical of this choice: for cases in which there is a strong permanent component, the fact that in each subsample the value is below average at the beginning and above at the end can create some spurious correlations between variables (Blanchard, Watson, 1986: 139-40).

39 The mutual dependence of trend and cycle is strongly argued by Goodwin: 'When output is falling it is falling, and that is all there is to say. There is no such thing as a trend factor which continues to rise right through the depression' (Goodwin, 1982b: 117). Nevertheless, the author presents some distinct versions of such mutual dependence: 'the cycle could exist without the trend but not vice-versa' (ibid.: 118), but also stresses that the cycles could not exist without growth and that 'technological progress, constructing a trend, produces booms and depressions ... [being] one and the same phenomenon' (ibid.: 123, also 112). The last insight is the essential point.

40 In his classical paper about growth, Harrod argued that 'Moreover it is possible, and this the following arguments seek to establish, that the trend of growth may itself generate forces making for oscillation' (Harrod, 1939: 15). Or, much later, Zarnowitz: 'In the present context, the critical question concerns the stationarity and predictability of the process observed during the business cycle. Their recurrent and sequential nature is indeed well established, but so is their lack of periodicity and the large inter-cycle differences in duration and amplitudes. The separability of business cycles from the long trends can by no means be taken for granted. These are arguments against the applicability of the Rational Expectations methods. In a nonstationary world with a mixture of random and serially correlated disturbances, uncertainty in the sense of Knight and Keynes is pervasive, even under the (empirically dubious) premises of no structural change and stable policy regimes' (Zarnowitz, 1985: 560).

41 See the following example from Zarnowitz: 'In reality, of course, the economy is always influenced by outside factors (for example, weather), so that a comprehensive explanation of its motion cannot be purely endogenous. But no outside influences can by themselves produce the recurrent sequence of expansion and contractions; in the first place, this presumably requires the particular dynamics of an interdependent economic system. A really satisfactory theory, therefore, should explain how business cycles are generated by the internal mechanism of the economy exposed to the impact of a great many potentially relevant external events. What matters, then, is the relative role of the inside and outside factors, not the extreme cases' (Zarnowitz, 1985: 544). No realist explanation of the cycles can avoid the consideration of both types of factors: the arbitrariness resides in their mechanical distinction.

42 'I would seem a better idea to redefine the role of exogenous phenomena as influences impinging upon the macro-economy which fall outside of the project of developing a model, but do not exist beyond the pale of explanation by economists ... . Non reversible and unique events which constitute the march of history should be included in explanations of macroeconomic instability ...' (Mirowski, 1985: 149).

43 Some decades ago, many economists would argue that governmental intervention and fiscal policies were not a matter for economics, but for law. But their incorporation in economic models is common-sense since the development of Keynesianism.

44 It is generally accepted that in a long run perspective variables of change and co-movement between tastes, technologies, social relations, etc., should necessarily be considered, which is rarely the case for short term analyses.

45 'With Marx he had one thing in common: a type of vision of the economic process ... one pure theory of economic change not merely supported in external factors to explain the transition of economic systems from one equilibrium situation to another. ... In effect, his conception

[Marx's] of the economic evolution of a differentiated process and created by the economic system itself is precisely what distinguishes from the economists of his time, as well as those who preceded him' (Elizabeth Schumpeter, in Schumpeter, TGE: 9). This is not quite correct about Marx, since he interpreted transitions between equilibria as a non-mechanic or non-deterministic process, but is certainly a fair although partial description of Schumpeter's ideas.

46 Kregel indicates that in an early draft of the GT, Keynes considered an explicit variable E in the consumption and investment functions, in order to represent the state of long run expectations, so that a stochastic change in E could change the behaviour of the function (Kregel, 1976: 210-11). The fact that Keynes wisely decided not to formalize the concept is rather indicative of its nature.

47 Previous concepts of 'autonomous variables' by Frisch and his disciple Haavelmo, implying exogeneity in the relation (Frisch, 1948: 368-9; Schumpeter, HEA: 1182 n.; Aldrich, 1989: 30). In later works, 'autonomous functions' have been defined as independent from time (Lorenz, 1989: 183; Medio, 1992: 27n.). A later concept of 'semi-independent variables', quite similar although independent from that of Kalecki, was developed by Mandel (1985; see Louçã, 1997b).

48 'In quantum mechanics and thermodynamics, and in many social science models, expressions in terms of probabilities have taken the place of completely deterministic differential equations in the relationship connecting the variables. However, if we adopt this viewpoint, we can replace the causal ordering of the variables in the deterministic model by the assumption that the realized value of certain variables at one point or period in time determine the probability distribution of certain variables at later points or periods' (Simon, 1953: 50).

49 Shackle criticized the 'Newtonian calculus' of Harrod, 'derived by analogy from physical mechanics' (Shackle, 1967: 251, 267), and ignoring some of the main insights from Keynes, namely the dependence of c, the capital-output ratio, on the rate of interest, which did not exist in Harrod's model (ibid.: 254; 1968: 115). Apparently, Shackle ignored that Harrod tried to prove the likelihood of instability and not the drive to stability.

50 'Statistics provide no direct evidence of the existence of "the" business cycle; what they provide is evidence of cyclical fluctuations in hundreds of time series. Indeed, it is difficult to construct from the data, or even to conceive of constructing any single index of the "general trend" in business activity. But the more thoroughly investigators anatomize the business cycle, the more they need a general term to designate the whole' (Mitchell, 1927: 454).

51 'We conceive of secular trends as drifts toward higher of lower levels that persist in a given direction for periods long in relation to business cycle' (Burns and Mitchell, 1946: 57).

52 This implies, for instance, that households have a maximizing behaviour and that the matrix of adjustment coefficients to the changes in conjuncture is the same for all agents, so that aggregation be possible (ibid. 174). This is quite unreasonable.

53 If, alternatively, the objective function is not the maximization of the vote but the maximization of the time of control of the government, then the result is closer to the optimum, as in the model of Frey and Ramser. The myopic result depends on the utility function of the party in government (Moura, 1981: 269).

54 This is, again, the reason why this type of model is not fully coherent General Equilibrium; if information is asymmetric and limited, there are transactions costs, and there should be a market for information. The supply in market z is in log terms: $y_t(z) = {}^p y_t + {}^c y_t(z)$, where ${}^p y_t$ is the permanent component and ${}^c y_t$ is the cyclical component. Since the trader in market z has lagged information about the whole economy, the cyclical deviations from trend are ${}^c y_t = b\{p_t(z) - E[p_t | I_t(z)]\} + \lambda\, {}^c y_{t-1}$, where b>0 and 0< l<1, $I_t(z) = \phi[p_t(z), I_{t-1}]$. The deviation from trend is a weighted sum of the differences of prices of market z to the general price based on available information in z at time t, and the previous cyclical component. Therefore, the cyclical component is serially correlated.

55 The total productivity factor is $\ln \lambda_t = g + \ln \lambda_{t-1} + \varphi_t$, where φ_t are the technological shocks, and g is the average growth rate. The logarithms of the (cointegrated) variables are considered to be the sum of a permanent component, a random walk, and a stationary serially correlated series (King et al., 1987: 11). Decomposition is therefore possible since the sources of the innovations are considered to be completely separate; the stochastic trend explains most of

the fluctuations, even in the short term (ibid.: 30).

56 Blatt proved that a series generated by this model, if analysed by the traditional tools of econometrics, would be easily confused with a stable linear model, with a very good fit and a correct Durbin-Watson statistics (Blatt, 1978: 294), concluding that the usual rejection of the nonlinear models and the persistent use of linear specifications is derived not from the empirical data but rather from the methods used to scrutinize the data (ibid.: 300). The question will be reconsidered in Part Four.

57 The evolution of the employment ratio is $(dv/dt)/v = [(1-u)/\sigma-\alpha-\beta]$ and the evolution of the share of labour in the added value is $(du/dt)/u = -(\alpha-\gamma)+\rho v$, where α and β are the growth rates of technical progress and the labour force, u is the share of capital, σ is the capital-output ratio, $f(v)=\gamma+\rho v$. Time is eliminated dividing both equations, and they are integrated. The result is: $v^{-(\alpha+\gamma)} e^{\rho v} A_1 = u^{\sigma-1-(\alpha+\beta)} e^{-u/\sigma} A_2$, where A_1 and A_2 are integration constants.
The graphical solution was suggested by Volterra, who studied a similar model for the evolution of the populations of predator and prey fishes in the Adriatic.

58 This is a characteristic of the current model rather than a desired quality of the theory. Goodwin stresses in the same volume that capitalism is inherently unstable and subject to morphological change, and not to steady growth (Goodwin, Punzo, 1987: 4). Therefore, the existence of the limit cycle as a form of equilibrium is also a (provisional) simplification: 'It is of use to know that an equilibrium solution exists, even if the system never gets there, but it is of limited use' (Goodwin and Punzo, 1987: 13; 108).

59 So did Rosenberg: 'But at last now we can understand why economists continue to lavish attention on general equilibrium theory. It is not because they think it can be improved in the direction of a descriptively and predictively accurate explanation of economic activity, but because they believe it is already part of the best contractarian argument for the adoption of the market as a social institution ...' (Rosenberg, 1992: 220).

60 In deterministic chaos, 'all sources of the system's dynamics are endogenized, so that the system itself produces its own dynamics' (Barnett and Chen, 1988: 201). 'Deterministic chaos' and 'chaos' are synonymous.

61 'The foregoing discussion shows that there is generally a large multiplicity of intertemporal equilibria with self-fulfilling expectations, in particular stochastic business cycles driven by expectations. And that we should expect a free market economy to perform rather badly under a laisser-faire' (Grandmont and Malgrange, 1986: 4).

62 'The difficulty of forecasting future values of a variable generated by chaotic systems seems to make necessary weakening the notion of perfect foresight' (Scheinkman, 1990: 38). Or, in the same sense, Jarsulic: 'the presence of chaos in perfect foresight equilibrium models raises an important philosophical question. As has been pointed out, chaotic systems exhibit sensitive dependence on initial conditions; and this means unpredictability in the face of measurement error. Since it is unlikely that anyone could maintain error-free measurement and expect to be taken seriously, what can perfect foresight mean? Chaos inducing nonlinearities appear to make the concept of perfect foresight a self-contradictory one' (Jarsulic, 1993: 359).

63 There are several methods to detect chaos and complexity in time series. Barnett and Chen list eight: (i) spectral analysis; (ii) Poincaré sectioning; (iii) subharmonic stroboscopy; and for series with less than 1000 observations; (iv) estimation of the maximum Lyapunov exponent; (v) inspection of the phase portraits; or (vi) of the auto-correlation functions; (vii) other time series frequency domain techniques, such as the bispectral test; and (viii) dimensionality measures (Barnett, Chen, 1988: 207 f.). The methods predominantly used in the papers here reviewed are (iv), (v) and (viii).

64 The economic interpretation of this condition is not obvious. No economic system can be conceived of as having so strict a dependence on initial (when is the beginning?) conditions; Schumpeter bitterly criticized Kalecki's model for its memory, so that something happening 'in the gardens of Paradise' could still influence our lives. But initial conditions must be conceptualized as the set of conditions which results from any significative change: as changes in parameters are possible and so is the change of the system itself, in a realist evolutionary context the initial conditions date from the last change occurring in the system. Furthermore, their influence depends also on the current state of the system: an appropriate metaphor is the

Freudian concept of childhood traumas: initial conditions really influence the adult's behaviour, but the conditions, relevance and evolution of this influence is also filtered, rationalized and submitted to change and is not strictly deterministic.

65 The authors are apparently unaware of the of the existence of a distinguished predecessor: in 1882, Maxwell discussed that class of phenomena such that a spark kindles a forest, a rock creates an avalanche or a word prevents an action (Maxwell, 1876: 443).

66 As proved by Dore, a large category of nonlinear models of cycles share this metaphor of forced oscillations, as represented by a Liénard-type of equation: that is the case of the models by Hicks, Benassy, Kaldor and Goodwin (Dore, 1993: 144 f.).

67 Darwin did not reject all teleological definitions. In one of the concluding pages of *The Origin of Species* he wrote: 'Hence we may look with some confidence to a secure future of equally inappreciable length. And as natural selection works solely by and for the good of each being, all corporeal and mental endowments will tend to progress towards perfection' (Darwin, 1859: 205). This may be understood as a purely rhetorical argument, in order to answer to the theological critiques which surrounded the scandalous publication of the book, since the whole natural selection process is indeterminate and therefore non-teleological in Darwin's work.

68 Bertalanffy defined his 'equifinality principle' as the equilibrium property of living systems which are able to contradict the second Law of Thermodynamics: 'A few main characteristics of open as compared to closed systems are in the fact that, appropriate conditions presupposed, an open system will attain a steady state in which its composition remains constant, but in contrast to conventional equilibria, this constancy is maintained in a continuous exchange and flow of component material. This steady state of open systems is characterized by the principle of equifinality; that is, in contrast to equilibrium states in closed systems, the open system may attain a time-independent state independent of initial conditions and determined only by the system parameters. Furthermore, open systems show thermodynamic characteristics which are apparently paradoxical and contradictory to the second principle' (Bertalanffy, 1962: 7).

69 'Furthermore, the concept of stability, based on a mechanical analogy, is inappropriate in economic analysis. For mechanical movements in space, there is no distinction between approaching equilibrium from an arbitrary initial position and a perturbation due to displacement from an equilibrium that has long been established. In economic life, in which decisions are guided by expectations about the future, these two types of movements are totally different' (Robinson, 1979a: 49).

70 'In traditional analysis, one may enquire into whether an equilibrium exists, is unique, is stable; and then the matter rests. The achievement of an equilibrium is the end of the story. In Keynesian analysis the opposite is the case. The achievement of full employment requires that a certain amount of net investment is undertaken so as to bring total effective demand to the level of full capacity utilization. But the very fact that the appropriate amount of new investment is undertaken comes to change the objective situation (that is, the existing productive capacity) on which equilibrium is based. In Keynesian analysis, therefore, the very achievement of equilibrium at one particular time, far from being the end of the story, opens up a whole new series of questions on how equilibrium is going to be maintained in the following period. A dynamic analysis becomes inevitable' (Pasinetti, 1974: 93).

PART THREE

Bounded Heresies

The pseudo-analogy with the physical sciences leads directly counter to the habit of mind which is most important for an economist proper to acquire. I also want to emphasise strongly the point about economics being a moral science. I mentioned before that it deals with introspection and with values. I might have added that it deals with motives, expectation, psychological uncertainties. One has to be constantly on guard against treating the material as constant and homogeneous. It is as though the fall of the apple to the ground depended on the apple's motives, on whether it is worth while falling to the ground, and whether the ground wanted the apple to fall, and on mistaken calculations on the part of the apple as to how far it was from the centre of the earth.

Keynes, xiv: 300

12. Introduction to Part Three: Disputed Interpretations

After the previous digression on business cycles theories and methods, Part Three returns to some of those themes by reassessing Schumpeter's writings on economic change and comparing his research program with Keynes's. In fact, Schumpeter presented the most systematic and extensive early work on the innovative creation and destruction under capitalism, and Keynes provided the first influential challenge of orthodox economics in the twentieth century. Both discussed thoroughly the place of metaphor in the scientific endeavour, and both commented on the constructive role of the metaphors from physics and biology. Equilibrium, certainty, rationality and expectations were discussed both by Schumpeter and by Keynes. One and the other presented their views on organic unity and evolution. And both argued about the cycle and the trend, the decomposition problem, the econometric procedures and the epistemology of economics and history. Therefore, the interpretation and evaluation of their works is an essential building block for the definition of modern evolutionary economics.

Keynes and Schumpeter are indeed at the roots of the contemporaneous critique of orthodoxy, although the impact of their contributions was quite different. The publication of Keynes's *General Theory* was immediately considered as a major scientific revolution in economics, and its incorporation in mainstream economics required drastic changes as those previously discussed about the IS-LM formalization; but Schumpeter's work, although widely known, did not have the same impact. The next chapters reassess those contributions in order to discuss the relation of both authors to the methodological problems previously identified.

The revival of interest in Schumpeter's contribution to economic theory, the first to be discussed in the next pages, was naturally followed by an intense dispute on classification and interpretation. Besides the strictly hermeneutic aspect of such debates, they still touch some decisive points for modern economics, namely its method and the identification of variables and processes of change.

The evaluation of such a contribution is controversial, and four main different interpretations are in dispute. The first, which is not considered now, is what can loosely and somewhat contradictorily be called 'Neoclassical evolutionism'

— the work relating the paradigm of neoclassical equilibrium to the study of economic change (for example Grossman and Helpman, 1991). The second and the third interpretations are discussed in the following pages: the second is defined by Neo-Schumpeterian contributions (for example Rosenberg, 1994; Andersen, 1994) and the third is constituted by the Institutionalist criticism of Schumpeter (for example, Hodgson, 1993; Mirowski, 1994). These debates are the subject of the next chapter.

The difference can be summarized along the following lines. While some Neo-Schumpeterians downgrade the contradiction between the Walrasian model, which was vindicated by Schumpeter during all his life and work, and the economics of change and mutation he argued for, the Institutionalist critique argues that Schumpeter was not able to emancipate from the neoclassical program, that is search for the rationale of change was condemned by internal contradictions and was therefore self-defeating. The first argument strongly denies the paradox (Rosenberg, 1994) or presents it as an exaggeration (Goodwin, 1982b: ix-x). Still, a somewhat more accurate version recognizes the paradoxical evolution of his thought towards a 'final thesis' to be reached a short time before the end of his paradoxical life[1] and defining a 'punctuated evolutionary' process which is historical by nature (Andersen, 1994: x, 1, 3, 40). The short periods of stasis in this punctuated evolution cannot therefore be confused with the Walrasian state, and history is considered as the main analytical tool for the study of change.

Hodgson presented an opposite argument: 'Schumpeter's theory is more close to that of Léon Walras than is often perceived' (Hodgson, 1993: 139) and 'Contrary to many admirers, however, Schumpeter provides neither a systematic theory nor an ideal epitome for a new evolutionary economics, if that is to be a precise and meaningful term' (ibid.: 151). Supporting his claims, the author presents several quotations of Schumpeter's texts idealizing physics as a model for science and lessening the role of biological analogies. But this entire assessment depends crucially on the assumption that evolutionism must be defined by the incorporation of the biological metaphor into economics (ibid.: vii) and on a somewhat unilateral reading of Schumpeter.

In the next chapters, the extreme version of the Neo-Schumpeterian interpretation will be rejected, as well as the strong version of the Institutionalist critique. Finally, a fourth interpretation of Schumpeter's work is that by his biographers, Allen and Swedberg, who devoted their work to the contradiction of a research defined in the framework of the dominant paradigm but permanently challenging its boundaries and even its metaphysics.

The next chapter is devoted to the discussion of such a paradox: the relation between Schumpeter's thought and the Walrasian system is investigated, namely in the framework of his deficient definitions of static and dynamics and, consequently, of equilibrium. The origin of Schumpeter's first theoretical sketches are then traced to two fundamental roots: (i) the first was the work by J.B. Clark, who was deeply aware but also unable to overcome the limits of the static equilibrium analysis: Schumpeter inherited his research program, developed it throughout his life, but was also unable to solve the enigma; and (ii) the second major influence on Schumpeter's theoretical evolution was the *Methodenstreit* debate, and the growing importance accorded to Schmoller's historicist arguments. Both these influences helped to establish a research program at the margin of the equilibrium paradigm.

The following chapter will then turn to the paradigmatic influences of physics and biology, defining Schumpeter's evolutionism. Chapter 15 will compare it to the development of the Keynesian theory. This procedure is essential to stress the difficulties of the two main research programs outside neoclassical economics in the first half of the century, in order to understand their compared achievements and failures. Such a comparison is natural and useful, given the fact that Keynes and Schumpeter had the same age, lived the same epoch, confronted similar problems even if they never really dialogued, and influenced generations of economists.

13. Schumpeter's Paradox

The present chapter discusses the place of Walras in the evolution of Schumpeter's thought, and consequently explores this relationship in more analytical grounds, namely in the framework of the central concepts of statics, dynamics, and equilibrium. The interpretation according to which the use and systematic reference to Walras was a mere academic strategy by Schumpeter is rejected, as well as the alternative presentation of his evolutionism as a mere *ad hoc* continuation of the general equilibrium paradigm.

Rosenberg argued that Schumpeter was the 'most radical scholar in the discipline of economics in the twentieth century' since 'he urged the rejection of the most central and precious tenets of neoclassical theory' (Rosenberg, 1994: 41). The author is very emphatic: 'Indeed, I want to insist that very little of the complex edifice of neoclassical economics, as it existed in the late 1930 and 1940s, survived the sweep of Schumpeter's devastating assaults' (ibid.). Based essentially on *Capitalism, Socialism and Democracy* (1942) and on the preface to the Japanese edition (1937) of *Theory of Economic Development*, Rosenberg identifies the alleged devastating assaults by Schumpeter against the neoclassical paradigm: since change is the decisive feature of capitalism and it means a permanent tendency to disruption, and since equilibrium has no welfare advantage because it means no progress, innovation is alien to rational-equilibrating decision making; in this context, the 'circular flow' described capitalism deprived of the essential movements of change and was therefore merely a simplification; as a consequence, Schumpeter committed himself to the historical analysis of this process of mutation as an alternative to the equilibrium paradigm (ibid.: 44-5, 48, 50, 56). But the whole case is based on partial and circumstantial evidence, given the fact that the same Schumpeter denied all these claims in other moments, including in posterior writings.

Certainly, one very essential point is clear: Schumpeter was opposed to the neoclassicals in the very definition of the research program — his explanandum was technological and institutional change, and these features are ignored and annihilated in the orthodox view by the *ceteris paribus* conditions (ibid.: 50-1). As a consequence, the real problem for the interpretation of Schumpeter is why did he not break with the neoclassical paradigm he could not use or follow.

Indeed, he held to the framework of the equilibrium paradigm through all his scientific writings. This paradox, not recognized by Rosenberg in his radical remarks, is the essential question or, as Allen puts it in the most complete and authoritative biography of Schumpeter:

> Paradox, failure, disaster, and disappointment were the keynotes of Schumpeter's life and work. He lived a paradoxical life and had a paradoxical career. He thought paradoxical ideas and wrote paradoxical books. Time and time again he failed as a scientist, scholar, politician, businessman, and even as a human being. ... Yet, paradoxically this career of failure was, in its totality, a success. (Allen, 1991-I: 4)

This paradox is never clearer than in regard to the Walrasian grandiose schema of general equilibrium.

13.1. Walras

Schumpeter's first book, *Das Wesen und der Hauptinthalt der Theoretischen Nationalokonomik*, was published in 1908 when he was 25 years old,[2] and it included a long appreciation of the Methodenstreit, the intense debate on method which opposed the Austrians theorists (Menger) and the German Historical School (Schmoller), between the 1890s and the 1910s. Although expressing his concern about the artificial separation between theoretical and historical methods, Schumpeter took sides with Menger, under whose influence he had studied at the University of Vienna. By that time, he was a supporter of the marginalist school and mainly of the Walrasian approach.[3] The book dealt with the general equilibrium and static analysis:

> In the centre of the book stands the problem of equilibrium, the importance of which is only slight from the viewpoint of practical applications of theory, but which is nevertheless fundamental for science .... The theory of exchange, price and money, and (...) the exact theory of distribution are based on it .... (DW, quoted in Allen, 1991-I: 61-2)

This presentation is very curious, since it indicates the limitation of the equilibrium analysis — its near irrelevancy for practical applications — but, in spite of it, its central status in the 'pure' theory, describing the 'changeless order and system in which everything fits together perfectly' (quoted in Allen, ibid.: 81). This is a precocious statement of the paradox, and indeed Schumpeter maintained the same attitude through all his life.

Both his general equilibrium framework and his doubts about its applicability were present in his discussion with Walras. In 1909, Schumpeter travelled to Switzerland and visited the ageing Walras, who received and praised the book that he considered a fair presentation of his own theories — although he thought until the very end of the interview that it was Schumpeter's father's (Swedberg, 1991:

31, Allen, 1991-I: 84). This visit was described by Schumpeter only in 1937, in the Japanese preface to TED,[4] according to whom Walras argued

> that of course economic life is essentially passive and merely adapts itself to the natural and social influences which are acting on it, so that the theory of a stationary process constitutes really the whole of theoretical economics and that as economic theorists we cannot say much about the factors that account for historical change, but must simply register them. ... I felt very strongly that this was *wrong* and that there was a source of energy within the economic system which would of itself disrupt any equilibrium that might be attained. (Schumpeter, 1937: 159-60, his emphasis)

There are substantial reasons for accepting this account.[5] First, it corresponds to Walras's approach of economic fluctuations, namely to his metaphor of the economy as the surface of a lake, permanently disturbed but always getting back to equilibrium (Walras, 1883: 207-8; Schumpeter, HEA: 999). Second, and much more relevant to the present purpose, it indicates the nature of the questions asked by Schumpeter to Walras and his rejection of the general equilibrium solution for those questions. Although this did not affect the reverence of Schumpeter towards Walras, it certainly suggested that the Walrasian scheme should be completed and that otherwise it would be wrong.

The following book, *The Theory of Economic Development*, published two years after the visit to Switzerland, presented his agenda for the study of development and change. As Schumpeter himself stressed in another preface to the English edition of TED, the ideas of this book were wholly formed in 1907-1909 (Schumpeter, 1935: 5), before the meeting with Walras, and did not change afterwards (Schumpeter, 1927: 289): the contrast between their theories is obvious and will be explored in a while. And the continuity in Schumpeter's own thought and writings was successively emphasized by the author: in 1941, in the Spanish preface to TED, Schumpeter argued that there was no change in his general vision of the capitalist economy from the first edition of that book (Schumpeter, 1941, in 1911: 9).

Yet, this consolidated theory of economic mutation did not challenge the eminent place of Walras in Schumpeter's Olympus. In 1910, Schumpeter published a biographic article on Walras, later reproduced in TGE: the general equilibrium theory was praised as being able to 'illuminate' the purely economic relations by 'one single fundamental principle' (TGE: 112), and the author was presented as 'an enthusiastic admirer of Walras' (ibid.: 140), 'the greatest of all theoreticians' (ibid.: 139), who defined 'the only truly general theory to be formulated in the whole history of economics' (ibid.: 442 n.). In 1935, Schumpeter stressed again his acceptance of the Walrasian concept of equilibrium (Schumpeter, 1935: 4); this was once more repeated in 1939 (BC: 45). In 1942, CSD presented the Walrasian system — and no longer the Physiocratic, as in EDM — as the foundation of economics. In HEA, he still stressed the General Equilibrium paradigm as the

'Magna Carta of exact economics' (HEA: 968) and presented Walras as the 'greatest of all economists' (ibid.: 827).

This clearly indicates the onesidedness of the previous argument by Rosenberg and constitutes the first approach to the paradox: Schumpeter formed a very early global sketch of his own theory, published and argued for it, acknowledging that the Walrasian system of stationary processes and static analysis was 'wrong' since incomplete and unable to deal with change and development. But, even so, he still considered Walras as the main modern economic theorist, the only one to create a science comparable to the achievements of physics and exact sciences. The compatibility of both sides of the story is the subject for the next pages.

13.2. Statics versus dynamics: positivist evolutionism

As previously indicated, the main reason for Schumpeter's feelings about the utility of the Walrasian system was its static nature, described as the first rigorous and yet partial analysis of the economy. This was considered rather a limitation than a mistake, since the dynamic side could always be added to this first picture of the theoretical system, as the 'classics' intended to do:

> The main concern of the classical system is to determine these constant rates, in other words to investigate the political economy in a state of equilibrium .... Yet this did not mean that the 'classics' followed the natural sciences either in form or in content. Thus they intended at first to present the 'static view' of the economy to which were later added certain statements about evolutionary tendencies — a 'dynamic view'. These expressions, as well as the actual separation of the two views, were introduced into economics by John Stuart Mill who derived the former from Comte. (EDM, 1914: 94)

All through his work, Schumpeter developed the same argument: since Stuart Mill took from Comte the distinction between statics and dynamics, and Comte was inspired by the zoologist De Blainville, the ultimate source of this metaphor was zoology and not mechanics. This was his argument in 1914, as indicated, but also in 1935 (Schumpeter, 1935: 6), in 1939 (BC: 37) and still in his last writings, in HEA:

> Adopting, as he [Comte] tells us, the terminology of the zoologist H. de Blainville, he called the former [phenomena] Statics and the latter Dynamics. J.S. Mill, the author who introduced these terms into economic theory, was well acquainted with Comte's thought, and it is natural to assume that he took them from Comte, though he did not say so. If this was the case, then Mill was wrong in speaking of a 'happy generalisation' of a 'mathematical phrase'. Since many people who failed to appreciate the importance of that distinction have tried to stigmatise it as an illegitimate derivative of a mechanistic way of thinking, it is time to state the fact that, so far as

> there is sense at all in talking about borrowing — as regards to words that is, not as regards the distinction itself which forces itself upon us in any case — the ultimate lender was not mechanics but zoology. (HEA: 417)

This argument is historically imprecise, theoretically misleading and therefore wrong.

13.2.A. Comte and De Blainville

The relationship between Comte and Blainville was complex and discontinuous. Comte attended a course on general physiology, by 'his friend' Blainville, at the Faculty of Sciences in Paris, from 1829 to 1832 (Comte, 1839, CW, iii: 209). Blainville was by then the main disciple and successor of Lamarck, and his work was well known, even if it did not make any impression on the future development of biology. Yet, it is very probable that it did influence the preparation of Comte's *Cours de Philosophie Positive*, which was to be published seven years after the end of that course on physiology. But shortly afterwards there was a violent rupture between the two men: Comte accused Blainville of 'brusque decadence' and 'fall' (Comte, 1854, CW, vii: 571; also 665), since he subordinated his vision of organic evolution to a theological principle. By that time, Mill's *Principles* had already been published (1848) and the correspondence with Comte developed for a long period (the first letter by Mill was sent in November 1841; Mill, CW, xiii: 488).

It is certain that the distinction between statics and dynamics was central to Comte's work, and Blainville might have had some influence on it. But this influence was probably felt mainly as a confirmation of Comte's vision, which was by itself defined according to the new trends of the science of the time and quite well established. Comte argued precisely this point, saying that both he and Blainville were inspired by D'Alembert, who treated mathematically the problem of reducing dynamics and statics to the same status (Comte, 1854, CW, viii: 442). And this was certainly the conception inherited by Comte, who argued that progress should be subordinated to the static conception of order, abstracting from time and representing the 'fundamental unit toward which tends our nature'[6](Comte, 1854, CW, iv: 3).

As Comte argued extensively in the book which was so influential in Mill's work, he derived his concepts of statics and dynamics from 'rational mechanics', treating the static case as a particular form of the dynamic one (Comte, 1839, CW, i: 480, 482, 491, 565). Of course, the physical metaphor makes possible this derivation of a static system as a particular case of the dynamic one but the biological metaphor does not, this crucial difference separating the concrete use of both inspirations.

From this conception followed that the 'social organism' was to be treated simultaneously from the point of view of statics (the fundamental nature) and dynamics (its evolution; ibid., iv: 430, 498; viii: 1). Anyway, progress should be subordinated to order and dynamics to statics (ibid., viii: 2; ix: 3). Blainville's and Lamarck's general views of zoological evolution supported this approach, since they were characterized by a strictly gradualist and therefore mechanical evolution through a defined and teleological scale of animals: the 'natural' evolution of the animal world is precisely the analogue of Comte's account of the stages of thought (theological, metaphysical, scientific), which was the basis of his theory of history.

The conclusion is by now obvious: the Comtian connection of Mill's distinction between statics and dynamics did not lead to any evolutionary conception but at most to a pre-evolutionary mechanical and teleological zoology or more exactly to the generally admitted scientism of the epoch, which was confirmed by the teleological version of zoology. Furthermore, this leads to a conception of the relation between dynamics and statics which states that both processes exist in nature and are closely related: in spite of Schumpeter's claims, mechanics was already ruling in economics by that time.

In this sense, the 'classical' program, formulated by Schumpeter and consisting of static analysis to which dynamical tools should be added later on, closely corresponded to Comte's program. Schumpeter himself accepted this program, and this was the substantial reason for his high praise of Walras. In this sense, he could not emancipate from positivism.

13.2.B. John Stuart Mill

Comte strongly influenced the evolution of John Stuart Mill, and this was indeed the main reason for the latter's rupture with the Benthamite School (Mill, CW, xiii: 488). It was in fact from Comte that Mill drew the distinction between statics and dynamics, the first being the study of the 'conditions of existence of the society' or the 'stationary society' and the second the study of 'the laws of its continuous movement' or the 'progressive state of wealth' (Mill, 1848, CW: xiii; 695-6; 1865: 88-9). Both types of analyses were considered to correspond to real processes, in this sense to 'natural' processes (ibid.: 100). In other words, Mill interpreted the Comtian conceptions of statics and dynamics as two simultaneous approaches to reality, order which subsumes motion and dynamics which includes statics: the progression of wealth should also lead to a final stationary society (1848: 746 f.).

But, as Mill argued in his *Autobiography*, the acceptance of Comte's gradualist law of social evolution was compatible with the pervasive and dominant scientism and with his own previous vision, which was inspired by physics[7] and not by biology:[8]

> This doctrine harmonised with my existing notions, to which it seemed to give a scientific shape. I already regarded the methods of physical science as the proper models ... . (Mill, CW, i: 173; also 75).

It is very important to note that Schumpeter acknowledged and stressed himself this passage in HEA, in order to underrate it and to defend his previous interpretation:

> Thus, most economists, J.B. Say and J.S. Mill in particular, thought altogether too much of the analogy with the physical sciences, which the latter declared to be the 'proper model' for economic theory — a point for critics to fasten on but actually irrelevant, since no practical use was made of it. ... We have already observed that the introduction of the terms Statics and Dynamics does not involve any such use, that is, any borrowing of a method from any physical science. Nor do economists borrow from mechanics when they employ the term equilibrium any more than does a bookkeeper who 'balances' an account. (Schumpeter, HEA: 537 and n.)

The nature of the physical metaphor is to be delt with later on, just like the new paradox introduced by this insistent argument about the 'zoological missing link' of the concepts — since Schumpeter did not hide in the same HEA his consideration of physics as the most exact and developed of all sciences and his lack of interest in the biological metaphor. What is essential for the moment is that this argument is wrong: the ultimate inspiration for Mill's distinction between Statics and Dynamics was not zoology but an extreme positivist version of nineteenth-century scientism, inspired by mechanics and which considered evolution as a special case of a gradualist and deterministic process. Mill, the 'classics' and Schumpeter received and accepted the *vision* that both static and dynamic processes are to be found in nature and that statics is a special case of dynamics. This argument is crucial to the comprehension of Schumpeter's acceptance and reverence of Walras, and in fact to the definition of his own research program with the aim of completing or constructing the dynamic side of Walras.

14. From Evolution to Evolutionism

Two instances were presented up to now of Schumpeter's relation to the physical and the biological metaphor. In the first, he insisted all through his work that the concepts of dynamics and statics he was endorsing were ultimately traceable to zoology and not to mechanics, in spite of the evidence and namely of the declarations of their creators. The second is a major case in the opposite sense: after acknowledging Juglar's role in the definition of a theory of the cycle, he concluded that

> And so we have reached a stage, perhaps for the first time, where facts and problems are before all of us in a *clear* and in the *same* light, and where analysis and description can cooperate in something like the spirit of physical science. (Schumpeter, 1927: 287, his emphasis)

The following pages are concerned with this obvious contradiction between the definition of physics as the authoritative model for sciences and the attempt to deny the incorporation of mechanicist influences into the province of economics.

14.1. Physics as the model science

For Schumpeter, Walras was certainly the major protagonist in the development of economics in the direction of the rigour of physics. His theoretical system of general equilibrium, the simultaneous system of equations determining all prices and quantities, was described as the perfect analogue for the exact features of physics:

> Walras is in my opinion the greatest of all economists. His system of economic equilibrium, uniting, as it does, the quality of 'revolutionary' creativeness with the quality of classical synthesis, is the only work by an economist that will stand comparison with the achievements of theoretical physics. (HEA: 827)

Physics was described as a paradigm for sciences in no other than in HEA, the last great text of Schumpeter, on which he worked for the last years of his life. Nowhere in his work was another indication of a different science having this central epistemological role. And physics was also mentioned when Schumpeter needed to present an argument of authority: since his critiques did not accept his views on the clusters of innovation, he argued in 1935 that these were postulated by the theory, just like hypotheses were formulated in physics, 'irrespective of what might be adduced for or against their objective truth' (Schumpeter, 1935: 6).

Finally, physics was also mentioned in order to present new methods (for example, being impossible to isolate a phenomenon, the simulation from a model in order to compare the result to the observation, ibid.: 3; or the use of simplifying schemata, HEA: 14) or to create or to incorporate analogies (BC: 12, 41).

These three types of instances — physics as the general paradigm, as the authority in science and as the source of pertinent analogies — were present in the whole work of Schumpeter. But this did not prevent him stressing several times what he considered the misleading character of the physical metaphor: 'Analogy with the entirely different problems of physics is much more apt to be misleading than helpful' (BC: 32).

In the same vein, he criticized Mitchell for his alleged overstatement of experimental procedures under the influence of the early physicist epistemology,[9] or Pareto for his illusions about the application of the method of physics (TGE: 189-90). He stressed the differences between the field of economics and that of physics, since the first is more complex, there is no possibility of experiments,[10] includes interpretative variables and the scientist is under pressure to get socially useful results (ibid.: 149, 189-90).

Schumpeter knew that this paradigm dominated the main works of the neoclassicals, from Walras to Pareto and from Edgeworth to Fisher, and that it defined the contours of the marginalist revolution, with the sole and relevant exception of his own teacher, Menger. This is why Schumpeter dealt with this point from the opening pages of HEA, considering the general criticism of Hayek — a representative Austrian — of 'scientism', that 'uncritical copying of the methods of mathematical physics in the equally uncritical belief that these methods are of universal application and the peerless example of all scientific activity to follow' (HEA: 17). Along his discussion about this point, Schumpeter was ready to accept Hayek's argument against the incorporation of such a 'scientist bias', but argued that this was not a general case in economics and that only words were being transferred from physics.[11] On the other hand, following Schumpeter's argument, there were two essential reasons why physical concepts — 'borrowing words and nothing else' — made their appearance in economics: first, mathematics was developed sooner in physics and exact sciences[12] and, second, the analogies were supposedly very useful for teaching since students understand a physical analogy more easily than its economic counterpart. The final argument by Schumpeter was that the presence of common concepts and methods only testified that economists and physicists have the same type of brains and act similarly faced with theoretical problems (HEA: 18), and consequently:

> This does not involve any mechanistic, deterministic or other '-istic' errors, or any neglect of the truth that 'to explain' means something different in the natural and in the social sciences, or finally any denial of the implications of the historical character of our subject matter. (HEA: 18)

The whole passage indicates very clearly that, while accepting the general scientific paradigm of positivism and consequently the role of physics both as forerunner of scientific rigour and as a model for pedagogic analogies, Schumpeter resisted the idea that economic concepts and methods were derived from physics. He did so because of the Austrian influence still present in his thought, because he was certainly aware of the uncritical exaggeration of the neoclassical tradition, namely by Fisher, and possibly because his own research indicated that the available physical concepts were unable to encapsulate the economic reality of movement and change. But the author also accepted important exceptions, since some concepts and not only mere words ('equilibrium', 'potential') were incorporated under autonomous contents for each science.

This interpretation accounts for *all* the instances previously identified: in the general epistemological stance, Schumpeter certainly praised the authority and clarity of physics; but when concrete economic concepts were at stake, he denied any significant influence from physics — even when such a claim was wrong, as the case of the Comtian definition of statics and dynamics exuberantly proves.

The argument about the concept of equilibrium is again very useful to illustrate this point. Schumpeter defined equilibrium by the twin properties of existence and stability: 'If the relations ... are such as to determine a set of values of the variables that will display no tendency to vary under the sole influence of the facts included in those relations per se, we speak of equilibrium' (HEA: 969). Schumpeter immediately denied that this definition was related to physics in any way (ibid.: 970). His definition of equilibrium was also presented as the core of the economic science:

> For our system is logically self-contained only if this is the case [being totally endogenously determined]: we can be sure that we understand the nature of economic phenomena only if it is possible to deduce prices and quantities from the data by means of those relations and to prove that no other set of prices and physical quantities is compatible with both the data and the relations. The proof that this is so is the Magna Carta of economic theory as an autonomous science, assuring us that its object is a cosmos and not a chaos. (BC: 41)

In other words, economics became a science after establishing its Magna Carta — the Walrasian system — but, what is still more relevant to the present purpose, equilibrium was defined as the creation of order, derived from a set of variables which exhibit no tendency to create change. Of course, if this was really a 'self-contained process', then the Law of Conservation of Energy could easily be translated into the system and we arrive in the wonderful world of marginalist economics, where all relevant information is included in the system itself. But then Schumpeter added that there was something else, since this real process was nevertheless incomplete: the main argument of BC was that there was also another set of endogenous variables, the innovative processes, that added movement and mutation to this orderly picture. In one word, we have two classes of endogenous variables, those that are well-behaved and those that are unpredictable, which create irreversible processes of change. Equilibrium exists, and so does disequilibrium: the system is intrinsically morphogenetic.

This is why Schumpeter did not follow the entire physical metaphor, although his determination to safeguard the notion of Walrasian equilibrium in the new synthesis also prevented him from avoiding or rejecting this metaphor. As a consequence, Schumpeter defined the two separate theoretical domains, one where the equilibrium paradigm was used and another where it was not relevant.

14.2. Biology and dilettantism

With his detailed discussion of the analogy with physics, Schumpeter proved to be aware of the importance and problems of the metaphoric redescription. This section deals with a second body of analogies used in Schumpeter's work, the biological concepts, which appear in two different frameworks: in the general sense of the development process and in the concrete sense of the analysis of the circular flow. They are obviously very different in nature and in scope.

In TED, Schumpeter commented on two comparisons, one with the circular flow and the other with development, and rejected both. First, the circular flow could be conceived of as the process of blood circulation, but no growth and decline was allowed in the case of circular flow, unlike the case of the organism (TED: 45). Nevertheless, three years later he came back to this topic, acknowledging that even for Quesnay — a physician — the process of circular flow could not exactly repeat itself every time, since there is a relation between society and the incorporation of energy and materials from Nature: 'This point of departure was in itself an obvious one, it was bound to be specially familiar to Quesnay because of the analogy to the nutritional process of organic bodies' (EDM: 54-5).

In Schumpeter's system, the analogue for this nutritional process was growth, under the impulsion of capital accumulation and population increase. The equilibrium can therefore be considered either as a stationary process or as an organic adaptive process (BC: 35), but does not explain the cycle.

Second, Schumpeter argued that the process of development is different from the growth of a tree, since it is not continuous (ibid.: 144; the tree metaphor was associated with Smith, Mill and Marshall, and rejected by Schumpeter, 1951: 233). The cycle — and the whole 'organic' process of capitalist development[13] — should be explained by another phenomenon,

> industrial mutation — if I may use the biological term — that incessantly revolutionizes the economic structure from within, incessantly destroying the old one, incessantly creating a new one. This process of Creative Destruction is the essential fact about capitalism. (CSD: 83)

This is indeed a rather exceptional statement[14] since, as a general case, Schumpeter rejected in the strongest terms any attempt to incorporate a biological metaphor into economics. In 1911, he wrote:

> Here [in the class of 'metaphysical' tendencies], too, belong all kinds of evolutionary thought that centre in Darwin — at least if this means no more than reasoning by analogy... . But the evolutionary idea is now discredited in our field, especially with historians and ethnologists, for still another reason. To the reproach of unscientific and extra-scientific mysticism that now surrounds the 'evolutionary' ideas, is added that of dilettantism. With all the hasty generalisations in which the word 'evolution' plays a part, many of us lost patience. We must get away from such things. (TED: 43)

Almost forty years later, Schumpeter still held the same opinion and expressed it in the same terms. Writing in HEA about the 1870-1914 period, he emphatically described biological evolutionism as 'a field infected by ideological bias and by dilettantism to an extent that surpasses anything that even we economists are accustomed to' (HEA: 788). In spite of it, Schumpeter considered Darwin's *Origin* as an important scientific achievement, comparable to the definition of the heliocentric system (ibid.: 445, 445n.) and his historical sketch of the previous biological theories as a crucial piece for the sociology of science, but did not indicate any possible kind of influence of theses texts on social sciences: his sympathetic references were probably mainly due to ceremonial reasons. Nevertheless, it is clear that his purpose was to attack the influential and wide-spread Spencerian type of evolutionism, which combined 'naive laissez-faire' with a simplified version of Darwinism, leading to conclusions such as the 'silly' suggestion for the abandonment of sanitary regulations or public systems of education and health (ibid.: 773). Schumpeter also cared to inform the reader from the introduction to BC that his assumption about the organicity of economic processes did not at all imply being a supporter of laissez-faire (BC: vi). Therefore, the concept of 'industrial mutation' was carefully chosen in order to emphasize the non-equilibrium properties of development and evolution. In this, Schumpeter was indeed closer to Marx than to Walras.[15]

14.3. Schumpeter's evolutionism

Schumpeter's evolutionism was not based on Darwinism or, in general, in the biological metaphor, which only played a minor role, if any, in his system. But it was still an evolutionary conception, since it was based upon two central concepts: first, the economy was defined as an 'organic'[16] whole, propelled by a process of development with mutations and, second, this defined a non-mechanistic and historical view of capitalism as creation and destruction. Since the concept of 'mutation', that change arising from innovation in the core of the system, was previously discussed, this section will now turn to the concept of the organic system.

Here is how Schumpeter presented the concept, criticizing the biological analogy:

> In the first place, we notice the idea that society, being an 'organic' system and not a 'mechanical' one, can be fruitfully analysed in terms of an analogy with biological organisms such as the human body. ... But the obvious puerility of this idea must not blind us to the fact that emphasis upon the 'organic nature' of the economic process may be but the means of conveying an eminently sound methodological principle — as it was, for instance, with Marshall. Theorists — specially of the 'planning' type — often indulge in the deplorable practice of deriving 'practical' results from a few functional relations between a few economic aggregates in utter disregard of the fact that such analytical set-ups are congenitally incapable of taking account of deeper things, the more subtle relations that cannot be weighted and measured .... 'Organic' considerations are perhaps the most obvious antidote — though in themselves hardly an adequate one — against such uncivilised procedure. (HEA: 788-9)

Besides the polemic bias — the 'theorists of the planning type' could be easily replaced by the 'theorists of the marginalist type', and the whole paragraph would keep its sense — this is a clear indication of the nature of Schumpeter's thought: organic considerations were supposed to be essential in order to avoid the useless biological analogies and hence to provide an overall method for the economic inquiry: the solution of a system of equations was unable to represent complex or 'more subtle' relations. This explains his approach to causality in economics (BC: 7) and to the analysis of its features (Schumpeter, 1949: 313). Evolutionism, then, was for Schumpeter simply the consideration of organic evolution in real time, or of historical and irreversible processes of change:

> Social phenomena constitute a unique process in historic time, and incessant and irreversible changes are their most obvious characteristic. If by Evolutionism we mean not more than recognition of this fact, then all reasoning about social phenomena must be either evolutionary in itself or else bear upon evolution. Here, however, evolutionism is to mean more than this. One may recognize the fact without making it the pivot of one's thought and the guiding principle of one's method. ... [James Mill's] various systems were not evolutionary in the sense that his thought in any of those fields turned upon evolution. And it is this that shall be the criterion of evolutionism for us, both as regards philosophy ... and as regards any 'scientific field'. (HEA: 435-6)

It is possible to conclude that Schumpeter defined the social process as an intrinsic dynamic disturbance of equilibrium through the creation of novelty

— the innovative mutation — and this was precisely what defined his evolutionary framework. It included stationary processes of equilibrium, the place of Walras, but also forces and processes moving towards disequilibrium, the place of Marx. And it was organic, since both processes were considered to be compatible and since all the relevant variables were considered to be endogenous to the system, which generates itself movement and change. Moreover, this particular combination was the very specificity of economics, and so Schumpeter believed that his general and historical approach was the only one able to integrate both the statics of general equilibrium and the dynamics of disequilibrating forces: in a superior synthesis, the unscientific bias of the physical and the biological analogies should be prevented, since those analogies took the part as the whole and thus developed dilettante or simplistic views.

14.4. Flirting with heresy

As previously indicated, Schumpeter considered that his own views were substantially unchanged from his first writings. This is why his early work is so important in order to understand the formation of the theory. In 1906, he prepared a review of J.B. Clark's *The Distribution of Wealth* (1899), generally considered as one of the important building blocks of neoclassical economics.[17] This was one of the first two papers Schumpeter published after concluding his studies at the university. Previous authors have indicated the influence of Clark's economics in Schumpeter (Streissler, 1994: 33), but the crucial aspect has not been acknowledged: it was from J.B. Clark that Schumpeter took the inspiration for his own definition of statics and dynamics, the first insights on the *entrepreneur*, the central concept of innovations as new combinations and, indeed, a whole theoretical program to be followed during the next decades. In fact, Clark's book represented a critical tradition within orthodoxy, since he was aware of the limitations of the static equilibrium approach and intended to overcome it: that was the aim and the proclaimed objective of the 1899 book, which was to be followed by another and more complete dynamic version. But the project failed, since the following 1907 book by Clark, *The Essentials of Economic Theory*, was merely a study on comparative statics.

Schumpeter's review was a short paper presented as a mere summary, 'abstaining from any criticism' (Schumpeter, 1906: 325). In fact, some of the main theses were presented without any detailed discussion: the 'natural' character of economic laws, the sea and tempest analogy for the equilibrium state and real economic processes, the social judgement implicit in the marginalist theory of distribution,[18] the concept of capital and the organic nature of social life. In general terms, the review was very limited and partial, and the book was clearly more influential on Schumpeter than his own text suggests — one of the crucial examples is the non-reference to Clark's treatment of the entrepreneur.

Five main influences will be now indicated, proving the importance of Clark's book in the formation of Schumpeter's research program.

Although concerned with standard marginalist theory, and therefore with static systems, Clark's book lengthily discussed the problem of the status of statics and dynamics, expressing the author's intention to develop in the future a new branch of dynamical economics. This was the first essential influence on Schumpeter's agenda: for Clark, the only real world was describable by dynamics — and this was a major departure from traditional marginalist theory, indicating how the young Schumpeter could have felt three years later when visiting Walras. But statics was 'imaginary' not because it did not describe real trends but because it was incomplete:

> All natural societies are dynamic ... . In the actual world unceasing changes thrust labour and capital, from time to time, out of one occupation and into another. In each industry they change, again and again, the modes of production and the kinds and quantities of the goods produced. Yet this does not invalidate the conclusions of a static theory; for static laws are nevertheless real laws. The forces that would work in a world that should be held in a fixed shape and made to act forever in a fixed manner still operate in a changing world of reality. (Clark, 1899: 30)

As economics was supposed to deal with this particular combination of static and dynamic phenomena, the author discussed carefully how do they match. Clark indicated that there are dynamical forces tending to change the whole society, 'changes that alter the mode of production and act on the very structure of the society itself' (ibid.: 31), but that the purely static state still describes a part of reality: 'The description of the purely static state, in fact, deals with realities. It is imaginary only by its omissions; for it presents an essential part of the forces that act in the real, dynamic world' (ibid.: 401).

Since 'all real knowledge of the laws of movement depends upon an adequate knowledge of the laws of rest' (ibid.: 442), the economy could be modelled as a set of forces of 'organization', defining the equilibrium prices and quantities, and of 'progress', impelling the system towards a new level of equilibrium (ibid.: 30, 32, 429). This was the second dominant influence of Clark's theory: after all, the tendency towards equilibrium still prevailed. This implied an important conclusion: statics was defined as a particular case of dynamics but, what is still more significant, it was supposed that dynamic processes converged towards equilibrium. In other words, equilibrium was supposed to be the final state of the system and, as a consequence, dynamics was reduced to statics. This crucial conclusion, which was opposed by Marshall and, later on, by John Maurice Clark,[19] was at the very centre of Schumpeter's paradoxical heritage.

The third influence concerned the definition of the dynamic change:

> New mechanical processes are coming into use. Machines supplant hand labour, and efficient machines displace inferior ones. New motive powers are taken into

> service, and new raw materials are used. Population increases and migrates, taking with it some of the increase of its wealth. Large industries grow up and crowd small ones out of the field. (ibid.: 31)

Those forces were distinguished in five categories: (i) increases in population; (ii) increases in capital; (iii) improvements in methods of production; (iv) organizational changes in firms; and (v) changes in the preferences of the consumers (ibid.: 56). Both (i) and (ii) were later classified by Schumpeter as the elements of growth, and indeed even Clark separated them from the others, since they were supposed to account for steady variations (ibid.: 416); the other factors were later on at the core of Schumpeter's concept of new combinations implying the displacement of the production function. Five years after his review, when writing TED, Schumpeter reconsidered those elements and explained the evolution of population and capital and even the changes in tastes as mere elements of perturbation, while the analysis of the changes in techniques and organization procedures required 'a new conception of the economic process' (TED: 45), which was the purpose of the book. The convergence but also the difference with Clark's definitions is obvious: while Clark studied the effect of a perturbation and the reestablishment of equilibrium, Schumpeter's main concern was the theory of the causes of perturbations, defining the 'new combinations' and the innovative processes as the motive power of capitalist development (ibid.: 48).

As a consequence came the fourth main influence by Clark: the notion of the role of the entrepreneur, since the 'mechanical inventions' imply 'new kinds of goods [and] call for new industrial groups to make them' (Clark, ibid.: 61). Entrepreneurship, as a function distinct from the ones of capital and labour, once again related dynamics to statics:

> Dynamic science deals with profits in their original state, as normally created by improvements in industry, in the proceeds of which the entrepreneurs have a share; while static science deals with them in their later and permanent state, as they are transmuted into increments of wages and interest. (...) Dynamic theory has to account for the whole of that friction of which the entrepreneur's share depends .... (ibid.: 410)

Later on, in HEA, Schumpeter emphasized this central contribution, which is completely ignored in his 1906 review: 'he made a great stride toward a satisfactory theory of the entrepreneur's function and the entrepreneur's gain and, in connection with this, another great stride toward clarification of all economic problems that must result from a clear distinction between stationary and evolutionary states. ... If his achievement fell short of Bohm Bawerk, Marshall and Walras in some respects, it rose above theirs in others' (HEA: 868). Moreover, 'Clark's contribution was the most significant of all: he was the first to strike a novel note by connecting entrepreneurial profits, considered as a surplus over interest (and rent) with the successful introduction into the economic process of technological, commercial or organizational improvements' (ibid.: 894). Clark — who was not one of the 'ten great economists' — was presented in HEA as the 'architect of one of the marginalist analysis' most significant theoretical structures' (ibid.: 868).

Last but not least, there was a fifth major influence: economic theory was really about the rate of progress induced by entrepreneurs through the profit obtained from innovations (ibid.: 411). This implied a specific scientific research program introducing history into economics — the conclusion which Schumpeter supported forty years after his paper on Clark. As Clark put it:

> Economic dynamics has a striking relation to those recent historical economic studies which have been so attractive and fruitful. ... Economic dynamics will, in its entirety, incorporate into itself historical dynamics. The changes that are going on in the world will in future be studied inductively as well as deductively; and it is the inductive part of the work that falls to the historical economist. (Clark, ibid.: 73-4)

These five influences synthesize the essential impact of the work of Clark on Schumpeter's future research: all the major questions were indeed present in this rough sketch by Clark, including an anticipation of the idea of long waves of approximately forty five years as specific periods of development of the dynamic forces of capitalism (ibid.: 429).

Schumpeter acknowledged explicitly some of these influences and, in general, presented a very positive appraisal of Clark's contribution to economics, although he never explicitly recognized the dimension of this global theoretical debt. Namely, Schumpeter indicated in HEA the relevance of Clark's contribution to the theory of the entrepreneur, comparing it to the presumed inferior notion of Marshall (HEA: 868, 894), and to the distinction between statics and dynamics, interpreted as descriptions of stationary and evolutionary states, but did not go farther. In reality, the review of Clark's work was the essential step for the early definition of Schumpeter's research program and he followed that agenda for all his life.

Although respected and recognized, Clark's work did not have the desired impact and the equilibrium paradigm developed along Walrasian lines. But his attempt drew the attention of some heterodox authors: Veblen's review of *The Distribution of Wealth*, written shortly after Schumpeter's, strongly opposed his 'hedonist' program and therefore the concepts of utility or the marginalist methodology, and namely challenged his equilibrium metaphysics, themes that did not provoke any opposition from Schumpeter. According to Veblen, Clark's book was just another effort in order to reduce dynamics to statics (Veblen, 1908: 189). From that point of view, Veblen argued that Clark's effort was a failure: 'All that it covers Clark's concept of dynamics is a speculative inquiry as to how the equilibrium restablishes itself when one or more of the quantities involved increases or decreases' (ibid.: 188).

Furthermore, according to Veblen, Clark limited the scope of his research to a trivial organicist conception (the economy was defined as an 'organism', Clark, 1899: 196), unable to account for change and mutation, that is, for dynamics:

> Economics of the line represented at its best by Mr. Clark has never entered this field of cumulative change. It does not approach questions of the class which occupy the modern sciences — that is to say, questions of genesis, growth, variation, process (in short, questions of a dynamic import) — but confines its interest to the definition and classification of a mechanically limited range of phenomena. (Veblen, ibid.: 192).

What Veblen charged Clark with was his adherence to the mechanical metaphor; once again, that did not affect Schumpeter, since he classified Clark's concepts of forces, equilibrium and statics as mere and inconsequential borrowing of words. Schumpeter fully accepted and followed, for all his life, Clark's program, which was at the same time criticized by Marshall and rejected by Veblen. In doing so, Schumpeter was exploring the boundaries of the orthodox paradigm and defining the daring — and impossible — task of completing Walras's general theory in order to include all phenomena, namely dynamics; as Clark, he did not want to abandon the general equilibrium paradigm — that was his bounded heresy.

*

The last sections followed the trajectory of Schumpeter's thought in two main directions: the definitions of statics, dynamics and equilibrium, and then the role and influence of the physical and biological metaphors in his definition of evolutionism. In both instances, the stability of Schumpeter's main concerns and definitions was acknowledged.

The following sections develop another aspect of Schumpeter's evolution, namely the important changes in his approach on the roles of theory and history in economics. This is the appropriate introduction to a comparison with Keynes, since the substantive difference between both authors was essentially one of philosophical stance.

14.5. From the Methodenstreit to the Sozialokonomie

The first book by Schumpeter (DW) was actually a large dissertation on the methodological debate opposing Menger and Schmoller, the 'theorists' and the 'historical school'.[20] Schumpeter, who entered the Vienna University just after the retirement of Menger and studied under the supervision of Menger's successor, Wieser, supported the 'theoretical' side of the Methodenstreit and praised the importance of the abstract and naturalistic approach in economics:

'From a methodological and epistemological viewpoint, pure economics is a "natural science" and its theorems are laws of nature' (DW, quoted in Swedberg, 1991: 28), having nothing to learn from biology (ibid.: 27).

Schumpeter was at the time considered to be a faithful supporter of the Mengerian side, although with strong Walrasian intonations: Hildebrand opposed in 1911 his appointment as teacher in Graz, because he did not want the University to be occupied by anti-'historians'. Later on, after the failure of his political career — as Minister of Finance in the republican government of Karl Renner from March to October 1919 — as well as of his banking career in Austria, the fame of the marginalist 'theorist' would precede him to Germany, the stronghold of the Historical School. Only the influence of his friend Spiethoff, a disciple of Schmoller, assured him a badly needed appointment at Bonn University (Swedberg, 1991: 69-70). Schumpeter considered himself as a marginalist by that time: in a note in HEA, he presented his positions as of a 'strong partisan of economic theory', just like Wieser (HEA: 819 n.). He was also, as throughout his life, an extreme positivist, supporting a strong demarcation between positive science and normative intervention. Under the influence of logical positivism, then dominant in Vienna, Schumpeter always defended this form of 'therapeutical nihilism'.

But his balance-sheet of the Methodenstreit changed during the years, and that was already obvious when, three years after the publication of TED, Schumpeter discussed again the problem in his EDM. By that time, his position was much more careful, even though still supporting the marginalist side: the whole debate was considered useless and exaggerated, and he presented 'an explanation for the controversy: it was a struggle between two methods of work, between people of different mental habits, who fought for elbow room or for domination' (EDM, 1914: 167).[21] It is hard to consider this as a compliment for either side.

The book was careful to insist on the elements of synthesis, namely on Schmoller's alleged acceptance of the similarity of the causal nexus in social and natural sciences and of the definition of laws as the aim of science (ibid.: 170). Schumpeter even argued that one of his main previous criticisms of Schmoller's insistence on reform policies could be dropped since the latter had changed his mind (ibid.: 175) — which was not at all evident.

Schumpeter indicated the six major innovative elements of Schmoller's contribution as the following concepts: (i) the relativity of theory; (ii) the unity of social life; (iii) anti-rationalism, which Schmoller was supposed to have abandoned; (iv) the evolution and the role of history, to be compared to Marx; (v) the affirmation of complexity; and (vi) the organic conception, as an analogy of society with a body (ibid.: 175 f.). The striking fact is that Schumpeter incorporated many of these features into his own research, namely ii, iv, v and vi, even if by the time of publishing EDM he was still fascinated by the marginalist revolution, a 'purer economics', 'incommensurably more firmly founded', 'more correct', 'simpler' and 'more general' (ibid.: 181 f.; 189-90).

But the very conception of EDM deserves some attention. The essay was prepared for Max Weber's *Grundrisse der Sozialokonomie*, a handbook which was intended to present a new methodology for a transdisciplinary social science. Weber was strongly opposed to the Methodenstreit, which he accused of having led to an artificial polarization of the statistical and theoretical methods against the historical method. The influence of this conception[22] on Schumpeter's thought was a lasting one.

In 1926, after a long period without any theoretical intervention, Schumpeter came back to the discussion of Schmoller's theories in an essay, *Gustav von Schmoller and the Problems of Today*, which represented a major turn, in the sense that it was a very positive assessment of the author and a formal endorsement of the Sozialokonomie, arguing for a fruitful combination of theory, statistics, history and sociology as the basis for a new economics. And this remained his consistent opinion from then on, insisting on the role of history for the understanding of capitalism.[23] In the opening of *Business Cycles* this was clearly stated: history has the 'most important contribution to the understanding of our problem' (BC: 13). This book is certainly a major piece of economic historical analysis.

And no other than his magisterial *History of Economic Analysis* is the most complete statement of the program of Sozialokonomie: the main techniques indicated for the research in economics are history, 'by far the most important' since 'the subject matter is essentially a unique process in historic time', then statistics, then theory and finally economic sociology (HEA: 12).

This methodological indication, combined with the definition of economics as the study of irreversible processes of change, plus the organic vision of evolutionary societies, defines the main conclusion of this section: since Schumpeter incorporated some essential traits of the Historical School in a very distinctive framework and since he invaded the new territory of historical mutation in the economies, Schumpeter was not a neoclassical economist, but at the same time he was not able to cease considering himself to be one since he did not wholly reject, and even tried to incorporate the paradigm of equilibrium along with the historical forces of mutation. This is more than the simple restatement of the Paradox, as it was met in the previous discussion about the role of the Walrasian system in Schumpeter's economics: it is also a programmatic conclusion, since the scientific viability of the modern evolutionary program depends crucially on the rejection of the Schumpeterian compromise.

This dilemma was obvious on several occasions in Schumpeter's work. The cold reception of BC in the scientific community was certainly one of the indications of the increasing difficulty for Schumpeter to maintain his profile as a mainstream economist while developing what was considered as an extravagant or esoteric research. As Kuznets put it, cycles are a quantitative phenomenon and should therefore be dealt with by statistical methods and not as a qualitative phenomenon as Schumpeter implied. Since Schumpeter had been the chairman of the founding meeting of the Econometric Society, he was at the time of the publication of BC its vice-president and was to become its president the next year, he was as a consequence expected to contribute to quantitative economics and to the mathematical formalism he praised so often. He did not and, in fact, in the whole scientific community only Frisch received the book with enthusiasm (Swedberg, 1991: 271 n.).

This contradiction was again evident in one of the last important scientific meetings attended by Schumpeter, the 1949 NBER Conference on Business Cycles. At this Conference, where 'historians' (the NBER researchers) and 'statisticians' (the Cowles Commission staff) collided, Schumpeter undertook the task of arguing for the historical method and to represent Mitchell, who had recently passed away. In his double and uncomfortable condition as the author of *Business Cycles* and as a distinguished member of the Econometric Society, Schumpeter began with a defensive declaration: 'I have no wish to advocate the historical approach to the phenomenon of the business cycle at the expense, still less to the exclusion, of theoretical or statistical work upon it' (Schumpeter, 1949: 308). But he then repeated his main definition: 'Economic life is a unique process that goes on in historical time and in a disturbed environment' (ibid.).

History is needed for the inquiry on exogenous, occasional events, but also and essentially on the very organism of the cycle:

> For historical research is not only required in order to elucidate the nature and importance of the non-essentials dealt with so far, but also in order to elucidate the underlying cyclical process itself. ... But it would not be quite correct to say that historical analysis gives information as regards impulses and dynamic [theoretical] models as regards the mechanism by which the impulses are propagated ... . Very roughly this is so and I should be quite content if my audience accepts the thesis that the role of the econometric model ... is to implement the results of historical analysis of the phenomenon and to render the indispensable service for describing the mechanics of aggregates. But the econometric models do more than this — they 'explain' situations which in turn 'explain' or help to 'explain' impulses. And the reverse is also true. (ibid.: 311-3)

This is a notorious argument, not only by its search for an incisive counter-logic pedagogy — the listeners should be driven to accept the historical method for the precise reason they were opposing it — or by the acceptance of some sort of Frischian formalism of cycles, but also because it indicates how far was Schumpeter engaged in the defence of the role of historical research and qualitative methods. And certainly the final advice by Schumpeter did surprise his audience:

> To let the murder out and to start my final thesis, what is really required is a large collection of industrial and locational monographs ... [including the historical change and the 'behaviour of leading personnel']. (ibid.: 314)

It is well known that his arguments did not change the course of history, and that the econometric revolution was already very well in its way. But his arguments surprised some of his colleagues — Samuelson[24] and Goodwin ('it was a great shock to me', in Swedberg, 1991: 176) — but neither did they stop the attacks by the econometricians against the historical method (Gordon, 1986: 27), nor could they prevent the rise of the new breath of equilibrium economics.[25] Schumpeter could not prevent it, and in fact could not challenge it, since it was too late and too little: he was not ready to emancipate from the equilibrium paradigm.

14.6. Equilibrium and the Cycle

In 1910, while preparing TED, Schumpeter summarized his own views in some short theses:

> First, the economic processes divide into two different and also in practice clearly discernible classes: static and dynamic. Second, the latter constitutes the pure economic evolution, that is, those changes in the model of the economy which arise from itself. Third, the economic evolution is essentially a disturbance of the static equilibrium of the economy. Fourth, this disturbance provokes a reaction in the static masses of the economy, namely a movement towards a new state of equilibrium. (Schumpeter, 1910, quoted in Andersen, 1994: 41)

For the current purposes, it is essential to emphasize that Schumpeter distinguished between statics and dynamics, as Mill previously did, as two real processes, related by the conception that without disturbances the system would be 'static'[26] but that those disturbances arise from inside the system itself. In 1908, Schumpeter argued that the central question for 'pure' economics was statics and equilibrium — which surely deserved the approval of Walras — and considered dynamics as a marginal phenomenon (DW, quoted in Bottomore, 1992: 171), despite some rhetorical declarations about dynamics as 'the land of the future' (DW, quoted in Swedberg, 1991: 29-30). But he quickly changed his opinion, as indicated by the 1910 thesis.

It is also certain that the 1908 book already presented some clues for the future discussion about the entrepreneur. Schumpeter was influenced by an economist of the early nineteenth century, Riedel, who stressed the role of innovations in economic life, by the previous work on the entrepreneur by Thuenen or Bohm Bawerk and particularly by his teacher, Wieser, under the ultra-romantic influence so important in Germany at the time: a figure of a 'great man' and some 'heroic individualism' was defined in economics, just like Spencer did in sociology and Nietzsche in philosophy (Streissler, 1994: 19 f., 34; Allen, ibid.: 107) and J.B. Clark in economics. Entrepreneurship was interpreted in 1908 as the function of carrying the adventuresome innovation (Allen, ibid.: 47).

In 1911, in TED, Schumpeter presented these conclusions as the distinction between the 'circular flow' and 'development', the main economic processes in action (TED: 145). The circular flow, the 'missing link' in economic causality (EDM: 43 f.) supposedly discovered by the physiocrats, described

> how each economic period becomes the basis for the subsequent one, not only in a technical sense but also in the sense that it produces exactly such results as induce and enable the members of the economic community to repeat the same process in the same form in the next economic period; how economic production comes about as a social process, how it determines the consumption of every individual and how the latter in its turn determines further production ... . (EDM: 43)

In other words, this is the stationary process or the condition for equilibrium, which are analytically equivalent (BC: 42 n.; also 68). On the other hand, development was defined as a *quantum* jump in the social conditions of the system, 'that kind of change arising from within the system which so displaces its equilibrium point that the new one cannot be reached from the old one by infinitesimal steps' (TED: 47 n.). But the 'static conditions' exclude the cycle but not growth: in fact, 'growth', defined as the combination of the evolution of capital accumulation[27] from savings and of the population, was included in the notion of static equilibrium (Schumpeter, 1927: 289f.). Equilibrium was thus defined as a 'shifting centre of gravitation' in a system which also generates the internal impulse for change, that is, for the rupture of the equilibrium conditions. While development accounted for the nature of the change (BC: 560 n.), equilibrium described the absorption of change (Schumpeter, 1937: 159), that is, was defined as the stability property of the system. This topic will be discussed further.

The real economic system cannot be understood without the integration of both processes: in fact, even if Schumpeter sometimes indicated that 'perfect' equilibrium was never really present (BC: 52) and that it was a 'methodological fiction' (ibid.: 964) or if he criticized Walras's and Clark's presentation of real prices oscillating around equilibrium (HEA: 999, 1000 n.), his general approach was to argue for an integrated account of the development process as including both change and equilibrium. In his 'first approximation' to the theory of the business cycle, equilibrium existed at the end of the depression and before the prosperity. In the 'second approximation', when the 'secondary wave' was considered and the cycle was described in four phases, equilibrium conditions were met at two of the inflection points,[28] namely when the recession leads to depression and when the revival leads to prosperity and a new cycle is supposed to begin.

Three main points should be emphasized. First, this schema considered the stationary process or the equilibrium conditions to be a special case of the dynamic movement, specifically that corresponding to the discrete points where the movement is null (BC: 70-1, 963). This quite closely matches with the mechanical Mill-Comte definition of the distinction between 'statics' and 'dynamics' and namely of the possibility of conducting a static analysis, the 'bare bones of economic logic', 'cleaning the ground for rigorous analysis' (BC: 68).

Second, the existence of equilibrium was stated and its stability was defined as the real processes of absorption of change and of disturbance:

> The thing that matters to us is nevertheless this tendency [towards equilibrium] considered as an actual force, and not the mere existence of ideal equilibrium points of reference. ... We wish to distinguish definite periods in which the system embarks upon an excursion away from equilibrium and equally definite periods in which it draws towards equilibrium. (BC: 69-70)

Or also, without room for doubt:

> Common sense tells us that this mechanism for establishing or re-establishing equilibrium is not a figment devised as an exercise in the pure logic of economics but actually operative in the reality around us. (BC: 47)

In this sense, the mechanism of equilibration provided the resistance to change in the economic system, namely the defence of established business and institutional traditions: it was the creation of order subsuming the creation of novelty,[29] namely imitation restoring equilibrium after innovation. Equilibrium or order would be the moment of the formation of prices, while development or disorder is the evolutionary process: in Schumpeter's emphatic words, 'fluctuations must be fluctuations around something' (BC: 69).

Third, this did not imply that equilibrium was considered to be the desirable situation. In the first approximation, it was considered to be the situation where the promises of the boom were fulfilled, that is, where the availability of consumption goods increased for the whole community (TED: 161). But in the second approximation this was certainly not accurate, since the system was described as in permanent turmoil, and its change — the disequilibrium processes — was the only form of progress. From this point of view, Schumpeter clearly opposed the 'classics' and the general equilibrium paradigm, and even condemned their incapacity to incorporate real economic evolution: as he stressed in EDM, at the very same time that the first modern industrial crises were exploding, the 'classicals' still argued for Say's Law and rejected the theoretical possibility of disequilibrium, against all easily available evidence (EDM: 150). The main achievement of Juglar, by contrast, was precisely to define a new agenda for a research indicating the problem, describing it empirically and presenting an explanation (Schumpeter, 1927: 287). In other words, the drive to novelty which moves the economic system forward depends on the ability of the entrepreneur to challenge equilibrium:

> What a miserable figure he is, this economic subject who is always looking so anxiously for an equilibrium. He has no ambition and no entrepreneurial spirit; in brief, he is without force and life. (DW, quoted in Swedberg, 1991: 29)

With these qualifications, Schumpeter's theory of the cycle can now be reassessed. The motion of the system was analysed under a steady state representation, then the possibility of change was introduced as an independent and separable dimension, since both correspond to social processes that can be isolated. In other words, Walras indicated a convenient approach to discussing one of the processes (BC: 47), but this was not enough, since evolution should also be explained: for Walras the needs were given, while for Schumpeter the real economic processes created new needs and lead to deep transformations. The Schumpeterian research program consisted of the bold task of providing the dynamic counterpart of the Walrasian schema, aimed at a truly general theory.

This implied that some sort of logical separability was possible between the problems of growth and cycle, since growth was reduced to the monotonic trend of capital accumulation through savings and to population increase, both being added to an equilibrating process. Of course, this did not solve the statistical problem of the assessment of the trend and cycle, since there was in this account no real trend of equilibrium — only a number of discrete equilibrium points, two for each cycle — and since the cyclical process by itself displaced the centre of gravitation upwards.

And, moreover, the three-cycle schema implied that the equilibria of the shorter cycles were defined in the artificial representation of the trend line of the larger cycles, and that the single true equilibrium occurred at the very beginning of a Kondratiev, when prosperity was to commence and the equilibria of the three types of cycles coincided. All other points are 'neighbourhoods of equilibrium', therefore unstable due to a new very structural reason: the dynamics of evolution in the larger cycles overdetermined the shorter ones even when they were in the neighbourhood of equilibrium in their own motion. This was a form of representation of the feedback mechanisms in action in real economies, but added singular difficulties to the mathematical treatment of the model — and Schumpeter certainly had these in mind when he acknowledged that his theories were very hostile to mathematical formalism.[30]

This permanent tendency to the dislocation of the centre of gravity of the system and the complex interaction of the different cycles explain an original form of instability, created by the system itself. Schumpeter's theory was a system of self-generating complexity and instability, where the equilibrium concept really played only a very subsidiary role. But Schumpeter was not prepared to break with the Walrasian part of his theory, for philosophical rather than for theoretical economic reasons. The rationale for this refusal can be discovered in his general view of science and the definition of his own place in economic theory.

15. From Bloomsbury to Cambridge and to Harvard

In the long *Biography* published shortly after Keynes's death by Harrod — a colleague and one of his closest collaborators — it was argued that he reverted after the war to the 'free trade' positions of Adam Smith. According to Harrod, whose evidence is the report of a long walk conversation near the Thames one afternoon, Keynes told him that it was possible to come back to the 'great truths which [Adam Smith] preached', and to accept his 'gospel', 'without sacrificing any of [Keynes'] cherished principles relating to employment and trade depression' (Harrod, 1951: 609; also 469), since the new conjuncture allowed for new policies. Haberler also supported the same interpretation of Keynes's 'reconversion to liberalism' (Haberler, 1981: 12, 13 n.). Although Harrod and Haberler eventually implied something else, when referring to Smith and not to Walras or to Say, Keynes was also vindicating a specific historical tradition which is distinct from neoclassical economics, the contemporary form of liberalism.

A hunch of the motivations for this reasoning can be found in one of Keynes's last public interventions, when he negotiated the US Loan after the end of the second World War. In fact, Keynes defended the agreement and the economic arrangements implicit in it for the new world order, stating that it combined 'the advantages of freedom of commerce with safeguards against the disastrous consequences of a laissez-faire system', and that this made possible the 'implementation of the wisdom of Adam Smith' (Keynes, xxiv: 611, 621-2). In another place, Keynes stated that Adam Smith's teachings had been overlooked since they had been wrongly associated with laissez-faire economics (Keynes, xxviii: 444, 824). It is therefore clear that Keynes's return to Smith was completely alien to any type of reconversion to liberalism or to the equilibrium paradigm.

Mini argues that the biography by Harrod was in fact designed to establish the closest possible continuity between Keynes and the Harvey Road or the Cambridge traditions of Marshall and John Neville Keynes, as some combination of a (mildly critical) version of the equilibrium paradigm and a recapitulation of the mainstream positivist epistemology (Mini, 1991: ix), just like the previous IS-LM formalization which attempted to translate Keynes's contribution in neoclassical language. This question is the theme of the following pages, which discuss some elements of the philosophical approach of Keynes's work, since this is the crucial element to understand both the nature of his opposition to the 'classicals' and his specific difference in relation to Schumpeter's scientific trajectory. The concept of 'organic unity', the discussion of equilibrium and then of the cycle are successively presented and commented.

15.1. Organic unity and the physical metaphor

Before studying economics with Marshall and failing to get a fellowship at King's College with a dissertation on philosophy which would later on become the *Treatise on Probability*,[31] Keynes's main areas of work were philosophy, namely logic, and mathematics. He was also interested in a vast range of problems, and was at the time mostly influenced by G.E. Moore, the dominant philosopher at Cambridge, who defined ethical principles as the basis of a theory of the right conduct[32] and who opposed Russell's positivism. The principle of 'organic unity' was one of the crucial elements Keynes took from Moore's philosophy. In *My Early Beliefs*, Keynes described this concept:

> Their value [of states of mind] depended, in accordance with the principle of organic unity, on the state of affairs as a whole which could not be usefully analysed into parts. ... I myself was always an advocate of the principle of organic unity through time, which seems to me the only sensible one. (Keynes, x: 436)

In Moore's tradition, Keynes indicated one telling example of this organic unity processes: the 'state of mind'[33] of 'being in love' depends on one of its parts (one's feelings) but is also influenced by all the other parts (by eventual reciprocity) and by the environment. In short, in the world of organic unity the whole is not the sum of its parts.

Keynes carefully considered the consequences of this anti-Cartesian manifesto for his analytical system. In the 1907 version of his philosophical dissertation, when dealing with rational conduct, he indicated that all individuals — the forces of action — are by themselves organic units, but that society as such is not and therefore society should be considered as the sum of its parts (O'Donnell, 1989: 128). This has two main consequences. First, it may make possible some last resort 'methodological individualism'. Second, social action may be predictable and the value and result of every individual action may be socially determinate, even if the rationality for the action itself is not fully known at the individual level: society may be analytically decomposed, even if individuals' actions and intentions cannot be. In that case, an overall mechanics of equilibration of organic units could rule.[34] But Keynes soon corrected his argument and increasingly stressed the organic nature of the social process itself at all its levels, or the deep rationality of social indeterminateness and complexity. This assumption had heavy consequences: it implied the rejection of the physical metaphor, related to the process of equilibration and, as in the case of Moore, it implied the rejection of a calculable future since we are condemned to the 'utter ignorance' about what will happen (Moore, quoted in Keynes, TP: 341). The challenge was frankly stated by Keynes against the forerunners of neoclassical economics:

> [Edgeworth's and the marginalists'] Mathematical physics has not, as a science or study, fulfilled its early promise.[35] ... The atomic hypothesis which has worked so splendidly in physics breaks down in psychics. We are faced at every turn with the problem of organic unity, of discreteness, of discontinuity — the whole is not equal to the sum of the parts, comparisons of quantity fail us, small changes produce large effects, the assumptions of a uniform and homogeneous continuum are not satisfied. Thus the results of Mathematical Physics turn out to be derivative, not fundamental, indexes, not measurements, first approximations at the best; and fallible indexes, dubious approximations at that, with much doubt added as to what, if anything, they are indexes or approximations of. (Keynes, x: 262)

Such a distrust of Edgeworth's mechanical equilibrium — which corresponded to the proposition that the economy is not an organic whole and is therefore calculable — was of course extensive to all its Benthamite foundations. Since his 'early beliefs' and under the influence of Moore's philosophy, Keynes presented himself as 'amongst the first of our generation, perhaps alone amongst our generation, to escape from the Benthamite tradition'[36] (ibid.: 445). This is how he challenged that tradition, in 1937:

> it was, I think, an ingredient in the complacency of the nineteenth century that, in their philosophical reflection on human behaviour, they accepted an extraordinary contraption of the Benthamite School, by which all possible consequences of alternative courses of action were supposed to have attached to them, first a number expressing their comparative advantage, and Second another number expressing the probability of their following from the course of action in question (...). In this way a mythical system of probable knowledge was employed to reduce the future to the same calculable status as the present. (Keynes, xiv: 124)

This rupture, since the early work on philosophy and on the grounds of the vindication of the organic view, was full of consequences: it meant the introduction of irreducible uncertainty and the rejection of the 'maximizing rationality' assumption of the neoclassicals, of their main procedures in economics and of the whole physical metaphor. This was clearly stated again and again: 'The pseudo-analogy with the physical sciences leads directly counter to the habit of mind which is most important for an economist to acquire' (Keynes, xiv: 300), since 'Unlike the typical natural science, the material to which [economics] is applied to is, in too many aspects, not homogeneous through time' (ibid.: 269).

The pseudo-analogy with physics was the main trend of the marginalist revolution, and of course Keynes was fully aware of it.[37] Rejecting this analogy, economics was defined as an intrinsically inexact science, including quantitative and qualitative changes (O'Donnell, 1989: 162). From Moore's tradition in Cambridge, Keynes derived a peculiar philosophy of organicism, which was supported during his whole life by his intellectual surroundings: the Bloomsbury group cherished this acceptance of distinctive individuality and of the psychological traits as a distinctive feature of a new vision of the world (Mini, 1991: 67). The novels by Virginia Woolf, the relation of the group to Impressionist and Expressionist painting, Moore's moral arguments, the influence of Ruskin, Carlyle or William Morris, the close relation of some of the mentors of the group with Freud and psychoanalysis, all contributed to and reflected this innovative trend. These influences illuminated Keynes's mature conception of the organic economy.

The organic conception meant that there is, in economic analysis, a general danger of the fallacy of composition: the paradox of saving, expressed as in the parable of the society of producers of banana, was certainly the most famous of such fallacies in the Keynesian schemata. But, of course, aggregate demand is still the sum of partial demands, and total employment is the sum of each sector's level of employment. This distinction between behavioural variables and accounting aggregates will be discussed later on.

15.2. Causality and organicity: the role of metaphor and analogy

Keynes was one of the most influential economists — if not the most — in this century; he was also one of the best prepared in philosophical matters. In fact, his whole approach to economics was based on solid philosophical arguments. His first major work, the *Treatise on Probability* dealt extensively with two of the most relevant epistemological categories that have been discussed in Part One: induction, or the logic of accumulation of knowledge, and analogy, or the logic of construction of knowledge. A short summary and discussion is in order, since this was his basis for the rejection of the positivist epistemology, contradicting Harrod's interpretation according to which Keynes followed Russell and Vienna's logical positivism (Carabelli, 1988: 10).

Keynes summarized his version of the Hume paradox in the following way:

> To argue from the *mere* fact that a given event has occurred invariably in a thousand instances under observation, without any analysis of the circumstances accompanying the individual instances, that is likely to occur invariably in future circumstances, is a feeble inductive argument, because it takes no account of analogy. (TP: 445; his emphasis)

In that sense, analogy was a precondition for induction (ibid.: 74), even if it was not demonstrative, as the example of the proposition 'the sun will rise tomorrow' proved (TP: 264-5). But analogy is widespread and indeed necessary, following Keynes, since it rules the positive and negative similarities and creates the structure of the measurement and theory: the new instances in an inductive process are important when and if they add some more elements of negative analogy, in the sense that they 'diminish the unessential resemblances' (ibid.: 259). Keynes's concept of the construction of knowledge was based on these two correlative operations: the non-demonstrative and non-conclusive induction, and the analogy which made possible a meaningful growth of knowledge from induction. Of course, analogy is an instrument for qualitative reasoning, which is the appropriate form of thought for most instances of an organic world.

Two examples of these qualitative and organic features are the definition of probability and of expectations. Probability itself was for Keynes one 'organic unit', and he denied that it could be reduced to a physical unit or to an empirical frequency:[38] 'A degree of probability is not composed of some homogeneous material, and is not apparently divisible into parts of like characters with one another' (ibid.: 32).

In that case, probability may not always be reducible to quantitative measurements, and qualitative analysis — of probability or of the 'weight' of the argument — is imposed by the organic nature of the process. And since one of the organic features of this world is the very formation of expectations (ibid.: 238), uncertainty is a building block of all types of economic action and process.

Given that no universal causation or perfect knowledge can claim to explain reality, Keynes abandoned the physicist concept of necessity in order to establish the 'moral' sciences' concept of possibility. Statements on causation are thus the cognitive syntheses of induction and analogy, obtained from qualitative analyses of cumulative processes. They indicate a sequential movement in time: in GT, the sequential causation was emphasized, so that the ultimate variable in the model is the exogenous one that defines the causation from the boundaries of the economic model; but it is sequential since it is organic, and as a consequence causality cannot be reduced to some unilinear evolution of impacts from the exogenous to the endogenous variables, since multiple interactions are causally significative and are the very reason for the non-additivity of the parts in the measurement of the whole. Since there are uncertainty and expectations in real economies, causality is both exogenous and endogenous: it is organic.

15.3. Whither equilibrium?

One of the main consequences of the organic vision of Keynes was his profound distrust of the mathematical methods as the last resort of truth, as they were introduced and generalized by the physical metaphor. Since society itself is an organic whole, and this implies irreducible uncertainty — namely under the form of the influence of expectations, which cannot be summed up or compounded

— a precise solution is never attainable in the relevant systems, in the sense of an optimal equilibrium. This was how Keynes answered in 1936 to a letter by Shove complaining about the difficulty of establishing very formal models:

> [You] ought not feel inhibited by a difficulty in making the solution precise. It may be that a part of the errors in the classical analysis is due to that attempt. As soon as one is dealing with the influence of expectations and of transitory experience, one is, in the nature of things, outside the realm of the formally exact. (Keynes, xiv: 2)

The mathematics of the econometric revolution of the 1930s was most suspicious to Keynes,[39] even if he was himself a distinguished member of the Econometric Society and its President in 1944 and again in 1945; the parallel with Schumpeter is quite obvious. The essential reason for his distrust was previously indicated: he could not support the omnipotent physical analogy and the presumptions of the Benthamite extensively calculable economics. Moreover, only breaking away from Say's Law he felt to be able to deal with the reality of social and complex phenomena:

> This [the rupture with Say's Law] could be described as the re-discovery of there being a problem with equilibrium of supply and demand of output as a whole, in short, of effective demand. It was an important moment in the development of my own thought when I realised that the classical theory had given no attention at all to the problem at what point the supply of output as a whole and demand for it would be in equilibrium. (Keynes, xxix: 214)

Therefore, equilibrium did not necessarily imply optimality, and so there were many possible equilibria with severely different social consequences: the selection among them depended on the unpredictable behaviour of the internal forces of the system.[40] As a consequence, the orthodox theory was considered to be inapplicable: 'Many people are trying to solve the problem of unemployment with a theory which is based on the assumption that there is no unemployment' (Keynes, ix: 350).

This concern with social intervention and the definition of economics as political economy were some of the reasons for Keynes's concentration on the short term problems, and for looking for a new terminology other than that defined for the conception of the long-run equilibrium of orthodox economics (Keynes, xxix: 35). But Keynes, in spite of the legend, did not completely restrict his analysis to the short term. Namely, he defined several possible processes of disequilibrium in long-run processes, including monetary factors (generating uncertainty), investment factors and a last category of 'changes in the industrial factors influencing volume of output and demand for money for income purposes' (TM: 231), interacting with the investment factors (ibid.: 233; also ix: 321 f.). Of course, those are innovations and changes in the industrial structure: TM, which is the most dynamical of all Keynes's major writings, included important insights on economic mutation.[41]

This concept of organic unity of the social movement explains one of the differences with Schumpeter, and not the minor one. Schumpeter, as previously indicated, tried to create a dynamic counterpart to the Walrasian grand schema of general equilibrium, through the introduction of endogenous innovations and change into the model: his ideal explanation was therefore a self-contained and complete model (BC: 41), and that definition was considered to be responsible for the emergence of the Magna Carta of economics. On the other hand, Keynes defined a dynamic and open model, a non-self-contained one, since expectations and uncertainty were not completely endogenizable, and therefore net investment was not fully explained by the other controlled variables.

This is indeed why Schumpeter was not able to formulate a mathematical model for his theory: it was simply impossible to do so, since the tool was completely inadequate to the purpose. And this is also why Keynes did not feel the need to formulate that type of model, since he defined a non-contained system which would allow for a distinct role of the historical variables. But he did not develop the historical counterpart of his theory,[42] since he was self-limited to the short term. Keynes was not interested in the newly opened door, while Schumpeter was unable to reach it.

15.4. Cycle and evolution

This brief survey of some of the foundations of Keynesian economics is now concluded: his analysis of the cycle, dominated by the fluctuations in the schedule of the marginal efficiency of capital, was the direct conclusion of the organic system, which generates by itself uncertainty — and in fact also complex and nonlinear relations.

As previously stated, the origin of this organic conception was Moore's philosophy, and was then supported by the Bloomsbury vision of the world: it was not connected to an evolutionary or Darwinian analogy. In fact, Darwinism was frequently referred to by Keynes, but only in two distinct frameworks: either as an example of the inductive and analogical reasoning that Keynes was arguing to be the genuine model for science and namely for his own theory of probability (TP: 5-6; 118), or as in the case of a perverse analogy used by the extreme Ricardians or liberals, presenting the survival of the fittest — the Spencerian summary of evolutionism — as the driving force in competition and in society (Keynes, ix: 276), analogy which was refused by Keynes (ibid.: 284). The first instance was a mere example and the second one a mere polemics: none of them was of particular relevance for his own philosophy.

In fact, Keynes's paradox was that his system was not evolutionary, since it was based upon an organic conception but no central role was assigned to historical evolution, but nevertheless he provided some of the most important insights for an evolutionary macroeconomics. Certainly, Keynes accepted a very peculiar form of Sozialokonomie, and his life reflected his own aphorism saying that economists must also be mathematicians, historians, statesmen and philosophers and, in a very Marxian fashion, that 'No part of man's nature or his institutions must lie entirely outside his regard' (x: 173-4). On the other hand, Keynes was aware and accepted Marshall's preoccupations with the long term dimension of economics, although no analytical solution was available to integrate both dimensions: 'his [Marshall's] views as to the perpetually changing character of the subject-matter of economics led him to attach great importance to the historical background as a corrective to the idea that the axioms of today are permanent' (Keynes, x: 209-10).

But, in spite of this preoccupation inherited from Marshall, neither of them worked out a satisfactory solution for the integration of the short term and the long term dimensions of the analysis. This, of course, demanded an evolutionary theory. And Keynes, who challenged orthodox economics more than any other of his contemporaries — denying the definition of economics as a science of scarcity dealing with choice, criticizing the assumption of maximizing rationality and harmonic aggregate behaviour assuring a calculable future, and simultaneously rejecting the physical metaphor — was fully aware of the lack of such a general and new theory. His dark conclusion anticipated the dissolution of the Keynesian program in the synthesis which defined the modern trend of mainstream economics:

> We have completely failed, indeed, to provide a substitute to these economic bogus-faiths capable of protecting or satisfying our successors. (Keynes, x: 445)

This global failure will be reassessed in the conclusion of this part.

15.5. Schumpeter and Keynes: the high theory of the first half of the century

Schumpeter and Keynes were born in the same year — 1883, the year of Marx's death — and were two of the most important economists of their generation. They faced similar theoretical problems but in very different environments: the decline of the British dominance in the world economy but still a somewhat stable political system for Keynes, and the desegregation of the Austrian empire and then the permanent exile to the United States, for Schumpeter. Their philosophical background was substantially different: Cambridge and Bloomsbury for Keynes, the Methodenstreit and the marginalist education for Schumpeter. Both were heterodox in specific ways: Keynes was in open conflict with the Benthamite school, Schumpeter was fascinated by Marx. And, finally, both intended to have a lasting influence in the renewal of economics; both finally concluded that their action was condemned to fail, although one had a large and lasting impact (Keynes) and the other was well known but by and large surpassed by the former's influence (Schumpeter).

Nevertheless, during all their lives there was between them no intellectual cooperation: in GT there was not a single reference to Schumpeter, and the existence of CSD was never acknowledged by Keynes, in spite of his deep interest in political matters; in BC and CSD the few references to Keynes are generally hostile and in HEA Keynes was presented as the 'father of modern stagnationism' or of the most successful 'static theory' (HEA: 1143). In fact, Keynes generally ignored Schumpeter, except in one relevant occasion, a passage of TM where it was indicated that the Schumpeterian explanation of 'major movements' could be considered as an essential factor for the behaviour of entrepreneurs and 'may unreservedly be accepted' as a cause for the intrinsic uncertainty of the system (TM: 85).

That book was sent by Keynes to Schumpeter, who kindly acknowledged the reception — this is one of the very few letters exchanged between both men — but it is known that the book had a devastating effect on him. Indeed, Schumpeter stopped the preparation of his own book on money and, in spite of several future attempts to finish it, was never more able do so. Swedberg indicates that he could have burnt the manuscript of his own book (Swedberg, 1991: 76), but Schumpeter just gave it to Spiethoff for safekeeping, while abandoning the entire project (Allen, 1991, i: 265). Of course, the very reference of Keynes to Schumpeter's theory as an explanation for the source of uncertainty indicated a potential dialogue between the two theories, but it was never further developed. On the opposite side, at least Schumpeter's stubborn hostility towards Keynes's theory of uncertainty prevented such a cooperation.

If Keynes generally ignored Schumpeter's contributions, the reverse was not true: Schumpeter had a profound hostility and a lasting jealousy about the success of Keynes's professional careers.[43] The content of GT was certainly the main factor of opposition between them: Harvard was suddenly converted to Keynesianism and Schumpeter lost a large part of his impact at the university. In a letter to Oskar Lange, in February 1937, Schumpeter bitterly complained against this turn of events, since Keynes was said to have missed everything in economics from the 1830s: 'The book could have been written a hundred years ago and skirts all real problems. It is the reverse of progressive ... it is the dying voice of the bourgeoisie calling out in the wilderness for prophets it does not dare fight for and shifts its ego to the real problems it does not face' (quoted in Allen, 1991, ii: 26).

The review of GT, published by Schumpeter immediately after the publication of the book, was a lively image of this bitterness. The short article was full of aggressive remarks: Keynes's GT was based upon 'artificial definitions', 'paradoxical-looking tautologies', 'treacherous generality', psychological laws of a 'bygone age', is 'Ricardian in spirit and intent', what certainly was not intended to please Keynes (Schumpeter, 1936: 792, 793). The final touch is a monument to perversity: challenging Keynes's conceptions of saving and effective demand, Schumpeter gave the example of Louis XV, who was supposed to have called Madame Pompadour in order to spend as much as possible as a means to avoid depression and to guarantee the well being of the people — of course, derided Schumpeter, if the whole story finished in a blood-bath it was by a mere coincidence.

Nevertheless, this review indicated two important themes. The first was the critique of the short term view ('a theory of another world') which did not make possible any change in the production functions since:

> reasoning on the assumption that variations in output are uniquely related to variations in employment imposes the further assumption that all the production functions remain invariant. Now the outstanding feature of capitalism is that they do not but that, on the contrary, they are being incessantly revolutionised. (ibid.: 793)

The second was the critique of the three *dei ex machina* of GT — 'there is a whole Olympus of them', wrote Schumpeter (ibid.: 794) — on the same grounds: expectations, the psychological law of consumption and the schedule of the liquidity preference, all three could not be part of an economic explanation, according to Schumpeter. These two main critiques are discussed in the following pages, in order to provide a brief comparison between the research programs of Keynes and Schumpeter.

There was also a third substantial critique by Schumpeter, which will not be developed in the present work: the acid rejection of all types of activist policies in economics, another 'Ricardian vice', as Schumpeter wrote (ibid.: 791). In fact, this point only emphasizes the political and philosophical distance between them: Keynes was a centre-left supporter of the Liberal Party in the United Kingdom, opponent to the Tories and vaguely sympathetic to Labour,[44] involved in avant-garde artistic activities, while Schumpeter was a conservative monarchist during the First World War, and politically a reactionary. The differences were indeed deep, not only in regards to their positions, but also in relation to the outcome of their political involvement: while Schumpeter had a brief and unsuccessful career as a minister in the republican coalition government of Karl Renner in 1919, Keynes had a long standing and influential intervention in political affairs throughout his life.

During the Second World War Keynes fully participated in his country's war effort against Hitler, while Schumpeter had an opposite stance: in fact, even if he generously supported colleagues persecuted by the German government, he still refused to condemn it. Schumpeter, who lived in Germany until 1932, even advised some of his students to join the Nazi party and, in his farewell speech to the University of Bonn, included the following remark: 'What enormous subjective individual possibilities there might be for a young man of today if there were any who, not deprecating economic techniques, felt like a National Socialist'. He also defined Nazism as 'a powerful movement which is singular in our history' (quoted in Allen, 1991, I: 284-5). Later, in his personal diary, Schumpeter asked himself why did he change his attitude towards Germany from the First to the Second World War: 'I cannot understand at all this revirement of my sympathies [for Germany] since 1916' (in Allen, 1991, II: 139). His diary includes many remarks supporting Hitler (in Allen, 1991, I: 288, II: 12-3, 58, 71, 92, 103). It must be added that none of his students rallied the regime and that his former secretary and mistress in Germany, Maria Stockel Bicanski, joined the underground and was shot by the Nazis.[45]

As far as activist policies were concerned, from the early days of the Methodenstreit Schumpeter criticized Schmoller's suggestions of political reforms; this was not considered to be the purpose of science. In his later reassessments of the polemics, as previously indicated, Schumpeter always argued that Schmoller was ready to abandon his reformist claims for the sake of the convergence of positions. In 1914, Schumpeter argued that the Physiocrats did not intend to give political advice, contrary to Steuart, and that it was the correct procedure for science not to do so (EDM: 48). In HEA, after praising Walras, Schumpeter lamented his reformist tendencies (for the nationalization of the land, for example; HEA: 827f.). And Keynes's anti-cyclical policies were certainly one of the main differences opposing them: Schumpeter was a fierce opponent to Roosevelt, since he considered that the New Deal was the wrong approach to the Depression and that the crisis should take its own way in order to raise a sounder prosperity.

There is nevertheless some indication that he later supported some anti-recessive spending measures, although thinking that it was not the task of the scientist to advocate those measures (Allen, 1991, I: 306-7), an extreme imposition of his positivist attitude. And in spite of his general position, in TED Schumpeter was ready to suggest some minor forms of governmental intervention for anti-cyclical purposes (TED: 166). These sharp differences are certainly part of the explanation of the opposition between the two scientists.

15.5.A. *Dei ex machina*: the semi-autonomous variables

In the 1936 review of GT Schumpeter rejected the use of expectations and of all psychological laws (such as Keynes's explanation of consumption and preference for liquidity) which were introduced as exogenous factors and ultimate causes for the economic behaviour. Those explanations were considered to be tautological,[46] as any purely exogenously driven form of causality.

The exact nature of expectations in Keynes's system is a matter of controversy. Mini considers that they are not independent variables, since they are part of the nature of economic agents (Mini, 1991: 179). But this is not completely convincing, since economics is not necessarily a global explanation of all the features of real life agents. O'Donnell, on the other hand, considers that long-term expectations, those concerned with decisions to invest and hence with transformations of the economic system, are typically independent variables, while short-term expectations, concerned with decisions to use the existing capital equipment, are endogenous (O'Donnell, 1989: 241, 236). In this interpretation long-term economic expectations determine the short-term expectations in the specific sector of capital goods, and thus the employment in this sector, while short-term expectations determine the employment in the sector of consumption goods.

Keynes's treatment of expectations was once again deeply rooted in his philosophy. From the first drafts of what would become the *Treatise on Probability*, Keynes insisted that logic should include uncertainty and not only deductive relations, which are certain or true if logically correct. This general logic implied the passage 'from the logic of implication and the categories of truth and falsehood to the logic of probability and the categories of knowledge, ignorance and rational belief' (TP: 62).

In this theory of rational belief, uncertainty can arise from three different sources: from the probability of an event (the measure of the degree of certainty), from the weight of the argument (namely, the nature of the available evidence) and finally from the unknown probabilities of events. But the traditional approach to mathematical expectations assumes the numeric and measurable nature of all variables and events, and thus ignores the weight of the argument and risk.[47]

In other words, uncertainty was to Keynes the reflection of a world where there are measurable and non-measurable qualities, or quantitative and qualitative phenomena. All these categories were already noted in the current inquiry concerning Keynesian theory, and it is time to schematize their relation. O'Donnell suggests that Keynes's interpretation of reality distinguished between those features of a quality which could be described by a degree and those which could not, including in the first category those objects or properties describable as the sum of its parts (weight), those for which the degree of the whole equates the degree of the parts (colour) and finally those for which the whole is independent of the parts (beauty; O'Donnell, ibid.: 62). In this framework, 'colour' and 'beauty' are the examples of organic units, while 'weight' is a non-organic one.

Expectation is typically one of such variables which can only be represented as an organic system, therefore non calculable and impossible to represent in the Cartesian world of purely deductive logic. According to the previous classification (see Part Two), this corresponds to the notion of semi-autonomous variables, which are to be represented as endogenous or exogenous depending on the scope of the model,[48] since they represent the crucial connection between distinct levels of abstraction in the context of a non-contained universe. The relevance of these variables flows from the fact that they are not compatible with the deterministic view of causality and that they represent the organic synthesis of network causality and complexity.

Rejecting this whole approach to expectation from his very first writings,[49] Schumpeter tried quite often to formulate an alternative one. He failed to do so. In BC, dealing with Knight's concept of expectations, Schumpeter reconsidered the stationary state in order to indicate that, even without omniscience, expectations are based on experience and perfect foresight is possible and indeed trivial in that case (BC: 52). But if the disturbances affect the system, consequently the role of the expectations may change: they may either preserve or prevent disequilibria (ibid.: 53). The drastic solution was to treat expectations as equilibrating features, and this was postulated in a very uncomfortable way, which is distinctively dogmatic and neoclassical:

> But although they [the disruptive effects of 'certain types of expectations'] may often temporarily counteract it, they do not in themselves disprove the existence of an equilibrium tendency or the proposition that at times it prevails in such a way as actually to draw the system toward equilibrium. The real trouble to the theorist comes from the fact that introducing expected values in his variables — we will now, on one hand, assume that they are expected with certainty and, on the other hand, also include past values — changes the whole character of his problem and makes technically so difficult to handle that he may easily find himself unable to prove an equilibrium tendency which, nevertheless, may exist, or even the existence and stability of the equilibrium position itself. (BC: 54)

In such a framework, expectations can be treated either as endogenous variables contributing to the equilibrating tendencies or as exogenous variables fully known, but both solutions are unsatisfactory. Schumpeter clearly preferred the first solution,[50] arguing that otherwise expectations would constitute a theoretical blank to fill another blank (HEA: 312). But even in that case there is no explanation for these variables ad hocly defined as endogenous: Schumpeter argued that the only available interpretation for expectations was to ignore them, since they are so difficult to handle (BC: 55).

In the previous chapters, it was argued that the only solution for this difficulty is that represented by the concept of semi-autonomous variables, those not wholly endogenously explainable by the system and whose behaviour is not autonomously determined by exogenous events in its full extent. In fact, they are not parameters, but the theoretical counterparts of the organic and complex realities, that is, they indicate the building nonlinearities of the system. What is striking is that Schumpeter in some way touched this exact problem when he left the domain of stationarity — where he discussed Knight's concepts of uncertainty and risk — and considered the notion of development.

In BC, he argued that there are three different types of variable: (i) *theoretical*, that is, those related to a law and, consequently, invariant in their behaviour;

(ii) *random*; and (iii) *historical*, defined as 'hybrid variables', since they represent the 'theoretical law in a process of change' (BC: 194-5) and 'Hence we may ... define a historical variable as a variable, the stochastic normal of which changes owing to a change of its theoretical normal' (ibid.: 196). And since 'the very concept of historical sequence implies the occurrence of irreversible changes in the economic structure which must be expected to affect the law of any given quantity' (CSD: 72 n.), it is referred to that very peculiar sort of variables which cannot be defined as endogenous and which cannot be simplified as exogenous (in the Schumpeterian sense of purely stochastic variables or otherwise as identified factors which are exterior to the scope of the theory). In other words, the morphogenetic process of mutation and evolution cannot be encapsulated in the strict formalism of the mathematical models of simultaneous and linear equations under the current qualifications — since these are not organic representations — and requires the inclusion of a new type of explanation, historical by nature.

Innovation, the key concept of Schumpeter, cannot be fully understood unless in that framework, and it is easily verifiable that all of Schumpeter's arguments against the Keynesian concept of expectation are directly extensible to his own case of innovation: its source (invention) is exogenous and therefore not explained, its diffusion is endogenous but it is simultaneously a source of disruption in the system; nowhere is it fully explained nor is it completely explainable by the system since it cannot be represented by the postulated relations, as it depends on singular decisions. Innovation is endogenous to the system, but it is finally determined by the entrepreneurial function, that unique capacity to make new combinations, which is clearly outside the domain of the model. And, of course, this boundary between endogenous and exogenous variables can change according to the purposes of each inquiry, as Schumpeter was aware, and is therefore irrelevant as a classification criterion.[51]

In a much more prudent and realistic way, Keynes introduced three main expectations-dependent variables: propensity to consume, marginal efficiency of capital and liquidity preference. He was right to do so and to stress the irreducible uncertainty in organic systems.[52] The reason why he could do so, without being entangled as Schumpeter in a self-contradictory net of explanations, was that Keynes's philosophy suggested the notion of organicity and therefore liberated him from the stringency of the concept of equilibrium, although he was not able to incorporate these notions in a dynamical approach.[53]

15.5.B. Between econometrics and history: the place for economics

The concept of equilibrium was at the centre of the econometric debate, in which both Schumpeter and Keynes strongly intervened. Once again, Schumpeter was close to an original solution for the problems detected in the analysis of the cycle: consequently, he was able to avoid the major dangers of the neoclassical theories, even if he did not draw an explicit alternative.

Keynes's intervention against the physical metaphor and the uncritical use of mathematical methods in economics was previously referred to. In fact, his concern about the misuse of statistics is as old as his studies on probability: it was one of the themes of the *Treatise on Probability*, which included a first sharp critique of 'inductive correlation' (TP: 444), and this had already been the theme of the 1910 debate, when Keynes discussed with Pearson in the columns of *The Times* and in statistical journals on the methods of correlation analysis: the research of the Galton Laboratory on the influence of parental alcoholism on children in Manchester and Edinburgh was the pretext. The discussion was prolonged in 1911, and Marshall joined Keynes in the criticism of the report, mainly concentrating on the representativeness of the sample and the methods of classification (Keynes, XI: 186-216). Keynes feared that the 'old Laplacean tradition has been followed' (ibid.: 186-7).

In GT, a whole section was dedicated to the criticism of mathematical models, namely the 'symbolic pseudo mathematical methods of formalising a system of economic analysis', leading the researcher 'to lose sight of the complexities and interdependencies of the real world in a maze of pretentious and unhelpful symbols' (GT: 297-8). Shortly afterwards, Keynes engaged into a polemic with Tinbergen on the method of multiple correlation, insisting on the points already made thirty years before.

His main contribution to the econometric debate — indeed, the starting point of that general debate — was this critique ofTinbergen's work on business cycles. Tinbergen (1939) tested the theories of business cycles earlier reviewed by Haberler: his results, published in two volumes, were based on a model of 22 equations and 31 variables for 1919-32 US data (the first volume considered UK, 1871-1910 and 1920-36, US, 1877-1913 and 1919-33, France, 1871-1908 and Germany, 1871-1912).This work constituted the first large applied econometric study with empirical data, and many problems of estimation were identified and discussed. Tinbergen used the multiple regression in order to study the influence of the variables, and correlation to verify the theory as a whole; after estimation, he tested every equation for structural stability in different sub-periods, and found first order serial auto-correlation of the residuals. Tinbergen admitted that the specification of the equations was somewhat arbitrary and that it could not encapsulate all types of causality: the 'proximate causes' defining a 'network of causal relationships', relations of definition and 'technical, institutional connections' (Morgan, 1990: 103), but argued that any way the rocking horse could provide a good estimation of the structure of the model and therefore allow for the comparison of the theories of business cycles.

Keynes's reservations and criticisms to the econometric methods[54] can be summarized in the following way: (i) the complexity of reality may be irreducible to the model: formal models tend to concentrate the explanation in a few variables and some of the relevant factors may be ignored or misspecified;[55] (ii) the interdependence in the real world can further complicate the attempt to define a model: some effects are also causes, what should not happen in a mechanical model; (iii) the specification problem also arises either from the fact that some of the possible influences are not identifiable, separable and measurable and we may be unable to separate the effects of each variable (for example, due to multicollinearity), or from the fact that they are not homogeneous through time, or from the fact that they are not representable in a linear universe (xiv: 319); (iv) the incommensurability problem remains: Keynes feared that the choice of the model depended on statistical availability of data, and that this manipulation could be responsible for whatever 'proof' desired; and (v) finally, the statistical method could hardly account for the qualitative influences in the economic system or for unobservable variables (expectations), and could lead to the confusion between statistical and economic significance.

As a consequence, Keynes was suspicious of these methods using non-experimental and unique sets of data to perform statistical tests originally designed for laboratories and physics. In short, whenever we are in the world of organic systems with structures which are non-homogeneous through time, then the correlation method may fail and, since this is the case for most of the relevant economic variables, no inductive claim is possible from this method.

The main issue was the problem of the application of the method of multiple correlation to non-homogeneous series in real time, and the consequent problem of misspecification: the method and the results are only relevant if the researcher is able to indicate all the possible influences on the endogenous variable, if the theory is previously established and is correct and if enough data is available to establish the correlation — a truly Laplacean set of requisites. Otherwise, spurious results may emerge, and the danger in fact is that the misspecified method makes possible any type of conclusion the researcher is looking for:

> It will be remembered that the seventy translators of the Septuagint were shut up in seventy separate rooms with the Hebrew text and brought out with them, when they emerged, seventy identical translations. Would the same miracle be vouchsafed if seventy multiple correlators were shut up with the same statistical material? And any how, I suppose, if each had a different economist perched on his a priori, that would make a difference to the outcome. (Keynes, xiv: 319-20)

On the other hand, since the method supposes homogeneity through time, the same structure must account for stable coefficients for the period under inspection, ten years in the applications by Tinbergen: Keynes argued that this is not conceivable and that there was a trade-off between the long series needed for the exercise of multiple correlation and the assumption of the stability of the coefficients, restricted to very short series (xiv: 294). The method was criticized in a letter to Khan as a 'mess of unintelligible figuring', and also as some sort of 'black magic', 'charlatanism', a 'nightmare', a typical product of 'alchemy'[56] (ibid.: 289, 305, 320, 315).

Tinbergen acknowledged and accepted some of Keynes's conditions for the use of the method arguing that:

> in so far as one agrees:
> a) that the explanatory variables chosen explicitly are the relevant ones,
> b) that the non-relevant explanatory variables may be treated as random residuals, not systematically correlated with the other explanatory variables, or
> c) that the mathematical form of the relation is given, certain details on the probability distribution of their 'influences' can be given. (Tinbergen, 1940: 141)

Tinbergen was obviously cautious about the misuse of the method, and he accepted that it could not provide a statistical proof for a theory; but still he maintained that empirical data could disprove a theory, what Keynes could not accept either (Keynes, XIV: 307). And, of course, the cursory critique of Keynes of the massive and innovative effort by Tinbergen mobilized all the econometricians, who did not ignore that Keynes did not follow any more the state of the art of the techniques, in defence of the method: the debate turned out to be a waste of arguments, since Keynes misjudged Tinbergen's work and mostly since Keynes's powerful critique was misunderstood by the early econometricians. But Keynes correctly anticipated the dangers of the unbalanced extension of the econometric program, which amounted to the extension of the ceteris paribus clauses as a condition for the regression and to the imposition of the epistemology of the rocking-horse.

Its deep reason was indeed the same that inspired the previous Keynes-Pearson debate, twenty-eight years before: the applicability of Bernoulli's Law of Large Numbers, that is, of the physical metaphor and of the laboratory methods designed to analyse fixed conditions and repeated samples. But, since in the social realm one cannot assume the 'principle of limited independent variety', the method fails and cannot be extended from the controlled experiments to the unpredictable and complex reality of social and economic life. Moreover, correlation proves little, since the *ceteris paribus* conditions — the analogue for the control in experiments — may lead to the fallacy of the *post hoc, ergo propter hoc* argument.[57]

Hendry and Morgan, who take sides with Tinbergen,[58] recognize that the crucial problems — the completeness of the set of causal factors, the inter-connection between variables, the homogeneity through time and the constancy of parameters — remain a 'greater threat' to the method (Hendry and Morgan, 1995: 55). Nevertheless, the econometric revolution which changed the daily methods of economic inquiry olympically ignored the crucial criticisms by Keynes.[59]

It is by now obvious that these themes of debate are related to the definition of the economic system following either the atomistic or the organic metaphors and using the Cartesian or Moore-Keynes's methods. In the first case, we are back to equilibrium. In the second, equilibrium and disequilibrium coexist; furthermore, if there are several equilibria, they are devoid of any general philosophical implication about the desirable state of society; on the contrary, different equilibria can only be compared through explicit political criteria, which are part of the action of coordination: choice is possible and, indeed, it is necessary. Keynes was not only conscious of the error of the mechanical metaphor but also engaged himself in the struggle to destroy it.

At the same time, Schumpeter — whose inability in mathematics is often offered as an explanation for the lack of formalization of his models[60] — was also aware of the limitations of the current methods. Not from reading Keynes, of course, but eventually from reading Mitchell: both insisted that theory was a precondition to the use of statistics since it ruled the architecture of the model, that the coefficient of correlation could merely indicate a spurious relation (Schumpeter, 1930: 151, 165; BC: 32), and that statistics was able to verify, but not to prove (Schumpeter, 1927: 296). In Schumpeter's words, any statistical procedure implies a theory:

> Even a mere arithmetical average, or its standard deviations is perfectly meaningless, unless we know beforehand whether there is some 'norm' in the set of data we have to deal with and what the nature of that norm is. (Schumpeter, 1930: 163-4)

In this context, Schumpeter criticized Mitchell's table of the 106 cycles of the US economy, one of his major results in the 1913 book, arguing that there was no statistical meaning for these cycles, since the structure of the economy should have changed during the period, and the results could not be fully compared.[61] Moreover, there can never be a complete theory of the cycle or a statistical representation of it, since there are factors which cannot be included in a general account.[62] This argument is obviously recognizable: it is no more and no less than the criticism formulated by Keynes.

And this is not the end of the story, given the fact that in the Schumpeterian explanatory system there is a further complication: he postulated the existence of a tendency to equilibrium, which was offset for some periods by a counter-tendency propelled by innovation, and insisted on the actual existence of some equilibrium points in every cycle. This suggests the application of the traditional statistical methods of analysis of the trend as the *loci* of those equilibria, and of the cycle as the deviations, as in the mainstream tradition. But Schumpeter did not accept that scheme, in spite of some rhetoric supporting Wicksell and Frisch's rocking horse metaphor, since his own impulses were defined as endogenous and therefore equilibrium was supposed to be by nature unstable: instability was structural, the disturbances changed the system and that was the condition for the progress of capitalism. As a consequence, econometricians reacted with great hostility to his formulations, which could not be reduced to a domesticated system of equations.

Tinbergen sharply criticized *Business Cycles*, considering that the book was 'alien to econometrics', since for Schumpeter the relevant variables were the shocks, and he 'belittles the importance of the mechanism', which for econometricians 'deserves the main attention': Tinbergen could not accept Schumpeter's theory of the impulses being endogenous to the system, since this was not compatible with the traditional cycle model and econometric schemata (Tinbergen, 1951: 59, 60). Schumpeter implicitly answered to this criticism, commenting that Tinbergen's model 'describe repercussions and propagations without saying anything about the forces or causes that put them into motion' (Schumpeter, 1937: 162): in other words, the mathematical formalism was empty and the concept of the shocks had no semantic value.

In fact, Schumpeter tried to save his allegiance to orthodoxy stating that the impulses could be of two kinds, both compatible with equilibrium:

> Now, what causes economic fluctuations may either be individual shocks which impinge on the system from outside, or a distinct process of change generated by the system itself, but in both cases the theory of equilibrium supplies us with the simplest code of rules according to which the system will respond. This is what we mean by saying that the theory of equilibrium is a description of an apparatus of response. (BC: 68)

Later on in the same book, Schumpeter compared the impulses with a water flow (BC: 179), which is close to Frisch's analogy in his 1933 paper. Nevertheless, there remain some remarkable differences: Frisch supposed a damping propagation mechanism, which is the only coherent way to reintroduce the notion of stability of equilibrium, while Schumpeter described a specific oscillator representing a cyclical and unstable form of growth.

In HEA, there is another metaphor for the explanation of this particular system — and, eventually, for the failure of the current statistical methods — when Schumpeter indicated that the economic system is a resonator for the impulses, just like a violin: the impulse and propagation autonomous systems were clearly stated, but neither the wooden box nor the movements of the fingers of the musician may fully explain the aesthetical pleasure of a concert (HEA: 1167).

Furthermore, the propagation and the cyclical mechanisms in Frisch's model may explain the cycle, but do not explain the trend-cycle behaviour. This is, of course, a major difference with Schumpeter's theory, which was instead concerned with the creative responses of the economic system to whatever impulses may exist (BC: 72). Indeed, when developing his theory, Schumpeter discussed this problem in detail.

The propagation mechanism was traditionally considered as describing the equilibrium and the Walrasian feature of the modelled economy, but the theory did not indicate how are these equilibrium conditions met. Schumpeter argued that the convergence process really existed, but he stressed that actual equilibria were only attainable at discrete and rare points, to be immediately abandoned by the motion of the system. Therefore, the equilibrium line indicated in the statistics was only an artificial representation:

> They [the neighbourhoods of equilibrium] are the most relevant items of a series .... A line or curve through those points, or a band or narrow zone through those neighbourhoods, supplies a trend that really has economic significance. ... We know ... that this trend does not describe a phenomenon distinct from the cycle. On the contrary, since evolution is essentially a process which moves in cycles, the trend is nothing but the result of the cyclical process or a property of it. ... Moreover, we also know that it carries realistic meaning only in discrete points or intervals. If we connect them by straight lines ... it must be borne in mind that the stretches between the neighbourhoods are nothing but a visual help, and devoid of realistic meaning. No fact corresponds to them. Real is only the cycle itself. (BC: 206-7)

Or else, criticizing the statistical methods:

> if trend-analysis is to have any meaning, it can derive it only from previous theoretical considerations, which must not only guide us in interpreting results, but also in choosing the method. Failing this, a trend is no more than a descriptive device summing up past history with which nothing can be done. It is, in fact, merely formal. (Schumpeter, 1930: 166)

Schumpeter acknowledged the efforts by Mitchell to solve the same problem of the relation between the trend and cycle through the 'reference cycles', 'a judicious compromise between eliminating trend and leaving it in' (Schumpeter, 1952: 339) and even defined the formal trend as those sub-intervals where the mean value is monotonically increasing or decreasing (Schumpeter, 1935: 3). But the economic meaning of the trend and the applicability of the 'statistical method' (of least squares) was supposed to depend on the interpretation of an economic mechanism explaining the monotonic variation. The only mechanism of that type is growth (savings and population), which constitutes the 'real trend' (BC: 201f.) but, as previously indicated, this was considered to be a minor influence in the overall behaviour of the system.[63]

In short, Schumpeter claimed that there was a real trend of growth which was nevertheless a secondary feature of the model, there were discrete points where the tendency towards equilibrium was achieved and there was a causal process unseparably explaining both cycle and evolution. Thus, the meaningful trend, the 'trend-result' describing evolution and the cycle, synthesized one and the same process, and therefore, according to Schumpeter, no multiple regression could be successfully applied to it since it required the decomposability of this process. But, according to the theory, there was no meaningful separation of the variables of impulse (innovation) and its propagation mechanism, since both were defined as endogenous, and every decomposition would be arbitrary:

> It follows that barring the elements of growth the trends of out times series are not due to influences distinct from those that create the cyclical fluctuations but simply embody the results of the latter. To these 'result-trends' ... it is entirely unwarranted to apply formal methods of the type of least squares. (Schumpeter, 1935: 6)

The same argument was given in other works (Schumpeter, 1930: 167; BC: 198). This was a substantial reason for the rejection of the Slutsky effect of the impact of random shocks on the propagation mechanism (BC: 180-1). This is the heart of Schumpeter's model: equilibrium was the reference to the trend, but this was without implication in the choice of the methods of analysis, since it could not in any case be meaningfully separated from the cycle itself. Furthermore, to add to the analytical difficulty, the theory provided an explanation for the inevitable disturbance of equilibrium and indicated the relevant process to be the disequilibrating and innovative process of creative destruction, the central feature of capitalism — and the very reason for its survival and adaptation.

This was in fact the consequence of the centrality of 'hybrid variables', whose 'theoretical norms' change was supposed to follow the irreversible process of mutation (BC: 196, 198): in that case, the historical approach was necessary to a general theory of the cycle, since the cycle is always a 'historical individual' (Schumpeter, 1935: 2). Equilibrium exists, but it is exiled to the domain of the secondary and artificial representations:[64] for Schumpeter, the cycle was the only persistent and meaningful reality.

16. Conclusion of Part Three: Split Heritage

When reviewing Mitchell's book, Schumpeter wrote that 'one of the best things' of the work was the implication that business cycle studies should help to reconstruct the whole economic theory, since dealing with the core of the capitalistic process (Schumpeter, 1930: 150-1; also 1952: 327). Mitchell and Schumpeter were right: the analyses of cycles and growth provided both the formalism that made possible the incorporation of the physical metaphor into economics, and the basis for some of the most consequential critiques against it. This part concludes with a brief comparison of the contributions to such a reconstruction of the economic theory by Schumpeter and by Keynes.

The fate of the scientific work of the two men was indeed very different. They were separated by their respective success or failure, or by their paradoxes — Schumpeter with general equilibrium and evolutionary dynamics, Keynes with disequilibrium and short term non-evolutionary processes — but what finally decided their theoretical impact was their philosophical attitude. Schumpeter's positivism,[65] as well as his program for the unification of statics and dynamics under a generalized Walrasian economics prevented him from establishing clear-cut differences between his theory of innovation and neoclassical economics. So, he rejected the most interesting and promising of Keynes's contributions, the ones that could influence the progress of his own ideas. That could eventually have been avoided: when Keynes recognized that the entrepreneurial function of Schumpeter was one of the main causes for uncertainty and changes in expectations, he was suggesting an important development to his own theory as well as to Schumpeter's. But he later on ignored the topic, while Schumpeter strongly opposed those concepts of 'psychological' schedules and Keynesian uncertainty, namely because their incorporation would force his own system to depart definitively from the general equilibrium paradigm. On the other hand, Keynes searched for a new general theory replacing the orthodox approach, but his stance prevented him from introducing history and evolutionism in his theoretical framework, even if he considered history as the arena of uncertainty. The non-convergence of these theories is partially responsible for the lack of a major challenge to the neoclassic synthesis in the first half of our century, and for the dissolution of some of the essential topics to the agenda both of Keynes and of Schumpeter.

Schumpeter and Keynes were confronted with many similar problems, and six major topics will be emphasized now in order to compare their programs. First, both discussed the hidden epistemology of the physical metaphor: Keynes rejected any borrowing from the physical methods, and discussed thoroughly its statistical and mathematical implications; Schumpeter denied that this metaphor was indeed influential in economics, and stated that no borrowing was taking place. He nevertheless accepted the use of some concepts whose generality he considered not to be questionable, which were indeed concrete expressions of that metaphor, such as 'force' and 'equilibrium'. Consequently, Keynes could easily contest the concepts of equilibrium and optimality, while Schumpeter accepted their relevance, in spite of all practical difficulties. Keynes therefore defined economics as an intrinsically inexact science, while Schumpeter hoped it could attain the status of the exact sciences. The difference can also be measured by the universe of references they used: the constitutive metaphors were the mechanic metaphor of the water deposit for Schumpeter, and the kaleidoscope or the organic metaphors and the complex inter-action of the banana parable or the widow's cruse for Keynes.

Second, both used the notion of organic systems, and both rejected the use of the Darwinian or of some sort of biological metaphor — but their conclusions from that were quite opposed. Keynes's concept of organic system was not at all trivial or merely descriptive, since it led to the definition of the nature of the variables, on the basis of a long standing philosophical reflection. This was not the case for Schumpeter, and his quarrels with the definition of the semi-autonomous variables witnessed his lasting positivism. As a consequence, his concept of innovation still has a somewhat ambiguous status in modern evolutionary models.

Third, the organic conception was completed by Schumpeter's dynamic notion of evolution in time, while Keynes, although acknowledging the constructive role of time — indeed, this was the decisive feature of his polemics against Tinbergen — did not study evolution and predominantly used a comparative statics approach. On the contrary, Schumpeter was more and more interested in economic, social and institutional history, and considered it as the central tool for a new theory.

Fourth, while Schumpeter's evolutionism was concentrated on the supply side, Keynes's theory was dedicated to the demand side. From this point of view, the desirable combination of these contributions, in spite of the distinctiveness of the analytical frameworks, could eventually improve both theories: innovation and entrepreneurship in order to understand the expectations and irreducible uncertainty in Keynes's model, expectations in order to understand the nature of the 'secondary wave' in Schumpeter's models.

Fifth, both described the nature of entrepreneurship as a function in the economic system, and not of a separate social class: the autonomy of this function explained disequilibrium in Keynes, since the decisions of investment are logically independent from those of saving, and in Schumpeter, given the fact that entrepreneurs decide to innovate and move the system away from equilibrium. Of course, this was a major departure from the Walrasian theory, which described the action of entrepreneurs as passive, since they were not supposed to take independent decisions (Walras, 1874: 380; or 1883: 207-8; Morishima and Catephores, 1988: 41). In this sense, both theories were essentially non-equilibrium accounts.

Sixth, both discussed the stability properties of the system. Keynes noticed that 'a profit-seeking organization of production is highly unstable in the sense that a movement from equilibrium tends to aggravate itself' (xiii: 394), while Schumpeter argued that 'under the conditions created by capitalist evolution, perfect and universal flexibility of prices might in depression further destabilize the system' (CSD: 95). But both conceded that the economic system has strong adaptive forces, and creates order as it creates mutation, that is, it is not violently unstable and it is bounded by institutional structures that control it.

In fact, both Schumpeter and Keynes were dealing with complexity in economic relations and trying to cope with its theoretical implications. What else were Schumpeter's entangled explanations about equilibria and that intrinsic drive to instability and mutation, or what else was Keynes's understanding of the role of small effects producing large effects (x: 362), but their recognition of complexity? They were not alone: so did Hayek, who provided a powerful early version of the economies as 'organized complexity',[66] although he counter-intuitively maintained his allegiance to the extreme *laissez faire* model as the sole possible and useful computational device to coordinate societies.

Both authors, who defined economics for the first half of the century, understood the problem. In a paradoxical fashion: Keynes defined the transformation in economics as an indeterminate evolution moved by expectations and kaleidoscopic movements, but analysed those features from the viewpoint of a closed universe; Schumpeter, who on the other hand guessed the importance of innovation and therefore of an open universe, rejected such an indeterministic rationality and tried to close his model in reference to general equilibrium. The domain where such a convergence was possible — and where it was rejected — was the concept of semi-autonomous variables: the development of that concept is one of the central theoretical conditions for the study of innovation and for the development of economics as a social science.

That implied a courageous charge against certainty and the 'principle of determinateness', as Schumpeter planned to do in his last days. In fact, when he was invited to give a series of Walgreen Lectures, Schumpeter prepared for the fifth conference a sharp critique of 'social determinism, [since] where it is non-operational, is a creed like any other and entirely unscientific'. The notes can be found in the Schumpeter Archive at Harvard University, but the conference never occurred, since the author died exactly ten days before it was scheduled.

The challenge was cyclopic. Schumpeter inherited from Walras the dream of a grand unification of equilibrium and movement, and from J.B. Clark the assumption that those real processes could be identified and combined: his inquiry conducted him to the foundations of mainstream economics and to express his own doubts and alternatives under the form of a specific historical method. Keynes, on the other hand, inherited Marshall's critiques against decomposition but, as Marshall himself, restricted his main research to the non-realist dimension of static processes.

Their whole inquiry was a failure, both came to conclude. But the identification of those historical processes of economic and social mutation, their challenge against the equilibrium assumptions and methods, their drive towards a realist science of economics, all those failures are still major inspirations for new generations of evolutionary economists. Midst failure, their success and heritage is immense. It is a program, rather than an alternative. It is a set of enigmas, rather than of solutions. It is incomplete, imprecise, contradictory, and yet it is an exacting agenda dealing with the constitutive problems of economics.

In fact, in spite of Schumpeter's claim that he was building a new general theory uniting Walras and the reality, Menger and Schmoller, statics and dynamics, theory and history, the evidence shows that such a theory required new foundations. And in spite of Keynes's designation of his theory as general, he was embracing only a very limited part of the problems of economic evolution.

The heresies can never be bounded, or they fail.

Notes of Part Three

1 Schumpeter's last years were marked by long periods of depression, probably motivated by the turn of world events, namely by the Second World War which was destroying Europe; by his political and academic isolation; by his progressive loss of touch with his students, in one university turned Keynesian; and mainly by the dramatic loss of his second wife in 1926 (Allen, 1991, I: 236).

2 Schumpeter never allowed a new edition of the book, of which only a thousand copies were printed in this edition (and was reprinted only in 1970 in Germany) and did not even take a copy with him when leaving Europe for the USA in 1932. He never explained the reasons for the rejection of this book (Swedberg, 1991: 30).

3 This gave him a very peculiar position in the economics of Austria. The 1908 book emphasized the importance of Pareto and Walras and Schumpeter's distance in relation to Austrian economics (Witt, 1993: xiii). Witt explained this feature by his early wishes of getting a specific standing: 'It is no secret, of course, that Schumpeter wanted to achieve a standing of his own and thus tended to distance himself from standard Austrian positions from the very beginning' (Witt, 1995: 84). Nevertheless, he was clearly on the 'theoretical' or

marginalist side: in 1906 Schumpeter published his first two papers on those lines. One was a paper on the role of mathematics ('pure theory') in economics, where he approvingly quoted Jevons: 'If Economics is to be a science at all, it must be a mathematical one' (in Allen, 1991: 56). The second one will be delt with later on.

4 Schumpeter's books are indicated, for simplicity, as DW (1908, *Das Wesen und der Hauptinthalt der Theoretischen Nationalokonomik*), TED (1911, *Theory of Economic Development*, using the revised edition of 1926), EDM (1914, *Economic Doctrine and Method: An Historical Sketch*), BC (1939, *Business Cycles*), CSD (1942, *Capitalism, Socialism and Democracy*), HEA (1954, *History of Economic Analysis*, posthumous), and TGE (*Ten Great Economists: From Marx to Keynes*, posthumous reprint of essays).

5 The reason why Schumpeter waited 28 years to describe this important meeting — why he did not do it in the biographic article about Walras the following year, or in 1911 when presenting his research or still in 1914 when arguing about the history of economic method, and did it in a preface to a book which was only accessible to Japanese readers at the time, was never explained but it is certainly a very curious detail. The reason why he did so in 1937, just when he finished the writing of *Business Cycles*, is nevertheless clear: by that time, the development of his system of thought had shocked several times with the limitations of the static analysis and the stationary Walrasian processes, as BC explicitly acknowledged.

6 Marshall criticized Comte's distinction between order and progress, equivalent to that distinction between statics and dynamics, in the 1907 preface to the Principles (Marshall, 1907 in 1890: 47-8). Schumpeter's attitude was consistently different: more on it later on.

7 Mill fully assumed the importance of the physical metaphor for the definition of the proper method in sciences: 'Now Induction is mainly finding the causes of effects; and in endeavouring to give an account of the manner of tracing causes and effects in the physical sciences, I soon saw that in the more perfect of those sciences we ascend, by generalisation from particular instances to the tendencies of causes considered singly, and then reason downward from those separate tendencies, to determine the action of the same causes when combined. ... My practice being to study abstract principles in the best concrete instances I could find, the Composition of Forces, in Dynamics, occurred to me as the most complete of the logical process I was investigating'. Nevertheless, Mill was aware that the principle of the addition of causal implications could not be extended to all sciences: 'On examining what the mind does when it applies the principle of Composition of Forces, I found that it performs a simple act of addition. It adds the separate effect of the one cause to the separate effect of the other, and puts down the sum of the separate effects as the joint effect. But is it a legitimate process? In dynamics and in the other branches of mathematical physics it is; but in some other cases, as in chemistry, it is not ...' (Mill, CW, i: 166).

8 Mill studied Darwin's *Origin of Species* and was impressed by the achievements of evolutionist biology, but considered that it provided no more than a fascinating hypothesis. In a 1860 letter, he wrote: 'It [Darwin's book] far surpasses my expectations. Though he cannot be said to have proved the truth of his doctrine, he does seem to have proved that it may be true' (Mill, CW, xv: 695). And nine years later he wrote in another letter: 'Darwin has found (to speak Newtonially) a *vera causa*, and has shown that it is capable of accounting for vastly more than had been supposed; beyond that, it is but the indication of what may have been, though it is not proved to be, the origin of the organic world we now see' (Mill, CW, xvii: 1553-4). But Mill considered that, since Darwin's theory was based on an analogy, it was not more than a new hypothesis.

9 The criticism did not challenge the importance of the physical analogies, but rather its general implication, that Schumpeter feared could launch a new and useless Methodenstreit: 'on those few and well-timed occasions when he is looking for formal analogy to the procedure of physical science, he seems to overstate the importance of the experimental, and to understate the importance of the theoretical side of their work' (Schumpeter, 1930: 152).

10 At least once, Schumpeter concluded that the impossibility of experimentation in economics might suggest another procedure also copied from physics, where such impossibility is sometimes the case: a model should be defined, in order to generate a series and then to compare it to the real observations (Schumpeter, 1935: 3). This suggested some form of proof by simulation. In spite of this and other evidence in the same sense, Schumpeter was

generally hostile to the physical analogies: in a letter dated May 1939, Roos criticized Schumpeter because 'you draw the line too sharply in your physics-engineering metaphor' and argued that economics should mimick the experiments of other sciences (letter in the Schumpeter Archive, Harvard University).

11 'As regards the question of principle, there cannot be the slightest doubt that Hayek is right ... in holding that the borrowing by economists of any method on the sole ground that it has been successful somewhere else is inadmissible ... . Unfortunately this is not the real question. We have to ask what constitutes "borrowing" before we can proceed to ask what constitutes illegitimate borrowing. ... Similarly, the concepts and procedures of 'higher' mathematics have indeed been first developed in connection with the physicist's problems, but this does not mean that there is anything specifically "physicalist" about this particular kind of language. But it also holds for some of the general concepts of physics, such as equilibrium potential or oscillator, or statics and dynamics, which turn up of themselves in economic analysis just as do systems of equations: what we borrow when we use, for example, the concept of an "oscillator" is a word and nothing else' (HEA: 17-8). Hayek's critique of scientism will be presented later on.

12 In another place in HEA, Schumpeter argued again: 'Finally, the reader should also observe that the conceptual devices sketched have nothing to do with any similar ones that may be in use in the physical sciences. ... Since the physical and mechanics in particular were so much ahead of economics in matters of technique, these conceptual devices were consciously defined by physicists before they were by economists so that the average educated person knows them from mechanics before he makes the acquaintance in economics, and hence is apt to suspect that they were illegitimately borrowed from mechanics. Second, such devices being unfamiliar in a field where a looser conceptualisation prevailed, some economists, I. Fisher in particular, thought it a good idea to convey their meaning to the untutored mind by way of the mechanical analogy. But this is all' (HEA: 965). Is this all?

13 In a letter written in the early 1940s, Schumpeter argued that the organic nature of his thought was responsible for the difficulty of formalization: 'there is nothing in my structures that has not a living piece of reality behind it. This is not an advantage in every respect. It makes, for instance, my theories so refractory to mathematical formulations' (quoted in Andersen, 1994: 2).

14 This 'biological term' was used for the first time in 1941, in the Spanish preface to TED (Schumpeter, 1941, in 1911: 15).

15 Schumpeter's main argument was that the nature of the economic reality was a disequilibrating process, just like Marx conceived it (HEA: 77, 774 n.; CSD: 83). In the Japanese preface, this was indicated when he argued that Marx was with Walras the main source of his thought, and that unlike the latter, he discussed the dynamical processes of change.

16 Of course, the 'organic' argument may be a trivial declaration of the self-containedness of a system, and in the previous chapters several instances of such a stance were found. In this case, the 'organicity' of the system is fully identified with its mechanistic character, that is, a 'natural' system excludes purposive action. The word is used in this chapter in a very distinct sense, still indicating the indirect influence of the biological metaphor: an organic system includes complex and indeterminate interactions and feedbacks including with the environment: it is an open system. This is the sense used in connection to the Schumpeterian concept of an organic whole.

17 See, for instance, Blaug (1986b: 51). After graduation, Clark spent three years studying in Germany in Heidelberg, under the supervision of Karl Knies, of the old German Historical School. But he studied also the works of Jevons and Menger, and inherited those combined influences of historicism and marginalism. Hollander described the evolution of the group of young US 'historical economists' returning from Germany, some of them, namely Clark, being 'more inclined to deductive analysis' (Hollander, 1927: 2-3). The cultural influence of Germany was obvious in his concept of entrepreneurship as adventuresome and heroic enterprise.

18 Many years later on, in HEA, Schumpeter defended Clark against the accusation that he was an apologetic of capitalism, saying that the marginalist theory did not imply any social philosophy whatsoever (HEA: 869-70). This is openly contradictory with Schumpeter's 1906

review. Mary Morgan, in her study on Clark, argues that he was a 'Christian of socialist leanings' (Morgan, 1994: 231), openly against *laissez-faire* capitalism (ibid.: 236), but still a marginalist, since he considered that the moral problem of distribution could be solved by marginalist economics (ibid.: 237-8).

19 Marshall polemicized against this conception, in a letter to Clark: 'What I take to be the Static state is ... a position of rest due to the equivalence of opposing forces which tend to produce motion. I cannot conceive of any such Static state, which resembles the real world closely enough to form a subject of profitable study, and in which the notion of change is set aside even for an instant' (1902 letter, in Marshall, 1925: 415). Later on, Marshall criticized again Clark's definitions, since 'an exclusive study of purely statical conditions must be unsatisfactory' and Clark's attempt to isolate statical forces was doomed to fail (1907 preface to the *Principles*, 1890: 51-52). In a note, Marshall added that the separation between statical and dynamical forces could only be accepted for short period analysis and just for 'illustrative purposes' (ibid.).

J.M. Clark wrote in 1927 an essay on his father's contributions, and argued that the solution of departing from static conditions and later on adding some dynamical premises was incoherent, and that a whole new theory was needed for the qualitative or 'chemical' change implied by dynamics, since society should be defined as an 'organic whole' (J.M. Clark, 1927: 46-7, 68-9).

20 Schmoller developed the historical method in economics, describing the successive stages of development of the societies with a combination of sociological, ethical and historical insights. His long time perspective was invoked by Marshall when he criticized the Comtian distinction between statics and dynamics: Schmoller's *Grundrisse* was 'an unsurpassed embodiment of wide knowledge and subtle thought' (1907 preface, in 1890: 48). Nevertheless, Keynes indicated that Marshall was all his life dissatisfied with the 'learned but half-muddled work of the German Historical School' (Keynes, x: 210).

21 Schumpeter maintained this opinion much later: in HEA he presented the whole polemics as 'a history of wasted energies' (HEA: 814 f.).

22 This was also another influence by Comte, which Schumpeter registered in EDM: the philosopher insisted on the 'altogetherness of social life and the need for an historical method for other problems other than the purely economic ones' (EDM: 96). Of course, later on Schumpeter developed this program much further, since historical methods were considered essential to account even for 'purely economic' problems.

23 One can hypothesize that the rejection of DW, the 1908 book on the Methodenstreit, was connected to this important change in Schumpeter's opinion and his later incorporation of essential elements of the Historical School in his own system. Simultaneously, important elements of differentiation with the marginalist school were developed by Schumpeter: unlike Menger, he did not consider that the value theory required any psychological foundation (Bottomore, 1992: 19), and he praised Pareto for getting rid of the concept of 'utility' and suggested that maximizing rationality was not a realist feature (TGE: 179, 192).

24 Samuelson's interpretation for this event is that Schumpeter loved to take the 'unpopular side' of the disputes (Samuelson, 1951: 49-50, 50 n.). Faced with the evidence, this is obviously a minor point. On this topic, also Machlup (1951: 95).

25 The impact of the Cowles Commission research program was by then dominant in the profession: the econometric revolution won the day. Friedman, by then a researcher at the NBER, argued at the conference that a final synthesis would be reached between the NBER method and the Cowles approach (Friedman, 1951: 114). But Koopmans was so convinced of the advantage of the econometricians that he could recommend, in an internal memorandum to the Cowles group with a balance sheet of the meeting, 'Let's not fight too much' (Epstein, 1987: 111).

26 In the first edition of TED (1911) Schumpeter used the distinction 'circular flow'-'development'; in the second edition, in 1926, these were replaced by statics-dynamics. But, since 1934, as indicated in the preface to the English edition of TED, 'in deference to Professor Frisch', and also in BC, Schumpeter used the distinction between static and dynamic *forms of analysis,* and stationary or development *processes in nature* (Schumpeter, 1934, in 1911: 6).

27 Schumpeter's concept of capital was defined as a flexible resource, distinct from the technical

structure of the production process (Oackley, 1990: 38). It belongs to the circular flow, and is 'that part of the social product of preceding economic periods which maintains the production of the current period' (EDM: 54). Thus, there are two sources of accumulation, one being the circular flow and another the development process, which is moved by innovation.

28 Formally, when $dx/dt > 0$ or < 0, but $d^2x/dt^2 = 0$. Thus, the upper and lower turning points, where $dx/dt=0$ — traditionally the centre of the polemics about causality in cycle theory — were not the main concern of Schumpeter.

29 Rosenberg interprets Schumpeter's position on the circular flow as a theoretical description, as opposed to the real processes of change in capitalism (Rosenberg, 1994: 43) and Swedberg interprets it as an ideal-type (Swedberg, 1991: 32). But the previous quotations refute this interpretation: for Schumpeter, the circular flow was a real process, simultaneous with development, and a complete theory should integrate both dimensions in the same framework.

30 One of the main reasons for the sense of failure Schumpeter felt in his last years was his incapacity to develop a formal model for his theories. His own diary proves that he worked almost daily and helplessly with systems of equations, at least since 1934 when preparing BC, and afterwards looking for a general equilibrium model accounting for the time path of the variables (Allen, 1991, II: 8, 142, 177, 190, 227). But he suspected that the available differential and difference equations were unsuited to define an evolutionary system including social relations and complex behaviours. Furthermore, his colleagues, such as Goodwin, witnessed his difficulties with mathematics.

31 Following the precedent method, the *Treatise on Probability* (1921) is indicated as TP, the *Treatise on Money* (1930) as TM, and the *General Theory of Employment, Interest and Money* (1936) as GT. Other texts of Keynes are quoted from the edition of his Collected Writings, and indicated by the volume (i to xxx).

32 Moore was not actually at Cambridge, where he took a teaching post only in 1911. He was nevertheless the most influential of the philosophers of what was to be called the 'trinity of Trinity' (Moore, Russell, Wittgenstein), and surely the most influential on Keynes's education. Moore's *Principia Ethica* was published in 1903, and this was the most important book in Keynes's life (Skidelsky, 1983: 119). From it Keynes derived a global conception of science and methodology, and namely an outstanding opposition to the Oxfordian mixture of Hegelianism and 'biological language', the statement that the good of society directly resulted from the good of the individuals (Skidelsky, 1992: 224), in the Mandevillean fashion.

33 In a 1905 letter, Keynes argued against Moore's inclusion of material things under the general category of organic units, which should comprise only states of mind (Skidelsky, 1983: 148).

34 In *Principia Ethica* Moore stressed the 'principle of organic unity': 'the value of the whole must not be assumed to be the same as the sum of the value of the parts' (Moore, 1903: 28). Or else: 'I shall use it [the term "organic"] to denote the fact that a whole has an intrinsic value different in amount from the sum of the values of its parts' (ibid.: 36).

Moore was strongly opposed to Hegelian idealism, namely to its dialectics, and condemned its formulation of a principle of 'organic unity' as a masquerade for a metaphysical approach: 'The principle of organic unity, like that of combined analysis and synthesis, is mainly used to defend the practice of holding both of two contradictory propositions [that is, that two things are distinct and yet identical since pertaining to the same "organic unity"], wherever this may seem convenient. In this, as in other matters, Hegel's main service to philosophy has consisted in giving a name to and erecting into a principle a type of fallacy to which experience has shown philosophers, among with the rest of mankind, to be addicted' (Moore, 1922: 16). Moore distinctly argued for a process of division of the whole into parts as the principle of definition (White, 1958: 73) but then denied the whole to be the sum of its parts.

35 In a biographical note on Marshall, Keynes quoted his remarks about Edgeworth's *Mathematical Physics*: 'It will be interesting, in particular, to see how far he succeeds in preventing his mathematics from running away with him, and carrying him out of sight of the actual facts of economics' (quoted in Keynes, x: 187). In another biographical note on Edgeworth, Keynes concluded that this actually happened: 'All his intellectual life through he felt his foundations slipping away from under him' (x: 262).

36 In fact, Moore accepted Bentham's utilitarianism, even if it was under the influence of his

philosophy that Keynes exuberantly rejected it. The contradiction was on Moore's side (Keynes, x: 436; also Carabelli, 1988: 242).

37 For instance: 'The Jevonian conception [of the "purchasing power of money"] would have been intellectually delightful and of great scientific convenience if it had been based on a true analysis. It is one of several quasi-mathematical economic conceptions, borrowing by analogy from the physical sciences, which seemed likely to be so fruitful when they were first derived fifty or sixty years ago 1860, 1870, but which have had to be discarded on further reflections in whole or in part' (TM: 78; also 71).

38 This was another difference between Moore and Keynes: the former supported a frequencist approach of probability (Skidelsky, 1992: 56-7), whereas Keynes argued that probability was mostly non-numerical and that many of the instances in science were qualitative and non-measurable.

39 'Mathematical economists often exercise an excessive fascination and influence ... . [They] introduce the student, on a small scale, to the delights of perceiving constructions of pure form, and place toy bricks in his hand so that he can manipulate for himself, which gives a new thrill to those who have had no glimpse of the sky-scraping architecture and minutely embellished monuments of modern mathematics' (Keynes, x: 186 n.). And in GT he wrote, in a very similar mood as Marshall did once: 'I do not myself attach much value to manipulation of this kind of formal models ... . I doubt if they carry us any further than ordinary discourse can' (GT: 305).

40 'The orthodox equilibrium theory of economics has assumed ... that there are natural forces tending to bring the volume of the community's output, and hence its real income, back to the optimum level whenever temporary forces have lead it to depart from this level. But we have seen ... that the equilibrium level, towards which output tends to return after temporary disturbances is not necessarily the optimum level, but depends on the forces in the community which tend towards savings' (Keynes, xiii: 406).

41 In fact, Shackle, one of the economists who vindicated a subjectivist interpretation of the Keynesian critique of orthodoxy, argued that GT is a contradictory book: its method includes equilibrium and its meaning is disequilibrium (Shackle, 1968: 5, 44; also 1974: iii). Concretely, GT is said to include a general equilibrium approach, a business cycle approach with a growth model, and a kaleidic method (comparative statics applied to sudden changes determined by the structure of the expectations; Shackle, 1972: 433).

42 This is how Joan Robinson defined the nature of the Keynesian way of incorporating history into economics: 'The GT broke through the unnatural barrier and brought history and theory together again. But for theorists the descent into time has not been easy. After twenty years, the awakened Princess is still dazed and groggy. Keynes himself was not quite steady on his feet. His remark about the timeless multiplier is highly suspicious. And the hard core of the analysis ... is based upon comparisons of static short term equilibrium positions each with a given state of investment going on, though it purports to trace the effect of a change in the rate of investment taking place at a moment of time' (Robinson, 1962: 78).

In relation to this argument, it is frequently ignored that Keynes flirted with some historical accounts of capitalist development, such as Commons': 'there seems to me to be no other economist with whose general way of thinking I feel myself in such a general accord', wrote Keynes in a letter to that author about his institutionalist theory of the stages of development (in Skidelsky, 1992: 229).

43 In her introduction to TGE, Elizabeth Schumpeter wrote that she could not understand the professional and personal distance between the two scientists (TGE: 15). So, no explanation was given for the outstanding fact. Smithies did not indicate any interpretation (Smithies, 1951a, b), and Heilbroner argued with the different cultural backgrounds and scientific interests (Heilbroner, 1986). Allen writes that Schumpeter was always extremely generous to all his colleagues when criticizing their work (Allen, 1991, i: 58); but the obvious and extraordinary exception was his review of GT, the 'strongest he ever wrote' (Allen, 1991, II: 24).

In the last years of Schumpeter's life, he worked intensely in a project of a new book, *The Theoretical Apparatus of Economics*, which 'should do from my standpoint what Keynes's General Theory did from his' as indicated in a letter to Smithies (Allen, 199, II: 227). In a letter dated May 1934 to Kaway, Schumpeter stated that the book was being prepared since his last years in Bonn, in the early thirties (Harvard Archives). He never succeeded to conclude this work, and apparently no manuscript was found indicating he was close to do so.

44 In 1923, in his first editorial to *The Nation*, Keynes presented his politics: 'Our own sympathies are for a Liberal Party with its centre to the left' (quoted in Skidelsky, 1992: 136-7); by that time, he suggested a Labour-Liberal government. He supported the miners during the General Strike of 1926, which caused a split in the leadership of the Liberal Party (ibid.: 223). In 1931, he was more inclined to support Labour than Liberal candidates; only with the outburst of the World War did he return to the Liberal party tradition (Moggridge, 1992: 465). But his public position was influential and consensual: in 1937, all the three parties offered Keynes a nomination as independent candidate for MP for Cambridge, which he refused after long hesitation (ibid.: 628).

45 In spite of his own anti-Semitism, Schumpeter rejected the discrimination against Jews: for instance he supported the appointment of Samuelson against the opposition of an anti-Semitic head of department, but his private diary included several anti-Jewish and very racist remarks. Anti-Semitism was, by the way, mildly shared by Keynes (Moggridge, 1992: 609), in spite of his irreproachable friendship with Kahn, Sraffa or Leonard Woolf. Stolper strongly denies this evidence about Schumpeter's anti-Semitic views (Stolper, 1994: 10-2).

46 'But expectations are not linked by Mr. Keynes to the cyclical situation that give rise to them and hence become independent variables and ultimate determinants of economic action. ... An expectation acquires explanatory value if we are made to understand why people expect what they expect. Otherwise expectations is a mere *deus ex machina* that conceals problems instead of solving them' (ibid.: 792 n.). Of course, this would be the relevant form of causality if Schumpeter still accepted his previous account in TED, where the cause was considered to be the first relevant exogenous factor for the system (TED: 10).

47 Risk was defined by Keynes as the mathematical expectation times the probability of failure (TP: 348) — this was written approximately one decade before the seminal book by Knight.

48 In GT, these variables (propensity to consume, marginal efficiency of capital, rate of interest) were formulated as independent, although the author recognized that it was an impure solution since they were influenced by other variables (expectations, etc.). The designation of the semi-autonomous variables as exogenous is the testemonium paupertatis of modelling.

49 In TED, before the publication of any of Keynes's writings, Schumpeter already stated that he was firmly opposed to the 'psychological prejudice which consists in seeing more in motives and acts of volition than a reflex of the social process' (TED: 43). But in TGE, Schumpeter credited Keynes for the dynamic feature of the concept of expectations (TGE: 381), even if this aspect was never developed in Schumpeter's work. As a matter of fact, Schumpeter was generally opposed to the Keynesian concept of expectations.

50 'Unless we know why people expect what they expect, any argument is completely valueless which appeals to the as *causa efficients*. Such appeal enters into the class of pseudo-explanations which already amused Molière' (BC: 140).

51 'What precisely is looked upon as inherent [endogenous] in it [the system] will, of course, depend on how we delimit it and which facts and relations we decide to treat as data, and which are variable' (BC: 7n.). Or, just at the end of the book, he indicated that variables such as 'mentality' could be considered as exogenous, as usually, or as endogenous, according to the researcher and to the research — BC is said to take normally a 'narrow sense' (ibid.: 1050n.). Of course, this created the greatest confusion among the historians of economic thought.

Hansen considered the exogeneity or endogeneity of Schumpeter's theory a 'perennial and inexhaustible subject for discussion' not clarified by the author himself. Here is Hansen's own interpretation, a Salomonic solution: 'It is exogenous in the respect that it places primary emphasis upon changes in the data. Yet it is also an endogenous theory in the respect that it runs in terms of an internal, self-perpetuating system ... whose impelling force, innovation, cycle after cycle renews the wave-like movement. ... [The business cycle] is an endogenous process determined by the inner nature of a dynamic economy, but it is exogenous in the sense that innovation is a change in the basic data' (Hansen, 1951b: 80).

52 In his GT, Keynes clearly stated that expectations could not be represented by a mathematical model (GT: 162-3).

53 Because of its static framework, in GT expectations were modelled as exogenous variables and the explanation abstracted from uncertainty, unless in the Chapter 22, in which expectations were introduced in order to understand fluctuations. For what matters for this book, cycles and irregular growth cannot be explained unless in a dynamical context.

54 Patinkin suggests that these criticisms were due to the failure of the previous mathematical formal models by Keynes: 'In fact, it may have been Keynes's lack of success with such formal model building in the TM that led him to the more critical attitude expressed in the passage from the GT just cited' (Patinkin, 1976: 1094). This is a highly arbitrary implication and, on the contrary, the future debate between Keynesians and Monetarists and the impossibility of getting conclusive results from the econometric methods that both sides fully used was a new confirmation of Keynes's point.

55 The misspecification can arise for instance from the representation of a model with a limited set of causes while in reality there is an infinite or complex causation, from the fact that the number of real regressors may be superior to the number of available observations, from errors in the formulation of the model, from the ignorance of the lag lengths.

56 This last charge was not so negative for Keynes, since he was fascinated by alchemy (he bought Newton's manuscripts on alchemy, Moggridge, 1992: 492 n.). His criticism ofTinbergen finished with the phrase: 'Newton, Boyle and Locke all played with alchemy. So let him continue' (xiv: 320). He did.

57 'In the case of the history of gold prices, to which defenders of the Quantity Theory usually refer with confidence, factors other than gold production have changed and fluctuated so hugely and so notoriously that the use of any apparent close coincidence between the level of prices and gold production in support of the Quantity Theory is a gross example of the *post hoc, ergo propter hoc* argument. Since other factors have not remained constant, the theory would lead us to anticipate coincidence between prices and gold production if the other factors happened to balance one another; and one cannot easily prove this without assuming the theory itself' (Keynes, xii: 765).

58 Morgan argued in 1990 that Keynes did not even read carefully Tinbergen's volumes and ignored the technical developments of statistics (Morgan, 1990: 121, 121 n.), and Hendry and Morgan added that 'Keynes might have been reading another book altogether' or even not read it at all (Hendry and Morgan, 1995: 54). It is certain that Keynes did not take the time to study the details of Tinbergen's method, and ignored for instance his concern about the stability of the sub-samples. But his critique was more general and still accurate on the main points: in 1980, Hendry accepted that many of these arguments remain unopposed forty years afterwards (Hendry, 1980: 402), and that the difference between statistical science and alchemy is still a narrow one (ibid.: 403).

The reason is of course the nature of the economic processes: 'Econometricians conceptualise this economic system as a complex nonlinear, interdependent, multivariate, disequilibrium, dynamical process dependent on agents' expectations and their adjustment, subject to random shocks, and involving many phenomena that are unobservable; relevant time series data are inaccurate, exist only for short periods and for a few major variables; economic theories are highly simplified abstractions usually of a comparative static form invoking many *ceteris paribus* clauses (with yet another implicitly required), most of which are invalid in empirical applications — little wonder our macroeconometric representations are less than perfect' (Hendry, 1980: 399). But isn't this the echo of Keynes's critique in modern parlance?

59 A very obvious example is the *Handbooks of Mathematical Economics* edited by Arrow and Intriligator: in the 2264 pages of the 27 chapters of the 4 volumes, only two include a (single) reference to Keynes. For mainstream economics, Keynes became a museum curiosity. And only when the problems mount in the paradigm there is again some attention to his critique.

60 Schumpeter compared himself to Moses, who knew about but could not reach the

Promised Land of mathematics and the 'foundation of a new economics' (letter to Haberler, March 1933, Harvard Archives). But the clearest text by Schumpeter on the difficulty to formalize his own model is the handwritten letter to Marshack about his review of *Business Cycles*: 'Now this is perhaps the most fundamental reason why I have never attempted to get my system into equations — except for individual bits of mechanics: if you have a system of interdependent quantities you will always be able to describe surface mechanical relations ... as disturbed by my process by partial sets of conditions which in the "partial" system may even be quite determinate. Therefore I do not feel any contrast between those schemata (Frisch-Tinbergen-Roos-Amoroso-Kalecki-Keynes and many others) and my way of thinking: they simply move on planes different from mine and I feel perfectly free to use any of them for those peripheric problems for which they are intended' (Harvard Archives). On his difficulty with mathematics, see also Samuelson (1974: 89).

61 'It has none [signification, for this description of the 106 cycles], because the norm of the phenomenon under consideration has obviously changed and the criteria as to what is to be called a "cycle" is neither definite nor uniform enough to warrant one in speaking in this case of a "statistical universe" of which the 106 cycles could be considered a sample' (Schumpeter, 1930: 164). The last point is essential and was already met in Part Two of this book. On that occasion, Schumpeter explained the problem as the result of the incorporation of the wrong metaphor: 'It is less universally recognized that this tool grew out of a soil very different from ours, and that by using it we are introducing the whole of the assumptions of the theory of errors — of "static" statistics — which may well reduce our results to meaningless' (ibid.: 165-6).

Frisch supported some of the criticisms by Schumpeter, namely about the tests of significance and their theory-dependent status: mathematical models are only tools 'subordinated to the general intuitional and philosophical interpretation' (Frisch, 1951: 9-10).

62 'Of course, there is no theory of the cycle, if we understood by this a complete explanation of all what happens. This can only be found in a reasoned history of industrial life. It is only the backbone of it' (Schumpeter, 1927: 298).

63 For the reasons for criticism of the correlation methods, Keynes shared such a suspicion about decomposition and the concepts of 'trend' and 'residuals' (Keynes, xiv: 319).

64 In realist accounts, the problem is further complicated by the abandonment of the hypothesis of perfect competition: under oligopolistic situations equilibrium becomes indeterminate (CSD: 79-80) and the same happens under monopolistic competition (BC: 57)

65 Once again, Schumpeter's positivism was itself paradoxical. In his major statement on the subject, the 1948 Presidential Address to the American Economists Association, he discussed the role of ideology and namely of the pre-cognitive vision in science. He acknowledged the danger of the ideological bias, and in a very positivist way indicated the solutions against it (the ideological neutral methods of economics and the public and institutional character of science) but added a surprising conclusion: the vision is therefore a condition for the success of science (Schumpeter, 1949).

66 Hayek fought against that 'imitation of physics' leading to 'outright error' and transforming sciences in 'cooking recipes' (Hayek, 1974: 23, 30), namely arguing that quantification arbitrarily restricted the domain of causation and ignored some decisive features of the economies: 'This brings me to the crucial issue. Unlike the position that exists in the physical sciences, in economics and other disciplines that deal with essentially complex phenomena, the aspects of the events to be accounted for about which we can get quantitative data are necessarily limited and may not include the important ones' (ibid.: 24).

PART FOUR

Dr. Pangloss Hunted by the Snarks

They sought it with thimbles, they sought it with care,
They pursued it with forks and hope;
They threatened its life with a railway share;
They charmed it with smiles and soap.

They Shuddered to think that the chase might fail,
And the Beaver, excited at last,
Went bounding along on the tip of its tail,
For the daylight was nearly past.

"There is a Thingumbob shouting?" the Bellman said.
"He is shouting like mad, only hark!
He is waving his hands, he is wagging his head,
He has certainly found a Snark!"

Lewis Carroll, *The Hunting of the Snark*

17. Introduction to Part Four: Turbulence Again

The previous chapters discussed the traditional methods for the study of the phenomena of cycle and growth, and considered in particular Keynes's and Schumpeter's critiques of those methods. The current Part Four deals in more detail with some of the assumptions and the elementary methods of traditional statistics and econometrics, which have been for a long period the basis of the ambitious and cherished research program responsible for important developments in the discipline. Their assumptions and implications are discussed, as well as the change of mood which is by now obvious: these statistical methods have been submitted to general criticism by some of its most distinguished and disenchanted practitioners or theoreticians (for example, Leontief, 1971; Arrow, 1984; Solow, 1988; Hutchison, 1992). Their arguments echo, several decades afterwards, the points made by Keynes in his polemics with Tinbergen, in spite of the fact that most of them defended the gates of the citadel from the day Keynes trumpeted his critique: the econometric applications are generally not based upon realistic assumptions, they support excessive conclusions, they are referred to opaque theories and ad hoc models and finally the pretension of axiomatic neutrality and objectivity of statistical methods is empty.

The large place of mathematical and statistical methods in economics is, of course, an old and necessary characteristic of the discipline. It is derived from its accounting function and reached a new stage given the success of the physical metaphor which defined the marginalist revolution and, later on, of the second resurrection of economics under the spell of the rocking horse. Frisch, Tinbergen, Koopmans, Haavelmo, the brilliant economists and physicists dedicated to the econometric program were responsible for a new generation of problems, of methods and of solutions. But they could not avoid a labyrinth of difficulties, since their method required a double decomposition between trend of growth and the cycle, and between impulse and propagation in the cycle, and the founding fathers of econometrics wisely maintained their reservations about the positivist requirement of an exclusive and exhaustive representation of social phenomena with mathematical symbolism. Such an appeal and those precautions were clearly expressed in the first issue of *Econometrica*, both by

Frisch in the editorial and by Schumpeter in the following paper, 'The Common Sense of Econometrics': 'Nothing is farther from our minds than any acrimonious belief in the exclusive excellence of mathematical methods, or any wish to belittle the work of historians, sociologists and so on' (Schumpeter, 1933: 5).

Some years after these warnings, Schumpeter defended the historical methods against the Cowles Commission advocates at the 1949 Conference organized by the NBER. By that time, as the econometric program was already much developed, his arguments were strongly contradicted and largely ignored and, given that he was not fully aware of the modern mathematical tools of the profession, Schumpeter was indeed accused of defending the historical method because of ignorance of more sophisticated instruments. The present chapters argue that, notwithstanding that justification, the utilization of historical insights in economic research is not only the possible last recourse for those mathematically deficient, but it is rather one of the very foundations of economics as a science and is therefore imposed to any competent mathematical treatment of economic data. It can be used because of ignorance of other methods, but it must be used because of sound wisdom. The historical methods are necessary simply because any economic explanation is devoid of sense without it: even if the technicalities of laboratory physics were fully applicable, still remains the certainty that 'Fluid dynamics do not depend on who is the Chairman of the Federal Reserve', as Brock, Hsieh and LeBaron put it (1991: 24); historical insights are necessary in order to direct the statistical research.

It is nevertheless well known that the historical approach was defeated at the 1949 Conference and thereafter was relegated to a separate and minor province in economic thought, and most of economic faculties are today reduced to departments of applied mathematics. But economics is consequently reduced in scope, since the science is submitted to the rules of the secondary subject, of the available physical metaphor: so it was while the general equilibrium paradigm maintained the static approach and the methods of maximization over a conserved field, or when the core concepts of the quantum revolution were introduced in economics.

That was the decisive contribution of Frisch, the introduction of the probabilistic approach in economics. With the rocking-horse model, he departed from the traditional equilibrium models, since two modes of oscillation were considered: the impulse and the propagation systems, the first creating the oscillation and maintaining its amplitude, the second one dampening the movement and establishing its regularity — the point of mechanic rest was replaced by a mechanism of equilibration. Of course, this completely depends on the assumption of the independence of the oscillators: if instead some oscillators are coupled (for example, by path-dependence), then turbulence is the probable outcome of the action of the system.

A very brief sketch of some of the debates about the econometric assumptions and methods is presented in Chapter 18. The purpose is to indicate the outstanding enigma: why do not the econometric methods detect the fluctuations and the distinct periods of economic evolution that the historians claim to exist — and why do they fail to do so, in spite of the rocking horse being just conceived of to study the cycles? In fact, traditional methods cannot but ignore the complex nature of economic series and, as a consequence, their spurious conclusions are imposed by the methods themselves and not necessarily by the structure of the real data.

But if statistical inference is suspect, will economics turn back to the pre-classic form of accountancy and lose all its modelling, explanatory and normative capacities? The answer is categorically no. New statistical tools and old historical methods can be combined in the evolutionary paradigm. Indeed, the next chapters present relevant cases for that argument: distinct methods are used in order to identify complexity in concrete series and to challenge the orthodox procedures. A simulation is generated by a nonlinear model and its series is studied with the help of some traditional methods, which are unable to discriminate the real nature of the data and therefore confidently accept the wrong hypotheses of the rocking horse: in that framework, the cycles are created by randomly distributed exogenous impulses impinging on a stabilizing propagation structure of the economy, which can be approximated by a linear specification. This view is strongly refuted and it is proved that if the cycles are generated by a complex nonlinear structure, then the standard methods and their conclusions are irrelevant and new tools are required to inspect the economies.

Some of these statistical tools are used in this part. One is taken from hydrology, and was used to identify the structure of a long series — the levels of the Nile — which apparently behaves according to the Gaussian law, but which really includes some hidden structures. Another is a non-parametric method recently developed by some economists studying chaos, and is used in order to detect one of the possible reasons for the confirmation of the mistaken hypothesis of linearity: the aggregation procedure generally used in economic measuring. Thus, two main possible sources of error are identified: the standard methods and the traditional ways of measuring the data. The pervasiveness of both procedures can explain the noticeable and strategic importance of the self-confirming hypothesis of linearity of the functional relations describing the relevant variables of the system; indeed, linearity and aggregation and inter-dependent and mutually reinforcing. Excluding interaction and complexity, these heroic assumptions are decisive building blocks of the mainstream epistemology in economics: time is conceived of as a continuation of space, dynamics collapses into statics and the general equilibrium paradigm is reaffirmed.

The hypothesis of complex relations in the reality of economic historical series is both more realist and more fruitful for applied research, since it establishes an alternative program avoiding some of the major flaws of the rocking horse, namely the attribution of causality to the additivity of elementary and independent factors, or the epistemic distinction between growth and cycle. Yet, this critical argument is not new: it goes back under distinct forms to Smith, Marx, Veblen, Kondratiev, Schumpeter, Keynes and so many others before the conceptualization of complexity or turbulence could even be thought about.

18. Uneasy Feelings

The history of econometrics could be written from the standpoint of the presidential addresses at the annual meetings of the American Economists Association. But the historian indulging in such a project would certainly be distressed by the shocking testimonies, in spite of the authority of the words and of the respect of the entire profession in relation to the respective presidents of the AEA. Although most of those were simultaneously members of the Econometric Society and leading practitioners of the statistical methods, several of the recent speeches were strongly opposed either to the general trends of abstract formalization in disdain of realist assumptions or to the excessive and implausible conclusions drawn from limited statistical inference.

Leontief's 1970 address was one of those assaults upon the 'palpable inadequacy of the scientific means' for solving practical problems, leading to an 'uneasy feeling about the present state of our discipline' (Leontief, 1971: 6). The balance sheet was sharp: 'Uncritical enthusiasm for mathematical formulation tends often to conceal the ephemeral substantive content of the argument behind the formidable front of algebraic signs .... In no other field of empirical inquiry has so massive and sophisticated a statistical machinery been used with such indifferent results' (ibid.). Shortly afterwards, Arrow's presidential address praised the 'rich formal development' of the general equilibrium paradigm, but recognized that unemployment and the real history of capitalism are 'scarcely compatible with the neoclassical model of market equilibrium' (Arrow, 1984: 154), and this theme has been repeated by the author ever since. Gordon had taken the same stance in his presidential address of 1976. In his presidential address in 1990, Debreu defended the achievements of econometrics arguing that they have been 'one of the prime movers in the transformation of our field', but still recognized their 'esoteric character' (Debreu, 1991: 2).

Simultaneously, Friedman rejected the new waves of formalism propelled by the general use of computers and simulations, since the economists are induced 'by the computer revolution to carry reliance on mathematics and econometrics beyond the point of vanishing returns .... More recently, the easiest way to avoid perishing by not publishing is to access an existing data base, download a batch of data to your computer and put the data through the econometric wringler' (Friedman, 1991: 35-6). Hahn called for a return to Marshallian 'biology' and to historical and sociological methods for the reorientation of economics (Hahn, 1991: 47-50). And so did many others, but the lock-in of the trajectory of the mainstream prevented any major impact of these expressions of uneasy feelings.

Obviously, something is rotten in economics.

18.1. The Panda Principle

Economics benefited from two types of resources from statistics: the methods defined in astronomy, namely the deterministic models and their errors of measurement, and the methods imported from biology, where observations are time dependent but eventually not unique and sampling techniques from large populations were developed. By the end of last century, economists were not only able to understand and to use those methods but also to improve and to adapt them: Jevons, Edgeworth, Slutsky were responsible for important developments in statistical inference and yet, although all of them were also major contributors to the definition of the neoclassical program, they did not attempt to transport the statistical and probabilistic theories to economics. Paradoxically, determinism came to dominate economic statistics, and that was not alien to the deep conviction of Jevons, one of the most influential scientist in the field, that the Laplacean theory was accurate and essential — a conviction he shared with many others.

The econometric revolution of the thirties introduced a gigantic change: the economic events were defined as mimicking the laboratory environment, where stimuli and response are controlled and repeated in order to get samples of series as the basis for inductive conclusions and tests of theories. The pendulum metaphor, or the simplest version of the rocking horse, dominated these years of high theory.

18.1.A. Correlation and causality

The notion of correlation was created by the French naturalists of the eighteenth century and was then appropriated by the British biometricians during the following century, who used it for confirmation of laws in experimental sciences. Pearson[1] generalized the concept for statistical inference, under the typical form: *the variation of Y is due to the variation of X* (Veuille, 1987: 45-6); so did Yule[2] (1897, 1899, Gutsatz, 1987: 72). Irving Fisher was one of the first economists to use the notion, for the analysis of the relation between trade and prices for 1915-23; he concluded that the cycle was a mere artifact, a 'dance of the dollar', since no evidence for it existed in the residuals (Fisher, 1925: 181 f.).

The acceptance of correlation as evidence for a causal relation implies, nevertheless, the assumption of several important premises. The first is that exogeneity is the necessary condition for causality but, since the functional relation does not unambiguously define the direction of those stimuli, and since the operation of correlation is reversible a new condition must be added. In that case, the adequacy of the relation must be established by further constraints and the causal ordering must be a priori defined. Furthermore, exogeneity of one variable requires its permanent and complete statistical independence from all disturbances of the system, as Haavelmo and Koopmans claimed (Christ, 1994: 37; Aldrich, 1989: 30-1). This is why correlationist causality requires the preliminary selection of variables by a criterion defining exogeneity, and consequently the whole operation amounts to a sophisticated version of the *post hoc, ergo propter hoc* syllogism (Sims in Gordon and Veitch, 1986: 282; Gutsatz, 1987: 83).

The second requirement is the acceptance of simple inductive inference, but the method is neither general nor demonstrative, namely because of the oblique transfer. This is why Keynes rejected the general implications of the inductive logic, since he could not accept Laplace's or Poisson's interpretation of the law of large numbers, according to which the increasing number of instances increases the stability of the inductive demonstration (TP: 368) — more on it in the next chapter. Third, correlationist proof requires a reducible, atomistic and stable system, excluding the notion of organic complexity since this includes non-definable relations and non-quantifiable probabilities. The social relations described by the theory and its model must be exhaustively computable, unidirectional, elementary and structurally permanent, if correlationist causality is meaningful: they must be 'legal atoms' subordinated to the principle of limited independent variety.

The limits of these assumptions were immediately evident to many researchers.[3] Thus, the development of statistical inference proceeded through regression analysis, based upon weaker claims about causality and allowing for other technical improvements: the most important was the ambitious program of the Cowles Commission for the conception of structural estimation, in order to solve the difficulties previously found in Tinbergen's research on business cycle models and in the early work on the simultaneous determination of demand and supply. The Cowles research program assumed a linear representation, the existence of observable systematic variables, that it was possible to know beforehand the list of exogenous variables, that disturbances could be approximated by normally distributed random variables and essentially that the system was dynamically stable (Christ, 1994: 48). Under those conditions, a concrete economic problem could be represented by a system of simultaneous equations explaining how the dependent variables were determined by the stable parameters, the exogenous and the predetermined variables and the well behaved shocks. In this sense, the Cowles program introduced and generalized the stochastic approach in economics and was the main driving force behind the econometric revolution of the thirties and the forties.

Hence, the rocking horse was transformed in a model of solution and in a privileged mode of theorizing about a typical economic problem: the structure of the impulses made possible the statistical inference, the mechanism was represented by the system of equations, simultaneity accounted for the time dimension and, moreover, the scope of the system was virtually infinite since all explicit and omitted variables were represented in the rocking of the horse. From that on, explanation concentrated on the horse itself, since the convergence to equilibrium was of course the most relevant of the properties of the theory. Yet, the model stands or falls on the acceptance of the epistemology of the double decomposition.

18.1.B. Experimental and organic environments

In this framework, traditional statistical inference depends on two families of properties: those of the mechanism which establishes convergence, and those of the stimuli which make possible fictionalizing the laboratory environment and its sampling conditions, an essential step for the appropriation of the experimental methods. The last one has been much criticized in social sciences: the analogy between time series and the urn, considering that each instance is a sample from a random process, has been frequently challenged, given the epistemological ambiguity of the concept. In fact, the early debates on the nature of the error divided those opinions describing the errors as random unobservable variables (errors of measurement) or as non-systematic and unobservable variables (errors to equations, also representing a composite of possible sources); if systematic observable (other explicitly omitted variables) and systematic unobservable variables (expectations) are added to the picture, the cocktail seems too dangerous to swallow. Of course, if all those variables are independent and of finite variance, then they can be represented by a single aggregate variable and the Central Limit Theorem provides the necessary reassurance about its obedience to the Gaussian law.

Nevertheless, three questions remain. Independence may not be sufficient: if the aggregate random term in the regression is conceived of as the combined effect of a large number of heterogeneous variables and of eventual misspecifications (lag structure errors, truncation errors, aggregation errors, omitted variables errors, unknown variables errors, etc.) and if these do not conceal each others, the error cannot be represented by a random term (Blatt, 1983: 343). Of course, no econometric model can claim to avoid all of these errors.

The second question is the epistemological contradiction: even if the virtually infinite component variables are aggregative under the Theorem, they are unambiguously incompatible, since some are described as residuals from the model (therefore causally insignificant) and others as (causally significant) shocks or innovations. Each interpretation excludes the other and causality is only meaningful in the second sense.

The third question is the implausibility of the urn metaphor and the derived randomness concept, since it includes the requirements of repetition of the experiment under perfect control and therefore in conditions of time reversibility. Yet, none of those requirements is met in economics, and that is why Haavelmo suggested the 'as if' methodology: economic series should be treated as if they were generated from a larger population of virtual series of the same type, for the same object and for the same period (the original paper is Haavelmo, 1944; for the debate, De Marchi and Gilbert, 1989: 2; Mirowski, 1989d: 229; or, supporting Haavelmo's assumption, Darnell and Evans, 1990: 146). But this urn metaphor implies at least two major conditions that defy logic:

a. Independence of the observations, which is assumed in the urn (with reposition) scheme, cannot be meaningfully assumed in economic series, unless the notions of causal sequence and cumulative evolution are abandoned.[4]
b. The definition of the procedure as the extraction of the series from a larger population, implying that there is an imaginary population of 'UK GDPs for 1801-1988' and that the current series is a representative instance from that universe is also implausible.[5]

The traditional testing strategies are based on the assumptions above indicated, namely on the assumption of exhaustivity of quantifiable variables accounting for an economic relation. If otherwise qualitative factors are considered, the conclusions from those tests are no longer valid or are not general. Furthermore, the *ceteris paribus* means that there is no possibility to refute the test: the Neyman-Pearson test is also inherently unfalsifiable, since protective hypotheses can always account for irregularities — an atypical sample as well as the falsehood of the null can both cause the large value of the t-statistics. And the significance of statistical tests is confined to the space of the model, since it depends on the specification of variables and of their functional relations; it does not make possible an independent inference about reality itself. The tests of significance and the tests of hypotheses are rhetoric tools.

The introduction of the stochastic theory into economics, indeed the most important contribution of Frisch's rocking horse, made possible an epistemological revolution in economics, although Frisch expressed doubts and differences with Haavelmo's general approach and subsequent practice.[6] It was at the origin of the research program of the Cowles Commission; it established the dominance of the Econometric Society in the discipline; it is still used in most of the concrete applications in economic statistics. Yet, it is epistemologically inconsistent since it requires some form of Newtonian atomism, the strict positivist quantification of the relevant economic relations, and therefore rejectes both the organic nature of social evolution and the simple existence of human choice. It may be argued that this was indeed a major success, given that atomism is the other side of the coin of Walrasian general equilibrium and neoclassical economics: the radical program of methodological individualism re-enthroned the virtues of *laissez faire*.

But the program collapsed, not because of the lasting heterodox critiques, but because of its own failure to deliver the complete solution to the structural estimation problem: by the end of the forties, the Cowles research program was dissolved and most of its mentors adopted an extreme axiomatic approach.

18.2. New problems from ancient times

Leontief accepted the most important of Keynes's points, the non-homogeneity of the series through time, and in that sense he described in 1953 the econometric guess-work with a vivid mechanical analogy: the task of the econometrician was as if he or she were told to draw the technical plans of a car motor on the basis of common knowledge of its working, without any other information but the noise of the motor and the indications of the panel of command. Furthermore, added Leontief, in economics we have permanent changes in the structure of the motor induced by its own evolution (Leontief, 1966: 60-1).

The homogeneity and empirical regularity assumed by traditional econometrics, the related assumption of the constancy of parameters, the hypothesis of a large universe of possible outcomes from which the inspected sample is obtained, the definition of probability as the limit of relative frequencies and the Neyman-Pearson strategy for hypotheses testing, all directly contradict Leontief's concern about the mechanistic metaphor: if the guess-work is to be successful, the motor is supposed never to change. Protesting against the excessive claims derived from models with a few dimensions, Leontief suggested instead the development of direct analyses.

The profound disarray of the program was noted by some authors who rejected the Cowles program: by Sims (criticizing the attempt to identify simultaneous equations), by Granger (criticizing simultaneity), by Friedman, by Wold (although he later on accepted structural estimation; Epstein, 1987: 6). But others went farther, challenging some of the decisive assumptions of mainstream econometrics and yet preserving the requirement of formal rigour in order to develop new statistical tools to inspect real series and pertinent problems.

Hendry assumed long-term disequilibrium processes, to be approximated by lag structures and dynamic adjustment. Another important trend in statistical methods is represented by the Bayesian approach: probabilities are subjective,[7] parameters are random variables, and new observations are used in order to correct the prior beliefs and the probability distribution of the parameters. In this framework, it is fully assumed that misspecifications may remain whatever the sample size, since there is no regularity and homogeneity in data. This obvious difference with mechanical economics does not preclude the possibility of a rigorous inquiry: sensitivity analysis and further supportive evidence may help to investigate the scientists' priors and, as a consequence, conditional probability openly incorporates the concept of uncertainty. Still, the treatment of subjective beliefs is a major difficulty in statistics.[8]

Finally, some authors, as Georgescu-Roegen, openly called for the disbandment of the program: his attack on the 'Walrasian organon', the 'mechanicist dogma' and 'arithmomorphism'[9] was led from the point of view of the biological metaphor and evolutionary economics. This alternative strategy based on the concept of organic evolution in historical irreversible processes has several major consequences for the current topics: (i) it establishes severe ontological limits to the traditional assumption of empirical regularities in the phenomena to be studied by econometrics: reality is complex, choice and change are pervasive in economic evolution and therefore no general deterministic law organizes the social life; (ii) therefore, it rejects the inductive proof of general laws, suggesting the adoption of weaker forms of implication; (iii) it establishes multiple forms of causal determination.

In that case, the analysis must reassess what traditional econometrics considers to be pathological cases, since they are the nature of most of the series: in the sequential processes, auto-correlation is not a nuisance but rather a current feature indicating path-dependency in social evolution, just as heteroscedasticity indicates evidence for structural change. This suggested change of paradigm is parallel to what is occurring in other sciences. When astronomy was considered as the queen of sciences, the first concept of random error was based on one single source of variability of data, namely unsystematic errors of measurement. Extended by Laplace to the realm of social sciences, recourse to probability was defined as the unavoidable consequence of the ignorance of the researcher. Either ways, the observations in economic series were considered as being generated by essentially deterministic processes, epistemologically autonomous from several sources of variation that could be described in one single aggregate 'error' variable. In this sense, the frequencies and modes of fluctuations were considered to be independent and additively superimposed, allowing for the distinction between growth and cycle, for the separate analysis of each type of cycle and for the clear distinction between an extrinsic form of causality and an intrinsic form of intelligibility.

But this account was challenged from physics itself by the introduction of irreversibility in Thermodynamics and by Maxwell's, Duhem's or Poincaré's preliminary definitions of complexity by the end of last century. One brilliant example is their lengthily discussed three-body problem, which was defined from Newton as a major enigma for celestial mechanics: as Poincaré suspected, it is now understood that if the trajectories of three bodies are interdependent, they cannot be fully predicted (Ruelle, 1991: 108). In contemporaneous terminology, that means that if at least three oscillators are coupled, chaos may arise, and indeed this was recently proved by Xia: three gravitationally interacting bodies in empty space exhibit chaotic properties, even if the mass of one of them is reduced to the smallest proportion (Xia, 1994: 289 f.).

In economics there is obviously a strong case for complexity, given the interaction of many different agents and institutions. But cycles were understood as the effect of action at a distance on Newtonian atomistic universes, that is, as the result of deterministic, independent and therefore superimposable modes of oscillation: our old friend, the rocking horse, is the eloquent paradigmatic example. If otherwise the modes of oscillation are related and are time-dependent, multiple cycles will exhibit sensitive dependence to initial conditions and create irregularity: that is turbulence in the economies and 'It is not unreasonable to affirm that we live nowadays in such economies' (Ruelle, 1991: 111).

If this is so, statistical inference as currently employed in time series is wrong and new methods must be used in order to cross that wild side of the street, economic evolution — not because the equilibrium models, linear specifications and probabilistic inference are technically incorrect but just because they are correctly applied to the wrong phenomena.

19. Order Out of Disorder

Reductionism was the common ground of Cartesianism, stating that explanation was equivalent to the partition of the problem into its smallest components, and of Newtonianism, which established the empirical confirmation of atomism. Such an epistemological strategy inspired the analysis of the economic cycles and took the form of the double decomposition: the discrimination between the trend, the loci of equilibria, and the cycle, the deviations from it, and then the discrimination in the cycle itself between the propagation system, once again the structure of equilibration, and the impulse system. The paradigm of equilibrium justified the decomposition and legislated about the explanatory value of the model: the systematic part represented the desired properties of convergence, but further assumptions should be assumed about the nature of the error term, so that statistical inference be possible.

Those assumptions are not intuitive, since they entail the aggregation of a disparate set of misspecifications and errors, known and unknown variables, quantifiable and unquantifiable entities. This is why the Central Limit Theorem is so decisive: it states a very powerful and general result and, under the appropriate circumstances, allows for the aggregation of distinct variables as a single random term whose distribution properties are well known.

The decisive conditions are therefore the independence and finite variance of the components of the aggregated term — the applicability of those conditions was the motivation for the debate within social sciences. Quetelet, the disciple of Laplace who formalized the Malthusian law of population influencing Darwin's acceptance of the notion of natural selection, stated that each individual is unpredictable, but that in the aggregate the population follows the law of errors in deviations from the average behaviour (Morgan, 1990: 7). So, when Chebyshev formulated the law (1887) and Pearson extended it as the basis for statistical inference, the concept was already firmly established, and it has been successively confirmed by a number of experiments of all sorts.

As a consequence, Galton argued that the 'law of the frequency of the error' was universal in social structures:

> The larger the mob, and the greater the apparent anarchy, the more perfect is its

> sway [of the normal law]. It is the supreme law of Unreason. Whenever a large sample of chaotic elements are taken in hand and marshalled in the order of their magnitude, an unsuspected and most beautiful form of regularity proves to have been latent all along. (Galton, quoted in Peters, 1994: 197)

In other words, the Central Limit Theorem describes the imposition of order out of disorder: if with Newton and Descartes we lived in the world of order and simple phenomena to be explained by universal laws, simplicity creating simplicity, the Theorem transported us to another world, where it is complexity that generates simplicity.[10] This very powerful result is the basis for statistical inference.

But is the Theorem so general?

19.1. Tales from Egypt

In 1906, H. Hurst, a physicist with an Oxford DPhil, was recruited to work in Egypt for the government. He spent there forty-six years of his life, and worked from 1913 on projects of dams in the Nile, directing the department responsible for the control of the river for most of the time. After thirty-eight years of work he knew the Nile quite well and published the results of his investigations, which are quite surprising.

The initial problem of Hurst was to compute the maximum capacity needed for the reservoirs of the dams. He disposed of an exceptionally good record of the levels and floods of the Nile, since the river has always been so important for the agricultural civilization of the region and a careful registration of data was available for centuries, and very accurately since 622 (Hurst, 1952: 259-60). From the inspection of those records, Hurst wanted to compute the storage of water needed in good years in order to provide for the agricultural work in the bad ones, and he wanted to know the maximum floods one could expect, in order to build the dams.

Very naturally, he adopted the hypothesis that the river discharges follow a random path, just like the pure games of chance such as the roulette. But soon he noticed that there was a pathology in the series: long periods of high floods and long periods of low floods could be detected. In other words, the series as a whole could be correctly described by the bell curve, as expected from the random hypothesis, but still some structure existed in it:

> it was clear that the storage required in the case of river discharges, rain falls, etc., was greater than that which would fit pure chance events. This is due to the tendency, already mentioned, of natural events to group themselves in runs in which high or low values preponderate. (Hurst, 1952: 298)

Hurst checked a large number — seventy five — of other natural phenomena from the thickness of the rings of the trees of California to the marks of mud in the river beds, from prices of wheat to sunspots, values of temperatures or pressure, and found the same features (ibid.: 297):

> In all cases so far these methods have assumed that the annual discharges are random that is that the annual discharge has no relation to its predecessors. The examination of some hundred of cases of natural phenomena ... shows that although one year is not directly connected with its predecessors, high or low years tend to appear in groups. (Hurst, Black and Simaika, 1966: 125)

If the order of the occurrences was ignored, the Theorem could be applied and the results fitted nicely. But if that order was considered, than 'something more complicated' emerged (ibid.: 59, 106). This suspicion was then checked with the distant help of Robert Brown, who discovered in 1828 that there was a relation between the time of the observation and the distance travelled by a particle following an erratic path: Einstein found the approximate values and, very simply, the equation is $d=t^{0.5}$. The Brownian motion was used in order to represent multidimensional and independent processes comparable to the games of chance. Inspired by this computation, Hurst calculated what he called the rescaled range (R/S): he normalized the series as deviations from the mean, in order to calculate the maximum range of the new series formed by the accumulated normalized values. This range indicates the distance travelled by the system for the available observations, and Hurst found that, after standardization, it could be described by the equation $R/S=cn^K$, where c is a constant and n the number of observations: the R/S value scales as time goes by.

If the exponent K, in Hurst's terms, or later on called H from his name, is 0.5, we have the form of Brownian motion, and this result was checked by exhaustive experiences with games of chance with cards. Otherwise, if $0<=H<0.5$, we have pink noise, the anti-persistent phenomenon of turbulence, for example, volatility in capital markets; if $0.5<H<=1$, we have black noise, the persistent phenomenon of trend-reinforcement. The system has memory of its past and there is no independence of successive observations. For the case of the Nile and the other natural phenomena under scrutiny, Hurst found values from 0.72 and 0.81: there was the 'Joseph effect', seven years of plenty, seven years of scarcity, a strong dependence on the past. The R/S analysis and the Hurst exponent can therefore be used in order to analyse a long enough statistical series and to detect the effect of memory and deeper structures that may be hidden by the Theorem or by the apparent randomness of the system.

Peters used the method to check the traditional assumption of efficiency and normal distribution of market returns: the standard theory says that any action is independent from the previous ones, that no investor learns from the past, and that the market is efficient provided that there exist investors with different time horizons (Peters, 1991: 14; 1994: 42) — order is created from disorder. Peters found that the financial series do not really follow the normal curve, as expected: some are black or pink noise processes, some exhibit kurtosis, fat tails which are aberrations in the Gaussian universe.

Now look at the following series (from Peters, 1991: 166). It is S&P 500 index returns from January 1950 until July 1988 and it presents the irregular features of the volatile financial markets, with several long trends for distinct periods. The graph is irregular: good years follow bad years, cyclic contractions and recoveries are successively indicated, the economy grows and changes. And, of course, there is an infinitude of factors explaining this global performance. If the decomposition method is accurate, we would be able to separate the trends of growth and the cycle, and to identify the mechanism of equilibration in the cycle — the constant parameters of the explanatory variables — and the aggregate effect of the random shocks, the residual or the error term of the equation. But, since we do not know the exact system, we are forced to proceed backwards and to use indicting methods, that is, to assume that we can use statistical inference since the random term must be normally distributed; in that case, the results are accurate. In fact, the proof is easily available: if we detrend the series with an exponential curve and then fit an auto-regressive process, we get a series of random residuals with no evidence of linear auto-correlation.

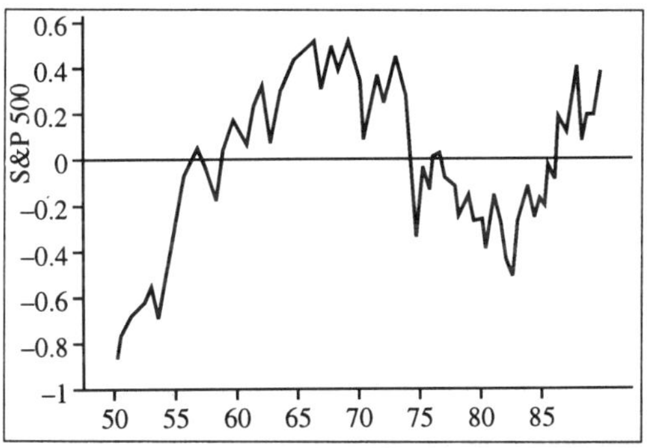

Figure 19.1 S&P 500 index from 1950 until 1988, detrended with CPI

But, as the Egypt tale just proved, the appearance of Gaussianity may hide important structures. If instead the rescaled range analysis is used, we find H=0.78: this is evidence for a Fractional Brownian motion. The series is generated by a black noise process, it is trend-reinforcing: after all, the conclusion is quite obvious, since the economy exhibits long memory. In the case of the

Nile, Hurst found H=0.72: if it was a white noise process (H=0.5), the probability of a year of high flood succeeding another year of high flood would have been just fifty per cent; instead, this event is more probable than the opposite and that was just what Hurst knew from his decades of observation and research on the available series. He had an explanation: complexity creates complexity. In our US historical series, a similar explanation is also obvious: depressions and expansions are not random events, but serially correlated and part of a structure. Again, that is complexity.

It can be argued that no proof is given, mostly if one standard well-established statistical procedure gives opposite results to those of one new brand of techniques. For the moment, just consider the economic explanation for the phenomena: interdependent events and complex relations are expectable after all. The whole idea of the evolution of the system being moved by random and independent shocks impinging on a marvellous device creating stability and convergence seems quite distant from reality. If just descriptive and theoretical reasons are invoked, the balance of evidence favours the complexity approach. Still, concrete statistical evidence is necessary.

19.2. Avalanches and other dangers

The results obtained from series of speculative prices are not the first exception to the rule of the Theorem. Many years before, Pareto had found that the distribution of wealth follows the normal curve, except for the upper 3 per cent of the population, which follow an inverse power law: a fat tail. Recently, Dosi and some colleagues found a stable Pareto distribution of sizes of firms (Dosi et al., 1994: 26-7). Many other observations confirm its pervasiveness: fat tails, again and again.

These pathologies were already studied a long time ago in statistics. Even without the necessary methods of nonlinear dynamics, both Maxwell and Poincaré touched at the heart of the problem: small causes produce large effects (Poincaré, 1908: 87) — that is the Noah effect, identified by Hurst and later on by Mandelbrot. In that case, complexity fails to deliver simplicity: the successive observations are not drawings from roulettes, but very concrete representations of sequential processes in social life. They are presumably not generated by independent processes and the small random shocks do not cancel each others. A structure is created: there is weather, but also a climate; there are crashes and booms, but there is an economy.

These were also the conjectures of Benoit Mandelbrot, an IBM scientist, when he studied in the early 1960s the series of prices of cotton at the Chicago market and some series of financial prices, where he found evidence against the Gaussian law: high speculative years followed more probably other speculative years and the roulette metaphor did not apply. In order to analyse these phenomena, Mandelbrot found and used the R/S method of Hurst: he confirmed deviations from normality. He was quite prepared for this conclusion: one of Mandelbrot's teachers, Paul Lévy, had argued long before that the normal distribution was just one of a family of 'stable distributions', from which we only know the density functions for the Gaussian, Cauchy and Lévy cases (Mirowski, 1989a: 462). Normality and the ruling of the Theorem can only be imposed under the condition of finite variance; if that is not the case and if long dependence exists, then the traditional procedures, OLS, spectral analysis, the ARIMA models, the maximum likehood estimation, all are suspect. The fact that they are nevertheless used is the ultimate proof of the allegiance to the omnipotent physical metaphor:

> The only reason for assuming continuity [instead of discontinuous and eventually non-Gaussian processes] is that many sciences tend, knowingly or not, to copy the procedures that prove successful in Newtonian physics ... . But prices are different: mechanics involves nothing comparable. (Mandelbrot, 1983: 335)

Hurst, Mandelbrot and many others detected this Noah effect: in the flow of the Nile, in traffic jams (Bak, 1996: 21 f.), in series of speculative prices, in economic and social evolution — fractional Brownian motion is everywhere in complex dissipative systems. In real series and in a new generation of models: following Bak and Chen, Scheinkman and Woodford studied some simulations of self-organized criticality, in which avalanches may be produced by small causes in unstable systems, and they found that the limiting size of the avalanches was a power law (Scheinkman and Woodford, 1994: 419). Those avalanches are in fact limiting cases of the structures created in long series: years of high floods follow years of high floods. That was just the same conclusion Mitchell had reached: when computing the normal curve for the percentages of changes in 5578 series of prices, Mitchell found a number of cases of fat tails (Mirowski, 1989c: 80). Fat tails, again and again.

19.3. General Bourbaki goes to battle

When he studied the raw materials' and speculative prices, Mandelbrot applied for the first time in economics the concept of fractal dimension. Previously in this book, his arguments for a second stage indeterminism were referred to: Mandelbrot's main point was that the first stage indeterminism, equating the study of economic trajectories to the stochastic behaviour learnt in the evolution of the particles of a gas, was irrelevant for most of social phenomena since depending on mild fluctuations and on Gaussianity (Mandelbrot, 1987: 118). In fact, economics moved from an Euclidean or Cartesian world of perfect order and geometry to the probabilistic approach and to the econometric program: that was first stage indeterminism, since extrinsic reasons prevented complete positive knowledge. Instead, Mandelbrot argued for a second stage indeterminism, considering nonlinear interactions and the emergence of complex phenomena, what today is called strange attractors, that is, intrinsic indeterminism.

When Mandelbrot's papers were published, they were the first important empirical inquiries to use Hurst's method and to criticize the well-founding of the general application of the Theorem in social phenomena. The implications of his research are summarized in the *Premature Fractal Manifesto* (1964, but only reproduced in Mandelbrot, 1987), where the author argues about the epistemological implications of the use of the physical metaphors — he was ignored. Economics was not prepared to abandon its dogmas, but still another reason intervened in order to create clouds of silence around Mandelbrot's work: by that time, the Cowles program had shifted away from the early econometric program, and endorsed an extreme form of Bourbakism, which avoided empirical analysis.

After the structural estimation project running into problems by the last years of the forties, Koopmans, Malinvaud, and most of all Debreu with his 1959 *Theory of Value* moved back to Cartesianism: in the new interpretation of the role of mathematics in economics, structures predominated over problems and theorems over empirical analysis. As Mandelbrot put it, this is the prototype of a 'fully self-referential structure', a 'top-down approach' which is alien to any concern about interpreting reality (Mandelbrot, 1989: 10-1). The program was originally developed as an ambitious project of a semi-secret group of French mathematicians, who took in 1934-35 the name of an obscure general of the previous century, Nicholas Bourbaki, to sign their collective textbooks on calculus, designed to reconstruct the science from a rigorous point of view. The work of this group was ruled by a rather bizarre contract: anyone reaching fifty years should abandon it. But it was quite disciplined and successful, and it became a driving force of the axiomatization of several sciences.

One of their disciples, Debreu, exported the program to the US and became a central figure in the transformation of econometrics and mathematical economics. The consequences were immense: 'The objective was no longer to interpret the economy, whatever that might mean, but rather to codify the very essence of that elusive entity, the Walrasian system' (Weintraub and Mirowski, 1994: 265) — under General Bourbaki and the logical positivist inspiration of the Cowles strategy, axiomatization became the new language for economics.

The emancipation of economics from the constraints of reality was indeed a necessary condition for the development of the general equilibrium paradigm, through extensive axiomatization. To the distress of Popperian epistemology, this also implied the rejection of refutationist strategies, as well as the abandonment of the previous efforts of identification, interpretation and explanation of real life phenomena. And, as the Panda Principle states, the following crises of the paradigm and the inability of its mentors to prove the simultaneous conditions of existence, uniqueness and stability of the postulated equilibrium were met with new waves of axiomatization and sophisticated Bourbakist developments.

It is therefore not surprising that Mandelbrot, according to his own words, left France because of the dominance of axiomatics in the academy or that he went back to Poincaré's work, the 'devil incarnate' for Bourbakism (Mandelbrot, 1989: 11), in order to understand the exceptional features of nonlinear dynamics and structural instability. It is even less surprising that he was ignored while doing so.

This mirrors one of the episodes of the last chapters: when Keynes polemicized with Tinbergen, the fundamental question — and the one most ignored by the subsequent debates — was the existence of evolutionary features dominating economic series, intrinsically creating heterogeneity and preventing the success of the methods defined by analogy to those of the laboratory experiments. In fact, both in the *Treatise on Probability* and in the multiple regression polemics, Keynes suggested the recourse to the descriptive method of Lexis, a distinguished member of the Historical School in Germany, in order to check the stability of the coefficients (TP, 434-5; xiv: 294); carefully enough, Tinbergen had proceeded to similar experiment, but he stuck to the hypothesis of the constancy of the parameters. Lexis's conclusion, on the contrary, was that the method of regression could not be applied in economics, given that it denied the evolutionary character of the phenomena and imposed a spurious approximation through the systematic part of the regression. This did not convince the statisticians: by the time of Lexis's critique, Edgeworth defended the generality of the normal law and of inference based on it (Mirowski, 1989c: 79). But this was not satisfactory to Keynes: as in his banana parable, small changes could ignite chaotic results, and the evidence of crises and structural unemployment of the thirties confirmed his mistrust of the alternative equilibrium accounts.

In the same critical vein of Mitchell, Mandelbrot recapitulated these critiques, like Keynes previously did. Their arguments about intrinsic complexity, irreducible to simplicity, are an impressive achievement by the current standards even if formulated in quite distinct terms. But they were buried until the very recent crisis in the neoclassical paradigm. It is now time to go back to the future.

20. The Crucial Dependency of the Answers on the Questions

In a paper published some years ago, John Blatt (1978) used a series generated by Hicks's nonlinear model of business cycles, defined by an explosive and deterministic oscillation and constrained by a ceiling and a floor representing the behaviour of full-employment resources and the level of investment. That series was then analysed by standard methods.

Blatt was able to prove that the estimated autocorrelations are indistinguishable from a stationary process, and to suggest that a scientist ignoring the original model could easily conclude that the series was generated by a linear structure with independently distributed exogenous shocks, that is, that a globally asymptotically stable system plus some exogenous shocks, small enough to be approximated by a linear specification, accurately represents the economic behaviour of that system. The author lately developed this conclusion into a general critique of the econometric methods (Blatt, 1983). That critique, substantially the same of the previous pages of this book, was not challenged by the mainstream econometricians and the experiment was even acknowledged by some of those scientists (Granger and Terasvirta, 1993: 23).

Blatt's point is very well taken, but in fact this 'billiard-ball' model is not satisfactory and was abandoned by Hicks himself shortly after its publication. The model does not generate growth, which is exogenously introduced by the growth rate of the boundaries, and discriminates by construction an asymmetric behaviour, the accelerator creating the expansion and the multiplier regulating the downswing. A more comprehensive approach is needed, and some recent developments in nonlinear dynamics may be of help. The same sort of experiment is here developed in a broader context, and the conclusion fully supports Blatt's findings.

20.1. A growth-cycle model

As in Blatt's paper, a nonlinear model is now used to generate an artificial time series which is then submitted to statistical tests. The model is that defined by Richard Goodwin in 1990, and uses a Rossler Band to control the chaotic structure and to assure dynamic stability. It is a five dimension model, which can be described by the following variables: rate of employment (v) above the permanent minimum v^*, rate of wages in the product (u), a control variable (z), the product (q), and a variable defining a logistic trend (k). It has the following form (in the current section, the subscript letters in the coefficients refer to each model):

$$dv/dt = -\alpha_{1A}\, u + \beta_{1A}\, v - \gamma_{1A}\, z \quad (20.1)$$
$$du/dt = \alpha_{2A}\, v \quad (20.2)$$
$$dz/dt = \phi_A + \alpha_{3A}\, z\, (v - \eta_A) \quad (20.3)$$
$$(dq/dt)/q = (dv/dt) / (v+v^*) + \alpha_{4A}\, dk/dt \quad (20.4)$$
$$dk/dt = \alpha_{5A}\, k\, (1 - \beta_{5A}\, k) \quad (20.5)$$

The model generates a growth cycle in the form of a Kondratiev wave, which is essentially determined by the logistic, with simultaneous Juglar cycles, in order to simulate the irregular behaviour of long economic series. It is also a very simple deterministic and nonlinear model, which is dynamically stable and structurally unstable, properties which presumably characterize the real economic systems.

In order to make possible a comparison, the parameters of the model were determined for this experiment[11] in order to generate a series close to the real data of the UK industrial production series (1801-1988, from Mitchell, 1992). The working of the model is represented in Figure 20.1, where the Rossler Band is drawn.[12]

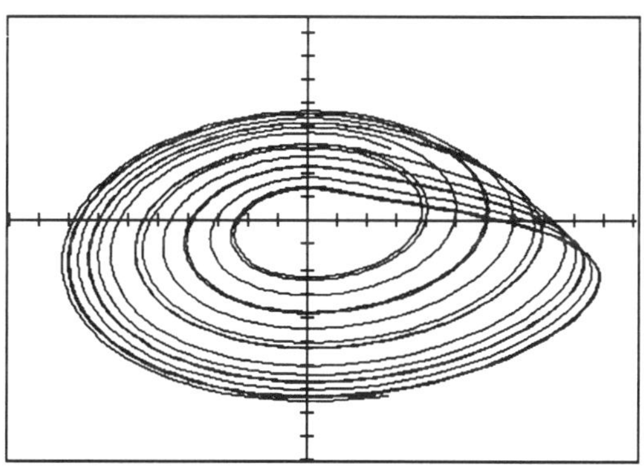

Figure 20.1 The Rossler Band in Goodwin's model

The simulated and the real series are compared in Figure 20.2.

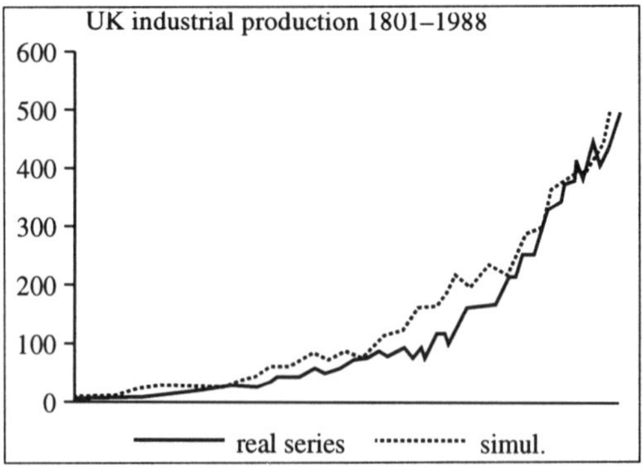

Figure 20.2 UK industrial production and simulated series

As indicated by the comparison in Figure 20.2, the simulated series acceptably represents the real data for the first, the second and the fourth 'Kondratievs', but it is far from it in the third 'Kondratiev'. Of course, the mathematical structure of the real series is not known, noise and multiple errors of measurement are also inevitable since it is an aggregated series and covers a long period with unequal quality of statistical patterns, but the simulated series is known by construction. In this simulated series there are of course no measurement errors.[13] Following the current methods, the series is detrended with an exponential curve and a linear auto-regressive process is fit to the residuals, that is:

$$ln\, x_t = \alpha_{1B} + \beta_{1B}\, t + ex_t \qquad (20.6)$$

$$ex_t = \gamma_B + \alpha_{2B}\, ex_{t-1} + \ldots + \beta_{s+1,B}\, ex_{t-s} + \delta_t \qquad (20.7)$$

The purpose of this model is to test the rocking-horse hypothesis of a linear structure of the stationary representation of the series, upon which exogenous random shocks (δ_t) are impinged. The statistical treatment of that historical time series is not simple, since both auto-correlation and heteroscedasticity are to be reasonably expected. The second problem will be discussed later on, but the first must be delt with right now.

The matrix of the regressors is stochastic since it includes lagged values of the dependent variable, and consequently the Durbin-Watson statistics will be biased upwards and cannot be confidently used as an indicator about linear auto-correlation. For that case, Durbin (1970) suggested a new statistics, which is computed from the regression:

$$\varepsilon x_t = a_{1C}\, \varepsilon x_{t-1} + \ldots + \beta_{s,C}\, ex_{t-s} + u_t \qquad (20.8)$$

with

$$u_t = \phi_C\, u_{t-1} + e_t \quad \text{and} \quad e_t \sim N(0,1) \qquad (20.9)$$

Under the null of ϕ_C=0, the statistics H is defined:

$$H = r \sqrt{[T/(1- T V(b_1)]} \qquad (20.10)$$

where T is the number of observations, $V(b_1)$ is the estimated sampling variance of the coefficient α_{1C}, and r is the estimate of ϕ_C from the OLS regression of $\hat{u}_t$ on $\hat{u}_{t-1}$, the $\hat{u}$s being the estimated residuals of the (20.8) regression. This test was used in the context of an AR(4) process fit to the simulated and the real series. The results, comparing the coefficients of determination and the standard DW statistics for each 'Kondratiev' period, are as follows:

Table 20.1 Auto-correlation in the simulated and real series

Simulated series	**R^2**	**DW**	**H***
I Kondratiev	0.98	1.9663	0.503
I+II Kondratiev	0.99	1.9014	0.187
I+II+III Kondratiev	0.99	1.9902	0.139
I+II+III+IV Kondratiev	0.99	2.0078	0.146
Real time series (T=188)	0.71	2.0167	0.292

* indicates the t-value for the coefficient: the H_0 of no auto-correlation is not rejected

The results indicate no evidence of a linear first-order auto-correlation in the simulated series or in the real series. The result is robust: the statistics for the four successive measurements of the simulated series confirm the same conclusion. The scientist using this type of model would presumably rest his case and note a new confirmation of the rocking horse: the structure of the horse is represented by the endogenous variables and their coefficients, the residuals are apparently well behaved.

Nevertheless we know that this is a gross misconception — at least for the experimental series — and that such a result is imposed by the theory and the statistical methods and is not derived from the data itself. Since for this case there are no random shocks, the origin of the apparent randomness must be intrinsic: indeed, it is the chaotic structure of the model. Of course, the inability of the traditional methods to deal with this series is due to the technical fragility and epistemological arbitrariness of the decomposition procedure.

The next sections will be used to criticize both premises of the method, the impulse system of the shocks and the linear system of the dampening propagation. Let's render therefore unto Caesar what is Caesar's and discuss the structure first: the argument is that linearity is not an useful and provisional simplification of the reality, it is indeed a wrong detour for science.

20.2. The nonlinear interactions and the complexity of real time series

A new hypothesis may be introduced now and stated in this form: in their inspection of the series, the previous methods hide the evidence of the pervasive nonlinear processes of interactions. Methods designed to look for simplicity cannot detect complexity: as Turing remarked, the justification of the assumption of linearity lies in the dubious claim that the topological patterns of a system are qualitatively similar when it is near the initial conditions and when it deviates from them (Turing, 1952: 66). Of course, the linear models can account for some forms of interaction between variables, but only for those additive factors and near-to-equilibrium situations. Reducing all other relations to such a status, linear econometrics plays tennis without a net. Alternatively, we are entitled to suspect that the real time series are dominated by changes of regimes, structurally unstable phenomena, creation of events (as represented by new variables or parameters) and complex interactions. Although the tools for this new approach are still in their infancy, some of them can illuminate our example.

20.3. The BDS statistics

In order to develop a statistical analysis of nonlinear and complex processes, one must deal with the problem of noise and non-stationarity, the notion of dimension and the concept of statistical inference. The second and the third problems are tentatively addressed by the procedure presented in this section.

In the theories of complexity, a crucial departure for time series analysis was the definition of the notion of dimension of the series $\{x_t\}$: the *correlation dimension*, as defined by Grasserberger and Procaccia, measures the geometric correlation of nearby points and namely indicates for trajectories generated at different initial conditions how close two points on those trajectories came to be in the phase space. This measurement may be viewed as indicating the minimum dimension or number of degrees of freedom which could generate the time series under inspection (Gabisch and Lorenz, 1989: 189, Brock, Hsieh and LeBaron, 1991: 2). The correlation dimension tends to a stationary and small value if the process is generated by a low-dimensional deterministic process, while it tends to infinite if the process is purely stochastic (Lorenz, 1989: 211).

The notion of dimension is intuitively perceived from the following illustration by Brock and Sayers (1988): a computer program generating random numbers, IID uniformly distributed in [0,1], is used to generate a series $\{x_t\}$. The distribution of the data in the interval is inspected in order to check if they fill it or if they cluster around some points. If the first case happens, then it is supposed that the dimension is >=1. If m-vectors (m-histories) generated by this program can also fill a m-cube, $[0,1]^m$, and do not cluster around some lower dimensional subset, then the dimension must be >=m. Of course, a truly random process should fill all the m-cubes for all m.

Brock, Dechert and Scheinkman developed in the late 80s a statistical procedure which is based on this notion of dimension and which deals with the questions of nonstationarity and nonlinearity. A m-geometric object, a *m-history* — that is, a possible subset $\{x_s,..., x_{s+m-1}\}$ from the original series —, is created in a m-dimensional space: it is proved that this reconstructed attractor is topologically equivalent to the object formed by the trajectory of the original dynamic system (Lorenz, 1989: 206). A correlation integral generated by the series $\{x_t, t = 1,..,T\}$ of T observations, is defined for $(t-s)$, the time distance between the elements of each pair of observations:

$$C(m,\varepsilon,T) = \#\{(t,s), \,||x(t,m) - x(s,m)|| < \varepsilon\}/ T^2 \tag{20.11}$$

where $x(t,m)=(x(t),..,x(t-m+1))$, $x(s,m)=(x(t),..,x(s-m+1))$, $||x(t,m)-x(s,m)|| = max\,|x(t,m)_i-x(s,m)_i|$, $i=1,..,m\}$ for a m-dimensional vector and $\#\{.\}$ indicates the number of elements (Brock, 1990: 432). In other words, for all possible m-histories $x(t,m)$ and $x(s,m)$, the correlation integral is that fraction of the pairs of data points (t,s) where the histories are closer than the radius ε. The measure of dimension of $\{x_t\}$ for the embedding value m which is used in this statistics is the limit of the correlation dimension, the elasticity of C with respect to :

$$d(\varepsilon,m,T) = dC(m,\varepsilon,T)/d\varepsilon \;\; \varepsilon/C(m,\varepsilon,T) \tag{20.12}$$

For a particular m-history $x(t,m)$, the correlation dimension measures the percentage increase in neighbours in the neighbourhood of when is increased by 1%. Brock and Dechert (1988) showed that if $\{x_t\}$ is IID, then it is true that:

$$ln\,[C(m,\varepsilon)] = m\; ln[C(1,\varepsilon)] \tag{20.13}$$

This is the basis for the BDS statistic:

$$W(m,\varepsilon,T) = T^{1/2}\,[C(m,\varepsilon,T) - C(1,\varepsilon,T)^m]\, /\, \sigma(m,\varepsilon,T) \tag{20.14}$$

where $\sigma(m,\varepsilon,T)$ is the consistently estimated standard deviation. The authors show that under the null that the series is generated by a linear system with IID innovations,[14] the distribution of W converges to $N(0,1)$ (Brock and Sayers, 1988: 80; Scheinkman and LeBaron, 1989: 216; Scheinkman, 1990: 111). Of course, if this is the case, W=0. The test is powerful against chaotic and other complex or nonlinear structures, although unable to discriminate them, and even for small samples the distribution is asymptotically approximated by the standard normal. It indicates the existence of hidden structures, that is, the presence of non-stationarity or of non-linearity in the residuals of the estimated models.[15] Of course, the case of the existence of non-linearity is the most relevant for the present purposes, and this is why the data are detrended and then linearly filtered as in the previous example in order to get stationarity. This is the non-parametric test used in the remaining parts of the chapter.

20.4. The UK industrial production and the simulated series, again

This test is now applied to the previous series, the 1801-1988 UK industrial production series and the simulation from the Goodwin model. The requirement of stationarity for both series, which is a pre-condition for the statistical test. is checked with Kendall's ι statistics for the null of stationarity in the mean: it is accepted that both series are stationary after detrending and linear filtering.

Since the series are very short, and one cannot be sure that the asymptotic approximation is valid, a method inspired in the *bootstrapping* (Efron, 1979: 17 f.; Freedman and Peters, 1984: 98 f.) was suggested by the authors of the statistic (Brock et al., 1991: 132): the residuals are randomly redistributed, and the statistic is computed so that the new values are compared to the original ones. The results of the tests are indicated in Tables 20.2 and 20.3, including between parenthesis the proportion of the bootstrapped values for the statistics whose absolute values surpass the BDS statistics for the original series:

Table 20.2 BDS statistics

Residuals of AR(4) of UK industrial production 1801-1988

ε/σ		**0.5**	**1**	**1.5**	**2**
m	**2**	10.32* (0.00)	6.87* (0.00)	5.84* (0.00)	5.27* (0.00)
	3	11.33* (0.00)	6.47* (0.00)	5.12* (0.00)	4.50* (0.00)
	4	11.68* (0.00)	6.04* (0.00)	4.93* (0.00)	4.16* (0.00)
	5	13.36* (0.00)	5.57* (0.00)	4.75* (0.00)	4.04* (0.00)

M, the embedding dimension, and ε, the given distance, are as specified. As there is no theory for the choice of both parameters (Brock, Hsieh and LeBaron, 1991: 169), a wide range of the admissible values is indicated in the tables.

Table 20.3 *BDS statistics*

Residual of AR(4) of the simulated series

ε/σ		0.5	1	1.5	2
m	2	12.81* (0.00)	7.28* (0.00)	4.52* (0.00)	3.44* (0.00)
	3	14.38* (0.00)	6.59* (0.00)	3.58* (0.00)	2.64* (0.00)
	4	17.60* (0.00)	6.16* (0.00)	2.81* (0.00)	1.94* (0.00)
	5	21.61* (0.00)	6.22* (0.00)	2.56* (0.00)	1.74* (0.00)

The values between parentheses indicate the proportion of values of the statistics for the bootstrapped series that are superior to [-*W*,*W*]; the bootstrap procedure was repeated 100 times; * indicates that the statistics is significant at the 5 per cent level and that the H_0 of IID is rejected

As the table indicates, for the relevant values of *m* and ε/σ the null of IID is always rejected. This supports the previous conclusions by Blatt: evidence of linearity is introduced by the traditional econometric methods, in spite of being absent from the data.

These are, of course, just conjectures and not definitive proofs, for several major reasons which must be pointed out. The first is that the original series is rather imprecise: for the first period, the quality of the UK statistical information is rather poor, and spurious conclusions can be introduced by errors of measurement. The second reason for calling for prudence about the interpretation of these results is however even more substantial: the applicability of the Grasserberger-Procaccia concept of correlation dimension is questionable. Ruelle (1990: 244-5) and Sugihara and May (1990: 741) share a clear lack of confidence in a direct estimation of the Grasserberger-Procaccia measure of low dimension attractors in so short series, namely if this low dimension is at odds with the theory.

In this sense, four major critiques have been addressed to the BDS statistics. Mirowski pointed out that the Grasserberger-Procaccia concept was transformed into something else, since the authors of the statistics 'ignore the usual interpretation of the correlation integral as an estimate of the exponent gauging the self-similarity of a geometrical object, and instead reinterpret it as analogous to a characteristic function in the case of independent and identically distributed random variables' (Mirowski, 1995: 593). Bausor argued somewhat differently that the BDS statistics are unable to interpret bifurcations, since all the data is presumed to 'arise from the same side of a structural instability' and 'in particular, they presume that the data follow one trajectory along a given manifold generated by an unknown process whose parameters are fixed' (Bausor, 1994: 120). Third, Mirowski criticized the assumption of the Neyman-Pearson strategy of tests of hypothesis, given its deficient epistemological foundations, already discussed in this book.

Although the last point cannot be by-passed, for the current purposes of this demonstration the acceptance of these critiques is not particularly damaging: the concept of a self-similarity of cyclical processes is doubtful and only interpretable in a very generic and qualitative way, and moreover the statistics were used essentially as an internal critique — it is because the results deny the methodology of linearity and equilibrium, the antinomy contingency-necessity and the procedures based on the estimation of fixed parameters, in spite of sharing a large part of the orthodox epistemology including the Neyman-Pearson formalism, that the method can be successfully used to reject the rocking horse and to highlight its logical and technical shortcomings.

Fourth, Mirowski argued that

> It would be better to retain the BDS statistics as a test for asymptotic absence of structure in a stationary ergodic situation in face of a non-specified alternative; but this would have highlighted the fact that it was rooted in a narrow hypothesis of randomness, and in fact had little or nothing to do with chaos. (ibid.: 593)

In the framework of this last argument, the BDS statistics were used in this book just to prove the error of the propagation and equilibrating plus non-correlated random shocks with no economic substance. No effort was made and none could be made in order to determine or to make assumptions about the precise trajectory of the system — but instead the nature of the motion and the general structure of the flow was discussed. The statistics were and will be used as a tool for identification of critical points in the evolutionary process, in order to elaborate conjectures about the turbulent interaction of political, economic and technological factors. This is why some precautions were adopted, namely resampling the data and checking the statistics for that case, as well as comparing the results with those obtained with the Hurst process.

With all these limitations, a provisional conclusion is possible, which is known to be accurate for the case of the simulated series, for which the BDS test gives the correct result. As in Blatt's example, the evidence of independence and of stationarity of the residuals after detrending and linear filtering might lead the traditional statistician to support the rocking-horse model, and he or she would be wrong. But it can be stated that neither of the series could have been generated by the linear auto-regressive process: the conjecture about the widespread character of nonlinearity in economic series is hereby supported.

21. The Aggregative Effects of Aggregation in Economic Time Series

The previous chapter argued for a new approach to the analysis of economic time series, on the basis of some evidence of nonlinearities in a real case and in a simulation. Nevertheless, it is well known that the cases for which that proof was presented are rather the exception than the rule. The most common examples for which nonlinearity or chaos have been proved are series of speculative prices of the stock market, with severe oscillations.

One of the reasons is that the adequate series are only available in those cases: the stock market, foreign exchange and financial daily series may be quite long, and in fact many observations are needed for the tests on nonlinear and complex structures; furthermore, the measurement of prices is exact. Comparatively, the best GDP or GNP series — and for a very restricted number of countries — have less than 200 yearly observations, some of them based on interpolation and guesswork, since the national systems of statistics are essentially a feature of this century. As a consequence, these series are noisy, incomplete and imprecise. But, a priori, the same possible sources of complexity can be accepted in real series as well as in financial series: the markets are formed by multiple agents, gifted with the capacity of choice and the determination of interdependent strategies and, furthermore, technology and the social institutions change frequently the settings of the system; these multiple interrelated oscillators endowed with free-will eventually create turbulence.

Under the indicated restrictions, the BDS statistics were used in order to study some macroeconomic series from different countries (United Kingdom, United States, Japan, Sweden). The original series were published by Angus Maddison (1991, 1993), and the previous series by Mitchell is also used. The series describe the GDP (at real prices and 1985 US dollars) and the volume of exports, investment and industrial production for periods between 120 (GDP, 1870-1989) and 212 years (Investment, 1780-1991).[16] The basis for all the series is the national 1913 level, so that a comparison of the performances of the different variables is easy from each graph; but they only make possible a relative comparison between countries, since the national levels of 1913 are obviously quite distinct. The periods of the World Wars are marked in each graph (values are in natural logs; abciss is time, 1 to 130):

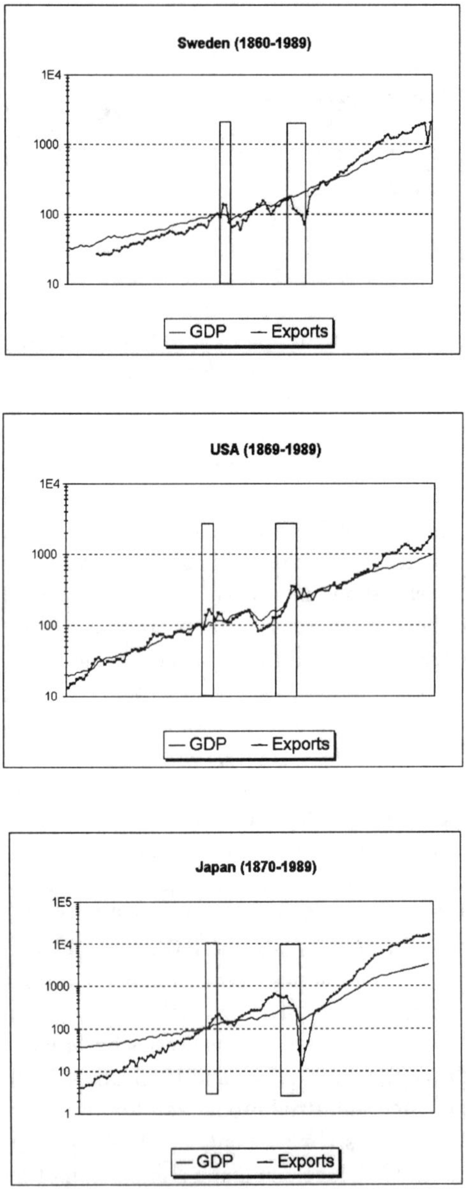

Figure 21.1 GDP and volume of Exports of Sweden, USA, Japan

There is a widespread *prima facie* appearance of nonlinearity in these series — note the different effects of the wars and the evidence for severe switches of regimes — which is undetectable by traditional statistical methods. The main hypothesis to be tested can be stated now: the aggregation procedure used in the national statistic systems is one of the major factors responsible for hiding the intrinsic nonlinearity and for the confirmation of the linear representation. This is probably not the single cause, but it is a crucial and widespread reason for the illusion of linearity.

Several authors were aware of this danger of a general statistical misspecification from aggregation. Gelderen argued that detailed statistical information is essential in order to assess historical processes of change, which are desynchronous:

> it will be necessary not to treat all commodities as a homogeneous mass, but to decompose the general price level into its components in order to distinguish a few large commodity-classes, each displaying their own individual price components. ... If, along this line of thinking, we consider the different stages of the development of prices as appearances of the different phases of capitalist development, then knowledge of the dearness will imply also knowledge about the core of capitalism. (Gelderen, 1913: 2)

Schumpeter was also aware of this problem and criticized the aggregation method:

> [aggregation] keeps the analysis on the surface of things and prevents it from penetrating into the industrial processes below, which are what really matters. It invites a mechanistic and formalistic treatment of a few isolated contour lines and attributes to aggregates a life of their own and a causal significance that they do not possess. (Schumpeter, BC: 44; also 463)

But Schumpeter's warnings, as well as those of Gelderen, were not taken seriously, given that the problem is not relevant in the universe of linear models, since they assume additive superimposition and consequently the decomposability of causes.

There are nevertheless very good reasons to consider the argument for desegregation. In 1957 Koopmans argued that econometrics could only progress under the condition of using detailed desegregated information, given his 'increasing awareness of nonlinearities' (Koopmans, 1957: 214), which was shared by Tinbergen (Epstein, 1987: 34), by Marschak (ibid.: 68) as well as by Frisch. Maddison discussed evidence about a smooth evolution of the capital stock in several countries, in spite of the obvious technological breakthroughs, and explained it by the combination of different generations of assets, so that even a sharp change in the current generation would have only a slight effect on the total (Maddison, 1989: 21). The same author compared the evidence of business cycles in the aggregate product of the countries and in the individual series, concluding that aggregation, since it includes compensatory movements, may hide the cycles (Maddison, 1977: 109-10). Freeman denounced the erroneous conclusions from the aggregation procedure when dealing with economic innovation (Freeman, 1987: 302-3). Goodwin argued on several occasions that aggregation 'usually masks the reality' (Goodwin, 1990: 7; also 1984: 67). Of course, some of the paradoxes of aggregation or fallacies of composition are well known in economics, such as Marx's argument about the typical behaviour of each individual capitalist, who does not expects his decisions on output and employment to affect his own demand, although the aggregate behaviour of all capitalists naturally does so, or Keynes's argument about the same relation between isolated entrepreneurial decisions and the employment level.

In fact, one may assume that the aggregation procedure which is the norm for measurements and economic studies affects the nature of information. This is to be proved; yet, it is a consistent idea: it is intuitive that the aggregation of different types of information in a single cocktail can produce indigestible results. Moreover, distinct researchers on business cycles stressed the importance of leading and lagging series: their combination in the same spectrum of time may hide and deform the nature of these relationships. On the other hand, the inspection of the historical aggregate must consider the essential fact of structural evolution and the creation of new industries: the continuity of the aggregate hides the discontinuity of the components, and the statistical sequence hides its own change. So, there are substantial theoretical reasons to look for evidence of the aggregative effect under the form of compensatory movements in historical economic series.

The proof for a similar reasoning was recently provided in zoology and epidemiology, where this aggregation bias was detected by some scientists applying the new tools of the theories of complexity. In a survey about childhood diseases in developed countries, Sugihara and May (1990: 739) found evidence of low-dimensional chaos in the measures of incidence of measles in British cities, but no such evidence could be found in the aggregate series for the country. Brock et al., when developing the BDS statistics, indicated the danger of aggregation (Brock et al., 1991: 187-8) and even that it could create spurious evidence of linearity (ibid.: 193), although they did not provide any proof.

A similar problem is now found in the UK series under scrutiny. The previously stated limitation of data severely constrains the generality of the proof, but the available series show nevertheless a very impressive feature. As previously, the GDP, volume of exports, investment in equipment and investment in non-residential structures series are detrended, and an AR is fit to the residuals. The results of this preliminary procedure are summarized in the following table:

Table 21.1 *AR processes on UK residuals from detrending*

Series/results	R2	D-W	Durbin's H
GDP (AR4)	0.91	2.041	-1.33
Vol. Exports (AR3)	0.92	1.970	-0.85
Ind. Prod. (AR4)	0.71	2.017	-1.23
Invest. Equipment (AR4)	0.68	1.99	0.12
Inv. in non-res. str. (AR4)	0.92	2.00	-0.32

NB: linear first-order auto-correlation is always rejected

Of course, all these series represent aggregates, and very large ones. The same procedure should be also taken for the cases of concrete production, investment, accumulation or export series in the cases of individual industries, but that historic data are not available for long periods. In spite of those limitations, the available series still present an obvious interesting feature: the GDP includes the value of exports, which is a function of its volume, and includes the industrial production as well as investment. Comparing the BDS statistics for these series, the test suggests that there is some strong evidence of nonlinearities in the volume of exports, in investment and in industrial production, while the null hypothesis of IID for the GDP could not be rejected.

The BDS statistics are now summarized in Table 21.2; the statistics for the industrial production series were previously presented. As the table indicates, the statistics always reject the null of IID in the cases of exports and industrial production, and do not reject it in the case of the residuals of the GDP for these countries. This is therefore an important piece of evidence for the hypothesis that aggregation at the level of the national product hides the nonlinearity of at least four of its components, for the period and the measurement considered. These results are the same as for the cases of Japan and Sweden, for the GDP and exports.

As the BDS is a portmanteau test, it does not give any indication about the hidden structure, and the alternative hypothesis is not specified. As a consequence, it works as a refutation tool, but does not contribute to the direct formulation of a new model. But a new point of departure is settled: the inquiry must account for the nonlinear structures which were detected and may be pervasive in real life economics.

Table 21.2 BDS statistics for UK historical data

UK GDP 1860-1989					
ε/σ		0.5	1	1.5	2
m	2	1.99	2.02*	1.94	1.18
		(0.08)	(0.04)	(0.06)	(0.34)
	4	1.74	0.84	1.19	0.85
		(0.14)	(0.46)	(0.30)	(0.44)
UK volume of exports 1870-1989					
ε/σ		0.5	1	1.5	2
m	2	5.92*	6.15*	4.99*	4.12*
		(0.00)	(0.00)	(0.00)	(0.00)
	4	6.60*	8.17*	7.59*	6.62*
		(0.00)	(0.00)	(0.00)	(0.00)
UK investment in equipment and machinery 1800-1991					
ε/σ		0.5	1	1.5	2
m	2	5.28*	6.73*	7.10*	6.46*
		(0.00)	(0.00)	(0.00)	(0.00)
	4	11.69*	7.04*	7.05*	6.35*
		(0.00)	(0.00)	(0.00)	(0.00)
UK investment in non-residential structures 1780-1991					
ε/σ		0.5	1	1.5	2
m	2	7.20*	7.08*	7.32*	7.31*
		(0.00)	(0.00)	(0.00)	(0.00)
	4	13.38*	8.40*	7.31*	6.79*
		(0.00)	(0.00)	(0.00)	(0.00)

* indicates that the statistic is significative at the 5 per cent level; values between parentheses indicate the proportion of values of the statistics for the bootstrapped series that are outside the interval [-*W*, *W*]

22. The Problem of Heteroscedasticity

The UK data are the most complete and, presumably, the most exact of all the available series. Furthermore, until now five different series were considered: 1860-1989 GDP and 1870-1989 Volume of Exports (from Maddison, 1991), 1780-1991 Investment in Non-residential Structures, 1800-1991 Investment in Equipment and Machinery (Maddison, 1993) and 1801-1988 index of Industrial Production (from Mitchell, 1992). After detrending and linear filtering, the GDP series indicates no evidence of nonlinearity, but the analysis of the major components leads to the opposite conclusion.

There is a possible economic explanation for the different behaviour of the GDP and export or industrial production series if the historical periods are concretely examined: the export series exhibits a much stronger impact of the world political events, namely the World Wars, of periods of international confrontation or of economic crisis and eventually of major technological changes. On the other hand, the series of industrial production indicates a more dynamic behaviour than that of the GDP. As a consequence, the changes in variance are more pronounced, and they can account for the rejection of the null: the presence of heteroscedasticity may indeed implicate the rejection of the hypothesis of linearity (Brock et al., 1991: 75; 139, 170). This is described in Figure 22.1, where the two world wars are indicated as previously.

The standard technique to deal with this suspicion is to test for heteroscedasticity, and if its presence is asserted to use for instance the Aitken estimator or other procedure exterminating its causes. In order to check this hypothesis and the effect of these techniques for the UK case, the Goldfeld-Quandt statistics were computed under a null of no change in variance, the alternative hypothesis stating that the variance of the later period is superior to the former's. The first periods to be considered were 1860(1870)-1913 versus 1914-1989. As expected, the null was indeed rejected both for exports and for industrial production, either for the residuals of the detrended series and for the residuals of the autoregressive processes, for both series. If the periods are instead 1860(1870)-1938 versus 1939-89, then the null is accepted, and no significative increase in variance is detected between the two periods.

These results suggest a closer look at the main periods of mutation in our

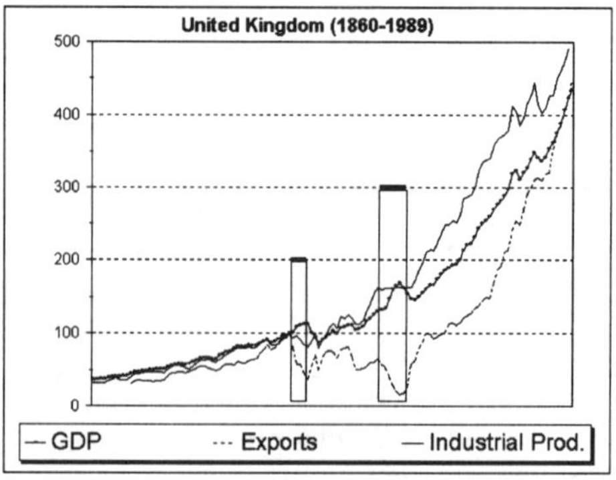

Figure 22.1 UK GDP, exports and industrial production (1860-1989)

series: consequently a third division was tested, considering now successively the years 1860(1870)-1913, 1914-1945 and 1946-1989. For both indexes and for both types of series, the null was rejected for the comparison between the two initial periods and not rejected for the comparison between the second and the third, as the following table summarizes:

Table 22.1 Goldfeld-Quandt statistics

Periods/series	Detrended Exports	Detrended Industrial Production	Lin. filtered detrended Exports	Lin.filtered detrended Ind.Production
1860/70-1913 vs. 1914-1989	36.777*	2.377*	13.146*	2.743*
1860/70-1939 vs.1940-1989	0.031	0.793	0.068	0.224
1860/70-1913 vs. 1914-1945	22.837*	3.027*	21.258*	4.721*
1914-1945 vs. 1945-1989	0.048	0.073	0.208	0.198

* indicates that the statistic is significant at 5 per cent level

These results suggest that a new pattern of evolution was created with the turning point of the 1910s, the period of the First World War, the maturation of the technological revolution of electricity and major changes in the world leadership, including the secondary role assumed by the UK from that period on.

Considering this evidence, several distinct procedures can be followed.

Testing the US real per capita GNP series (1872-1988), Scheinkman and LeBaron used both OLS and GLS for their regressions, but could not reject nonlinearity in any of the cases. The authors then suggested a different approach, the introduction of dummy variables for the periods supposed to create most of the nonlinearity: the Great Depression (1930-1939) and the Second World War (1939-1945). This procedure eliminated the evidence of nonlinearity in the series they were inspecting.

Both methods have severe deficiencies and introduce strong theoretical assumptions and implications that cannot be ignored. The GLS procedure implies that distinct periods are considered and, as a consequence, are treated as separate entities under the appearance of a single series. The method and the series which are finally explained by the model are incoherent, since the original AR process implicitly looks for a sequential explanation, but this is of course abandoned by the procedure. In other words, the technique is correct and necessary so that a least squares regression is possible, but it is inadequate to the current problem since the enigma remains unsolved, as the variation of variance is a priori annulled from the series and as a consequence it cannot be explained.

The second procedure also implies a heroic assumption about the nature of those impacts, which is quite common in cycle analysis: several authors dealt with this type of problem introducing dummy variables or simply eliminating the period of 'excessive' variation, through interpolation. Both versions are now used to test the robustness of the previous results.

The method of interpolation was successively used for the exports series, the one evidencing a larger historical mutation. First, the values for the years of the World Wars were replaced by those computed from a linear interpolation, and the new series are described by the following figure:

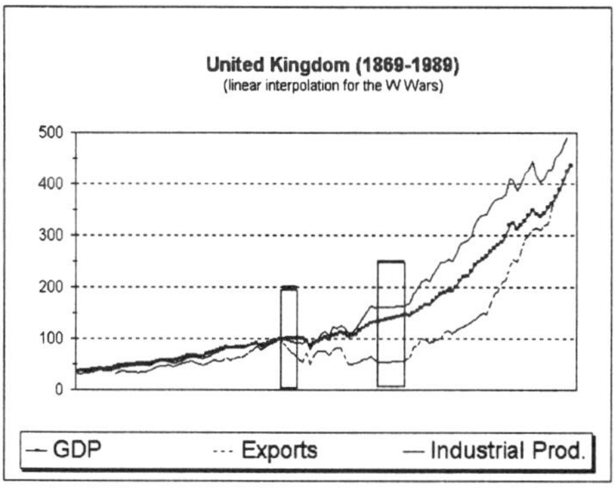

Figure 22.2 First scenario of interpolation for the War years

Second, the interpolation was extended to the years 1914-21 and 1939-47, considering Maddison's evidence about the prolongation of the economic depressions in the UK during those years. Third, the years considered were extended to account also for the Great Crisis of 1929-33. Then, the interpolation was prolonged from 1939 until 1953 (when the UK product reached the peak level of the war). For the four scenarios, the BDS statistics were computed for the series, and still rejected linearity for the first three of them, accepting the null of IID for the fourth one (with the series of industrial production, such a result was attained with the third scenario). The statistics are as following:

Table 22.2 BDS statistics for the interpolated UK export series

Scenario (years of interpolation)	(1) (W. War years)	(2) (1914-21, 1939-47)	(3) ((2) + 1930's crisis)	(4) (1914-21; 1929-33; 1939-53)
BDS value ($\varepsilon = 1.5\sigma$, m=2)	5.53*	11.49*	4.68*	1.86

* indicates that there is a prima facie evidence that the statistic is significant at 5 per cent level

A second procedure is the introduction of dummy variables, a technique used by Scheinkman and LeBaron. The result in this case is even more conclusive: in the three series, dummy variables were introduced for the periods of the World Wars, and then the procedure was extended for the same years as the last scenario of the interpolation procedure (for the years 1914-1921, 1929-1933, 1939-1953). For both cases, the conclusions remain unchanged: the BDS statistics for GDP do not reject IID and the BDS for exports and industrial production very clearly rejects the null. The results are robust: even under the fire of powerful punitive methods, the analysis of the series indicates evidence of complex nonlinear structures.

23. Reconsiderations on the Rocking Horse: The Vanishing Property of the Errors

So far, the equilibrating structure was discussed. It was noted that it cannot be isolated, given that in order to prove its properties the traditional methods are forced to add new assumptions about the supposedly independent impulse system: the nature of the shocks becomes the condition for inference and for the theory. This is the theme of the current chapter.

23.1. Errors of errors

The weighting statistics, the dummy variables and the interpolation procedures were unable to deliver the tranquilizing confirmation of a stable structure under the stress of well-behaved random shocks. They are technically inadequate in this case, but they are also doubtful on the theoretical level.

The first problem is a methodological one: the procedures are incoherent with the model which is assumed in cycle theory. Indeed, they eliminate the major shocks and its economic influences, in the context of an inquiry designed to study the cycles as created by exogenous shocks impinging on the linear structure. In other words, there are only two possible strategies to contest the results presented so far: either the traditional econometrician argues that the evidence of nonlinearity is an artifact from heteroscedasticity — and consequently eliminates the source of the variance, eliminating as well the major shocks and consequently her or his own cycles — or she or he argues that those shocks are indeed the driving forces of the cycle, and therefore must accept their presence and the implication that they may be responsible for or symptoms of nonlinearity. The difference between both strategies is not a minor one: in the first case, one can say that heteroscedasticity is a statistical nuisance to be corrected, whereas in the second the implication is that it is the relevant fact for the inquiry. In both cases, the traditional theory collapses.

The second problem concerns the exact nature of these 'shocks'. In order to eliminate heteroscedasticity from the exports series with the interpolation procedure, it was necessary to eliminate two different types of shocks, the first ones being those related to the World Wars, and the second ones to the prolonged periods of their economic impacts as well as the Great Crisis of the 1930s. In this case, the consequences are immense and the very purpose of the inquiry is questioned, since there is no economic rationale for eliminating the economic crises in order to explain the cycle. That strategy is self-defeating: the evidence of nonlinearity can only be eliminated if we abandon any economic theory at all — and the inquiry is reduced to the manipulation of a simple statistical artifact, deserting any attempt to produce realist explanations of real events.

Third, the nature of the wars and their economic impact must be carefully evaluated. Economic theory normally considers the wars to be exogenous processes defined in the political sphere, and many researchers feel quite at ease when they eliminate these 'outliers' from their series. Again, the paradox is quite difficult for the mainstream paradigm. Either those shocks are considered to be strictly exogenous, and so they are the best candidates for the causal force creating cycles and consequently cannot be ignored by the statistician, or they are internally caused by the economic structure itself: both ways, the shocks cannot be eliminated. Either case implies that those periods and factors cannot be ignored by statistics. Indeed, there is a major argument in favour of their consideration: simply, the wars effectively happened and, whatever the implicit historical theory of the scientist, the research should consider these periods since they are part of the reality, as a cause or as a consequence of the economic endeavour.

The series under consideration show the evident influence of those big exogenous impacts, but also indicate something else: it is obvious that there is a long decline in the UK since the outbreak of the First Wolrd War, and that for at least forty years the exports lagged in relation to the GDP or industrial production. The level of the volume of exports of 1912 was only overtaken in 1955, and still for some twenty years the evolution of the series did not reach the levels of growth of the national or the industrial product. The main factor for the detected behaviour is therefore a long tendency of decline in international competitivity for the British economy as a whole, in which war impacts, international leadership evolution and economic performance are mixed as causal relations. And we may guess that these historical facts are indeed the major cause for the evidence of nonlinearity.

There is a possible strategic recourse for the econometric standard argument, and indeed it is implied by some scholars, based on the distinction between two different types of exogenous shocks, namely between those which are considered as non-economic, rare, very strong and to be despised, and those which are supposed to be truly random and small. The log-linear macroeconometric models are generally based on the notion of a stable system affected by these small random shocks, and other more significative impacts are excluded from the analysis. This may be so for toy-models, but definitely cannot be accepted for the analysis of real time series.

23.2. All that is solid melts into air

One major piece in the building up of the econometric program was the seminal 1927 article by Slutsky,[17] which did not present a general theory of the cycle but suggested an important conjecture: 'it is possible that a definite structure of a connection between random fluctuations could form into a system of more or less regular waves' (Slutsky, 1937: 106).

Slutsky used series from the Russian lottery, and computed a moving average in order to prove that 'the summation of random causes may be the source of cyclic, or undulatory processes' with 'quasi-strict periodicity' (ibid.: 114, 120). This is a very powerful result: Slutsky proved that the cycles could be mere statistical artifacts imposed by the computation of moving averages. Alternatively, the results can be interpreted in a distinct way: if the propagation mechanism is considered to be a real feature of real systems, the random shocks can create cycles, provided that propagation part is designed to operate in order to combine, to sum and to average their effects. This interpretation was at the core of Frisch's theory of the cycle, in which the random shocks were defined as 'the source of energy in maintaining oscillations' (Frisch, 1933: 197), impacting on an economic structure modelled as a system of mixed linear differential and difference equations with complex conjugate roots. This provided a global theory for the cycle:

> The most important features of the free oscillation ['the majority of the economic oscillations'] is that the length of the cycles and the tendency towards dampening are determined by the intrinsic structure of the swinging system, while the intensity (the amplitude) of the fluctuations is determined primarily by the exterior impulse. An important consequence of this is that a more or less regular fluctuation may be produced by a cause which operates irregularly. (ibid.: 171)

This is, as the reader knows, the rocking horse.

The crucial point of the Slutsky paper was the conclusion about the effect of the random shocks: this hypothesis was discussed here under the form of the null of the generation of the process by a linear system with independent and identically distributed random shocks, and was rejected in several instances. It should be once again noted that a linear system can only mimic the behaviour of a real series if exogenous stochastic shocks are added in order to create the cycles.

But, for one of the cases in which the IID hypothesis has not been rejected, UK GDP (1860-1989), the moving average of the residuals of the series seems to substantiate the Slutsky effect. The following graphs show some evidence of cycles in the pattern of evolution of the residuals if a moving average of 10 years is computed:

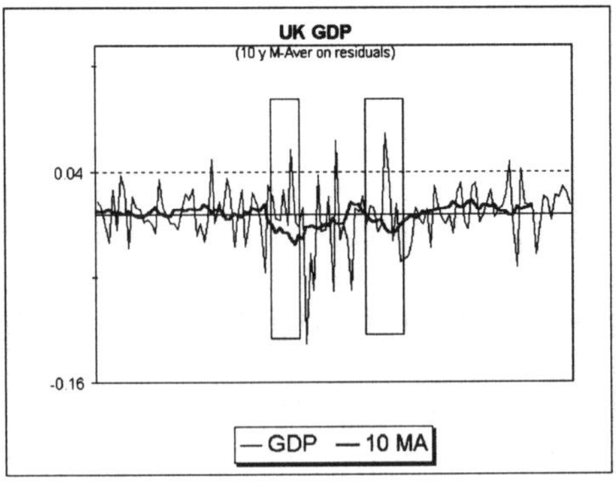

Figure 23.1 *A Slutsky effect on residual of UK GDP? (residuals of an AR process on GDP, and a 10 years moving average on the same residuals)*

Indeed, this seems to support the argument that the summation of random variables (what one could argue to be the case of the GDP residuals) creates the *artifact* — in a strict interpretation of Slutsky's article — or the new *fact* — if one follows Frisch's interpretation, of the cycles. Moreover, against Kondratiev, this graph suggests that a moving-average procedure may create the illusion of long cycles since three patterns of large waves become evident. The next graph follows closely Kondratiev's method for the treatment of the series, and shows some evidence of three long cycles in the UK GDP, which are even more striking than the previous ones (as previously, the two World War periods are indicated; the abciss is time, 1860-1989):

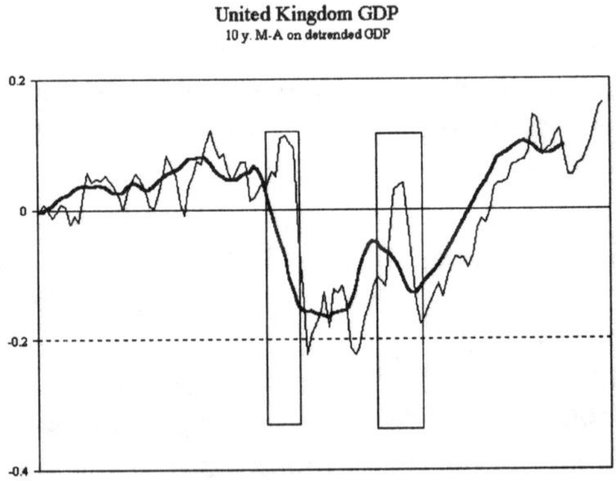

Figure 23.2 *Long cycles in UK GDP: a Kondratiev artifact? (detrended GDP and a 10 years moving average on the residuals)*

If the residuals are indeed IID, this evidence of long cycles may be legitimately interpreted as an illusion caused by the summation procedure, which is more or less highlighted according to the filtering procedure adopted. If Slutsky was right, then the cycles may indeed be spurious effects of the random shocks on the propagation system. Nevertheless, all these conclusions strategically depend on the basic assumptions about the independent nature of the exogenous shocks on a linearly structured propagation system. Now, as shown by the earlier discussion on aggregation, as far as the current case is concerned the appearance of the IID distribution of the residuals of GDP is itself an artifact, since at least several of its main components are presumably generated by nonlinear processes. This nonlinearity and the presence of a structure in the residuals, which as a consequence are not IID, is hidden by the overall aggregate index of the GDP: the multiplicity of structures obscures the existence of a structure. This is very important, since it falsifies the whole conclusion: it is because of complexity that simplicity is wrongly assumed.

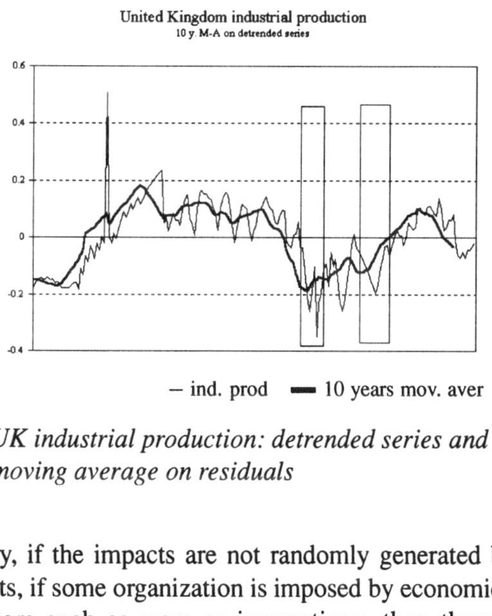

Figure 23.3 UK industrial production: detrended series and a 10 years moving average on residuals

On the contrary, if the impacts are not randomly generated but clustered in certain moments, if some organization is imposed by economic or non-strictly economic factors such as wars or innovations, then there is a plausible explanation to each switch of regimes.

A final point to be made is about the comparison between these results and those of the simulated series, whose structure is known by definition. The moving-average processes on the detrended series and on the detrended-filtered series indicate both the existence of four Kondratiev waves, which of course we know to exist by construction. The single important difference of this

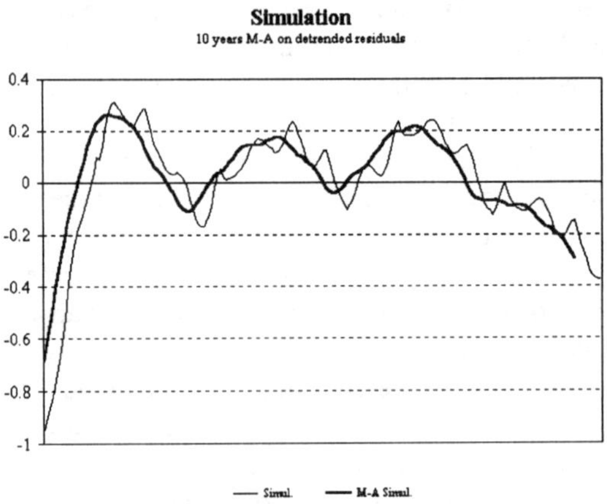

Figure 23.4 *Results of detrending the simulated series and applying a 10 years moving average on the obtained residuals*

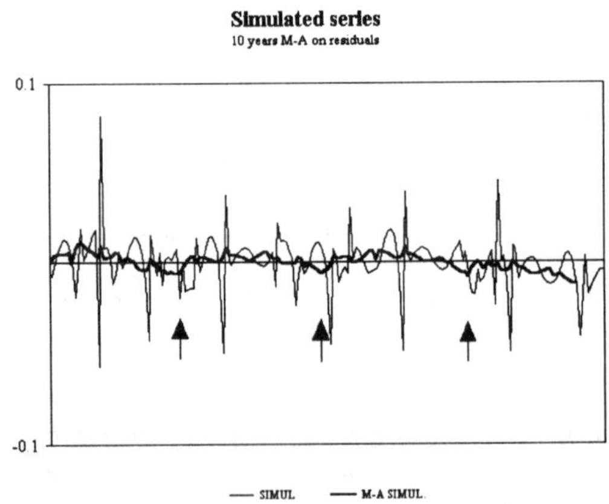

Note: the turning points are indicated by arrows

Figure 23.5 *Results of an AR process in the simulated series and a 10 years moving average on the residuals*

representation in relation to the original series is that the fourth Kondratiev is scarcely detected, as a consequence of the shortness of the information available for that period. The turning points are clearly marked in the second graph by strong oscillations, which indicate a change of regime.

These effects are created in the series which is known to be generated endogenously by a nonlinear deterministic system, that is, there are no random shocks at all. This proves that the Slutsky effect does not prove any sort of causality: a low-dimensional chaotic system, as the Goodwin growth-cycle model here used, can also generate a series simulating randomness and whose moving average will exhibit the same effect, even if the residuals of a linear filtering are not IID and in fact indicate the presence of a hidden nonlinear structure. In other words, the Slutsky is true, but not general; it is a sufficient, but not a necessary condition for cycles. Moreover, some series which are considered by the standard techniques as randomly generated may result from an effect of aggregation hiding their true nonlinear nature. These series may also simulate a Slutsky effect which is really meaningless: it is no more than the illusion of an illusion.

24. Conclusion of Part Four: The Snark was a Boojum

Among the precedent pages, a number of issues on the analysis of the evolution of historical economic series was discussed. The most important are the nature of the generating process and the possibility of modelling it through a traditional impulse-propagation system. Linearity and the crucial existence of independent random shocks are two definitional conditions for this scheme, and both were rejected. It was also proved that such a story is incompetent to defend the theoretical coherence of the paradigm whenever challenged by the concrete problems of data analysis.

One of the major problems is the interpretation of the exogenous shocks, given that their relevance for the orthodox theory depends on their elimination, as the previous pages have proved. To take one step further: the sole solution for the paradox is to take real history into consideration for the analysis of historical series, and this is quite evident in the current case. If there were periods of major turbulence such as the 1929 Crisis or the World Wars, the economic series reflecting those events should be discussed as such: there is no point in an explanation which eliminates its object, or in a theory which disintegrates its evidence. Therefore, if the variations in variance may cause the statistics to refute the linearity and the IID hypothesis, all the efforts to save those assumptions against stubborn reality are pointless and pernicious. Indeed, such a refutation is relevant by itself since those changes are a real feature of the economies and a symptom of a possible nonlinear process. The switch of regimes is a crucial theme for economic theory and must not be avoided. Furthermore, it was also proved that the evidence of nonlinearity can only be destroyed for these cases with the application of technical procedures arbitrarily eliminating not only the periods of wars but also some of the major economic crises, and in that case it is useless to claim that the theory is studying real cycles existing in real economies.

The case against the double decomposition strategy can now be summarized. The first decomposition is designed to bound the variance of the time series and to annihilate its long memory — the Joseph effect — but it leads to spurious results if the series is generated by some low frequency dependence, if there is history.

The second decomposition is designed to whiten the series, so that the normality of the residuals makes possible the statistical inference, that is, to eliminate the Noah effect or to reduce the structure to the propagation system, described by the systematic part of the linear regression. The first decomposition imposes the finite variance property, the second one imposes the independence of the variables, so that the statistical journey may proceed in the domains of the Theorem.

Otherwise, if the sequential explanation and the notion of cumulative evolution are to be conserved since corresponding to one of the features of the historical nature of the time series, one must look for alternatives. As a consequence, when arguing for the presence of nonlinearities and complexity in economic series, and thus for the essential unpredictability of their evolution and for the condition of structural instability as a consequence of the sensitivity to the initial conditions, one must evaluate the historical methods: the scientist must explain but cannot predict; the model must explain the past and the present, while predicting the future requires a metaphysical trip which is alien to science. It is much more fruitful, as in the current example, to use historical knowledge and methods in order to scrutinize the time series than to manipulate them statistically in order to obtain implausible and non-theoretically based results, such as those derived from interpolation or dummy procedures. Moreover, these results are conclusive but perfectly indeterminate: the scientist cannot produce meaningful propositions about the historical period referred to by his time series, since its artificial representation was successfully designed to eliminate the time dimension.

The complexity approach developed in this book suggests therefore a different strategy to deal with heteroscedasticity: important variations in variance are presented as evidence for (nonlinear) switches of regimes in economic time series, and this historical evidence becomes the relevant enigma for the research. This implies a strong plea for the development of historical methods in economic analysis, like Adam Smith, Marx, Schumpeter and so many others argued for — not for the pleasure of wild speculation, but just because of the need of formal rigour.

The choice of the impact of the wars[18] for the argument of this book is not innocent. In fact, war is the historical event which is easier to present as a 'totally exogenous' factor, therefore suggesting a strict separability.[19] This assumption is implicitly or explicitly stated by several researchers who use the interpolation procedure or introduce dummies for these periods. Yet, some theoretical problems remain even after the discrimination and elimination of the 'irrelevant' facts: are the war periods in fact totally autonomous, as to be treated like separable factors? The positive answer supposes that the economic evolution did not have an impact in the political developments, that is, that the new international order of the Versailles Treaty did not affect the German society, that the increase in unemployment and the collapse of the Weimar Republic were uncorrelated to the rise of Hitlerism, or still that the international conflicts had no relation to trade, technical innovations, capital accumulation and market domination, and vice-versa.

This is obviously untenable, and only in scholar blackboard exercises can one maintain the pretension of pure endogeneity, ignoring the 'exogenous' political factors. Even more so, because of the outstanding fact that most of these explanations require random shocks — which are exogenous, unknown and indefinable — as the very condition for the cycle and for showing some similitude with reality. What the paradox reveals is that most definitions of exogeneity and endogeneity are logical implications of the mathematical formulation of the model, which implies by itself an arbitrary discrimination: exogeneity is defined in order to exclude the complex factors and endogeneity is conceived of in order to represent the equilibrium mechanism. In other words, the antinomy is a property of the model and not a general feature of reality, and imposes a drastic choice between the endogenous mode of explanation which supposes history without events, and the exogenous mode supposing events without history.

A major reason for the use of historical methods in economics is the necessity of reincorporating combined explanations in order to understand social phenomena, which would otherwise be absurdly disintegrated in a nonsensical puzzle. However they are defined within each theoretical universe, the 'exogenous' and 'endogenous' factors must be integrated in the same concrete economic analysis. It is because of history and not because of mechanical representations that we understand and may eventually explain the turning points, the structure of the epochs in economic evolution, the social constraints, the main innovations, the rise and fall of institutions.

There is still a further reason for the reconsideration of the historical methods as a complement to economic theory. As indicated, the BDS statistics detects hidden structure, be it nonlinearity or nonstationarity. In the calculations for this part, the procedure of detrending and then filtering our time series was used for the indicated technical reasons. But these procedures cannot be taken for granted, since we know how damaging they may be: the traditional detrending method is indeed a major concession to the standard techniques, and a most pernicious one, since there is no clear theoretical basis for the elimination of the trend, or in general for the double decomposition. Detrending is a mathematically arbitrary procedure, and it implies the abandonment of any evolutionary perspective since the object of the research becomes the reversible fluctuations around a theoretical line, instead of the irreversible evolution of the historical process. This is the price for the divorce between the growth theories domain and the business cycle domain, which still dominates the formulation of techniques and theories — and it is the ultimate victory of Maxwell's demon or of Laplace's infinite intelligence.

The procedure of detrending was nevertheless used in the previous pages since the BDS test cannot be applied to monotonically increasing (decreasing) or non-stationary series. In the case of the computation of the Hurst exponent no detrending against time is absolutely necessary, although log series or first differences are typically used— and the range represents the changing average level for each subset; consequently, the technique is only able to identify the nature of the historical structure in the case of major variations in the series.

The simultaneous and combined inquiry into nonlinearity and nonstationarity is still a task for the future, and this is why the reader must be alerted about the limitations of the current statistical instruments. Once accepted the agenda of complexity, the old wisdom is challenged: nonstationarity may be a major influence creating the nonlinearity, and vice-versa — the variation in variance may be the evidence for the changes of the 'trend', or in other words of a nonlinear process. The solution here adopted, to investigate changes in variance as symptoms of nonlinear relations and to use both techniques, is just a preliminary step.

The procedure could therefore be charged for accepting the same arbitrariness as the neoclassical paradigm, if it was the case of producing conclusive affirmations about some precise structure of the macroeconomic series depending on the previous decomposition. But, on the contrary, the current argument is that economics must eliminate that type of explanation and look for a new mode of theorizing, based on non-parametric methods in order to study non-specified nonlinear and complex relations. As on the other hand it is known that a fundamental nonlinearity cannot be hidden by the traditional procedures of filtering and extracting the trend if these new methods are used, these results are robust.

The complex approach excludes the procedure of extraction of the trend and the impulse-propagation model in order to test hypotheses or to establish precise parametric specifications in the Laplacean mood. In the same sense, it challenges the epistemic distinction between endogenous and exogenous variables and criticizes aggregation: the rocking horse is indeed a toy. In fact, there is no point in detrending the series whenever there is no economic explanation for the procedure and for its results: Schumpeter suggested the trend to be the incorporation of 'growth' effects from population and capital expansions, but he was wise enough to argue against any general procedure of trend decomposition. Mitchell did the same. On the other hand, detrending may create meaningless entities, such as cycles with no real time dimension. One may indeed suspect that the Slutsky-Frisch paradigm has been hunting the snarks just to find a Boojum and to vanish with it, as in Lewis Carroll's 'agony in eight fits'. Indeed, the presumed linear structure of the models is contradicted by a tenacious and badly-behaved nonlinearity.

In order to deal with the nonstationarity of the series — with the whole original series as the very subject of the research — it is necessary to reconsider the historical insights as the best reliable instrument to complement and to lead the statistical inspection. However, this is not generally accepted, primarily at the methodological level: the dominance of econometric confirmationism in economics and the unending quest for a laboratory type of control and production of experiments vitiated the debate about the interrelation with other social sciences. One very impressive piece of evidence is the general acceptance of the explanatory capacity of closed and reversible models; this in fact implies that a theory should be able to explain exhaustively all the social processes, or that economics should be a global science able to measure, to explain and to predict all human actions. This design implies the reincarnation of praxeology, which was endorsed by the Austrians and indeed at one time by all the neoclassical current: the newly unified social science should be defined by the computation of utility, all humanity reduced to the measurement of pleasure and pain.

After the positivist attempt in the same direction, scientists should wonder if this all-inclusive theory is possible and even if it is prudent. It is therefore necessary to reverse Comte's strategy for the unification of social sciences: he thought he could eliminate metaphysics in economics, replacing it by a sociology based on strictly inductive, fact- and law-finding knowledge according to the positivist patterns; it is now obvious that such a type of unification is neither possible nor desirable, and that multiple forms of cooperation from the specific dimensions of each science are needed, in the sense of a Sozialokonomie articulating historical, sociological and economic researches.

In this framework, it is indeed possible to progress in the explanation of complex social phenomena without simultaneously abandoning the formal rigour which is so badly needed in economics. And it is also possible to deal with the relevant phenomena not as pathological aberrations to be treated by punitive methods, but as the pertinent questions: cycles in irreversible time, switches of regimes in economic evolution or disturbances representing in history the relation between the economic and the political spheres. The analysis of an economy in historical perspective is just that: there are no cosmological essences or cyclopic regularities moving the universe like a clock, and any effortto find such creatures is wasted time. Hicks once stated that 'As economics pushes beyond "statics" it becomes less like science, and more like history' (Hicks, 1979: xi); on the contrary, the argument of this book is that the traditional distinction between nomological (natural sciences) and ideographic (history) sciences must be abandoned and dissolved in a global approach to evolution and, indeed, that historical knowledge is indispensable to economics, and that is the condition for theoretical precision and scientific exactness.

As Schumpeter argued in the 1949 NBER Conference, the construction of economic knowledge is essentially needing the contribution of other social sciences, namely history and sociology. These last chapters argued that common sense favours such an inter-disciplinary approach, necessary to cope with the complexity of the real world. And, if the proof was convincing, so does statistics itself.

Notes of Part Four

1 Karl Pearson, a disciple of Mach — the empirocriticist philosopher who influenced the future Vienna Circle — introduced the concept of correlation in statistics. Under the influence of Mach's dismissal of causality, Pearson reduced the notion to the conceptual limit of the probability correlation. Pearson, a Baconian, was engaged in the 1910 polemics with Keynes, described in Part Three.

2 But not the last ones, in which Yule identified the dangers of spurious correlation (Yule, 1921, 1926), namely criticizing the acceptance of non-existent sampling conditions, since each economic observation is not drawn from a population and it is not independent of the previous ones. In this framework, Yule even argued in 1926 that the 'usual conceptions' of statics were not applicable.

3 The debate between the Neyman-Pearson (Egon Pearson, the son of Karl Pearson) strategy of test of hypotheses and R.A. Fisher's strategy of tests of significance is understandable at this light. Fisher privileged tests of significance and analyses of variance for small series, while Neyman and Pearson argued for an infirmationist action. In fact, only a weak and indirect confirmation is produced, since the alternative and not the null hypothesis is tested, and the test is unable to prevent or to detect eventual errors in the definition of the variables. The N-P strategy is nevertheless a hallmark of the HDM in economics.

4 'When we cannot accept that the observations, along the time series available to us, are independent, or cannot by some device be divided into groups that can be treated as independent, we get into much deeper waters. For we have then, in strict logic, no more than one observation, all of the separate items have to be taken together. For the analysis of that the probability calculus is useless; it does not apply. We are left to use our judgement, making sense of what has happened as best we can, in the manner of the historian. Applied economics does then come back to history, after all. I am bold enough to conclude, from these considerations, that the usefulness of "statistical" or "stochastic" methods in economics is a good deal less than is now conventionally supposed' (Hicks, 1979: 121).

5 Hicks again: 'Thus it is not at all sensible to take a small number of observations (sometimes no more than a dozen observations) and to use the rules of probability theory to deduce from them a "significant" general law. For we are assuming, if we do so, that the variations from one to another of the observations are random, so that if we had a large sample (as we do not) they would by some averaging tend to disappear. But what nonsense this is when the observations are derived, as not infrequently happens, from different countries, or localities, or industries — entities about which we may well have relevant information, but which we have deliberately decided, by our procedure, to ignore. By all means let us plot the points on a chart, and try to explain them; but it does not help in explaining them to suppress their names. The probability calculus is no excuse for forgetfulness' (Hicks, 1979: 121-2). Or in the same sense: 'The national income and electricity consumption of OECD countries in 1970 is not a sample of anything, and therefore the calculation of the "significance" of the relation between them lacks point: the relation is what it is' (McCloskey, 1985: 167).

6 Curiously, Frisch, who was aware of the difficulties involved in the traditional methods and clearly distinguished between substantial and statistical significance, approved Schumpeter's claim in that matter. Significance tests must be 'subordinated to the general intuitional and philosophical interpretation', wrote Frisch (1951: 9-10), who strongly advised his colleagues about the intrinsic limits of the proof by the statistical tests: 'They [the significance tests] have a clearly defined meaning only within the narrow confines of the model in question. I wish it were more clearly and more commonly recognised by all model builders that all the shrewd mathematical tests are of this relative sort' (ibid.). Neither Frisch nor Tinbergen followed the probabilistic revolution they had introduced in economics: Frisch moved in the thirties from econometrics to an important and passionate work on planning, and so did Tinbergen after the Second World War.

7 This is again based on a critique of the metaphor from physics: 'Economists have inherited from the physical sciences the myth that scientific inference is objective, and free of personal prejudice. This is a utter nonsense. All knowledge is human belief; more accurately, human

opinion. What often happens in the physical sciences is that there is a high degree of conformity of opinion .... The false idol of objectivity has done great damage to economic science' (Leamer, 1983: 36). The extreme relativist opinion expressed by Leamer is not a precondition for accepting his point on the subjective component of probability or on the specification of the model.

8 'We comfortably divide ourselves into a celibate priesthood of statistical theorists, on the one hand, and a legion of inveterate sinner-data analysts, on the other. The priests are empowered to draw up lists of sins and are revered for the special talent they display. Sinners are not expected to avoid sins; they need only confess their errors openly' (Leamer, 1978: vi).

9 'Indeed, any system that involves a conservation principle (given means) and a maximizing rule (optimal satisfaction) is a mechanical analogue' (Georgescu-Roegen, 1971: 318). Georgescu-Roegen criticized the Cowles Commission epistemology, namely the assumption of the constancy of the parameters, the 'as if' sampling process and the deterministic conclusions, and suggested an alternative dialectical approach (Mirowski, 1992: 91, 93 f.).

10 In that regard, in the first issue of a new journal dedicated to *Complexity*, the editorialist John Casti (1995: 12-3) argued that there are three general laws in the emergence of complexity: that described by the Theorem, the Fourier representation of functions as infinite series of sines and cosines, and the Power Law, namely Zipf's Law of the frequency of words, of the relation of dimensions and occurrences of falls of meteorites, earthquakes, speciation, dimension of cities, etc. (Krugman, 1996: v, 39).

11 The PHASER software was used to generate the series using the Runge-Kutta algorithm. Other software used was TSP, SPSS (to analyse the time series), Excel (to prepare the graphs), BDS and DMC (to compute the correlation dimension).

12 The initial conditions and values of the parameters are fixed at moment 0, and then only three parameters are corrected every 50 'years', just as the initial condition for k. So, the simulated 'Kondratievs' compared to the real counterparts for the accepted dating (in parenthesis, the date for the end of long wave) are 1801-1850 (1848), 1851-1901 (1893), 1901-1951 (1945), and 1951-1988 (...).

13 Even if no measurement errors were possible, a complete precision is never available, including for a controlled experiment. The degree of accuracy of the measurement of the residuals is a choice of the scientist in the case of the computer experiment: the conclusions will reject nonlinearity very easily if a gross measure is used, ignoring the irregularities that characterize chaos or nonlinear series by the procedure of rounding. For the present purposes, measures with 17 digits were used in the computation.

14 The sampling distribution of the statistics is not known under the null of nonlinearity, but the asymptotic distribution is known under the null of IID.

15 Brock and his colleagues proved that if the series is generated by a low-dimensional chaotic process, then the residuals of the linear filtering will also be low-dimensionally deterministic, will have the same dimension and the same largest positive Lyapunov exponent. So, the pre-whitening procedure does not affect the dimension of the series if its dynamics is chaotic (Brock, and Malliaris,1988: 301; Mullineaux and Peng, 1993: 70). The correlation dimension indicates the minimum number of degrees of freedom needed to simulate the process and to describe the data, or the number of lags in a nonlinear system (Brock and Sayers, 1988: 86). The fractal dimension obtained is typically smaller than the real one (Barnett and Choi, 1989: 151). In this framework, a deterministic system implies a low dimensional process, whereas a purely random one implies an infinite dimension. In practical terms, it is impossible to distinguish between high dimensional chaos and a stochastic process for a given series (Brock and Malliaris, 1988: 322).

16 Maddison indicates that some data are interpolated from fragmentary information for 1871-85 and 1945-6 for Japanese GDP.

17 The article was published in English only in 1937, but its first version (1927) included several pages of an English summary, and it was widely known in the Western countries. Frisch wrote to Slutsky in March 1932 asking for permission to publish it in one of the first issues of *Econometrica* (which was still in preparation); by that time, a translation had already been prepared under the supervision of Schultz (correspondence of Frisch, University of Oslo Library). But the publication was successively delayed until 1937.

18 The same can, and must be said in relation to other processes of social change, such as

the diffusion of new constellations of innovations, new modes of organization of the working force, changes in institutions and, in general, all the processes occurring in the socio-institutional sub-system. The example of the wars was chosen among them because there is no difficulty to date it very exactly.

19 Still, if this was the case, the example discussed in this Part provides further arguments against any simplification about the effect of the 'war' factor. The countries previously considered suffered the wars in very different conditions: one was defeated in the devastating war (Japan), others won but were severely affected (UK) or benefited considerably from it (USA), another was neutral (Sweden).

CONCLUSION

Complexity, The Condition of the World

No doubt, if our means of investigation should become more penetrating, we should discover the simple under the complex, then the complex under the simple, then again the simple under the complex, and so on, without our being able to foresee what will be the last term.

Poincaré, 1903: 132

25. The Complex Under the Simple

This book told the story of a general failure: the epistemology and the techniques of the positivist approach in economics, based on closed models and assumptions such as reversibility of events, decomposability of processes, atomistic and deterministic causality, have been incompetent to analyse real oscillations, irregularities, interdependency and complexity. Equilibrium became a theory of theories and a paradigm of paradigms — and it is a wrong assumption, leading to inappropriate methods and to false conclusions.

Three dimensions and critiques to the general equilibrium paradigm were discussed in this book, in order to establish the scientific conditions for the development of the research program on the historical evolution of capitalism.

I.

The epistemological conditions were delt with in Part One, discussing three key questions: (i) can the positivist epistemology be improved or should it be abandoned?; (ii) are metaphors acceptable tools for the progress of science?; and (iii) can the evolutionary metaphor be the inspiration for an appropriate alternative to positivism?

Positivism, the triumphant religion of modernity — 'order and progress', says its motto — looked for a long time for the methods, the language and the verification procedures that could support its trust in the exclusivity of science as the rational representation of a knowable, appropriable and reproducible universe under a single causal principle. The quest for this unique ensemble of essences has lasted at least from the seventeenth century, attempting the bold task of explaining order as a sequence of separable, contiguous and fully identifiable causes and effects. The absolute failure of that project is proportional to its ambition: it has neither been possible to determine a form of cognition above any suspicion, since the Paradox of Hume challenged the legitimacy of the conclusions from inductive inference, nor has it been possible to define a demarcation criterion in order to definitively exile metaphysics out of science, since the inefficacy and logical flaws of Popperian infirmationism have been noted. The crisis is general and has been deeply aggravated by the fall of the insular boundaries between academic sciences, since the moment when mutual and metaphoric fertilization invaded the territories where old style physics, exhaustive quantification and strict empiricism previously ruled as illuminated despots.

But, as always happens, the ivory tower of epistemology is so far from the daily life of scientists in the battle field, that in the trenches nothing or almost nothing transpired about the crises in the court. The majority of the courageous positivists in social sciences went on looking for simple and universal cause-effect relations, reversible movements, experiments in close laboratory environments, confident deterministic predictions with the future defined in analogy to the present, the elimination of normative science in the name of positive science and the end of ideologies, as they still go on looking for Maxwell's or Laplace's demons, the Invisible Hand or the post-historical *homo economicus*. As a consequence, the 'normal science' born in the decade of 1870 with the Walrasian general equilibrium paradigm has been developed for generations with the help of bright tactics of immunization, such as a powerful axiomatization protecting the core hypotheses from tests, since it was based on non-observable entities (the information and rationality of agents), and on a forest of *ceteris paribus* clauses which forbids empirical analysis.

Yet, not all scientists followed that path: some resisted in small villages and looked elsewhere. In social sciences and in economics in particular, some of the most distinguished practitioners of orthodoxy slowly tried to overcome the classic positivist creed and to realign their methods and beliefs using some more sophisticated alternatives, which were more limited (infirmationism) or less vulnerable to the pitfalls of reality since being wholeheartedly relativist (instrumentalism). Some even tried to overcome the neoclassical heritage.

In spite of these intellectual *partisans*, the impressive success and the duration of the paradigm prove its ability to overcome the frequent crises, provoked either by external challenges (answering and defeating the institutionalist and the Keynesian critiques of the twenties and thirties, during the 'years of high theory') or by internal difficulties: in fact, the winners of the theoretical crusade of the first half of the century quickly changed their opinions, moving from structural econometrics to an abstract axiomatics which prevailed ever since.[1] But this process of self-reformulation seems to be exhausted today faced with the development of the complexity paradigm. As a consequence, the answer to the first question was that positivism should be replaced and that other concepts and tools are necessary.

One of these old and new concepts is that of metaphor. Metaphors are literary tropes, traditionally discussed and classified by rhetoric: they belong to the realm of common language and are one of the dominant modes of predicative reference and learning. In communication, metaphors are general and essential tools for the transference of meanings; this is the case of *substitution* and *comparison* metaphors. Furthermore, they are also part of scientific creation, and an important one: *interaction* metaphors constitute the basis for abductive formulation of new conjectures, introducing the necessary semantic impertinence under the form of new hypotheses which abolish the static character of the traditional correspondence rules — in particular, models are active metaphors from theories.

The properties of inductive inference were here discussed from the point of view of their constitutive metaphors: Laplace, and the most Laplacean of all economists, Jevons, were fully aware that inductive inference, as defined, depended on similitude relations to be associated with causal sequences, thus restricting the scope of the model to processes which are structurally stable and homogeneous through time - the oblique transfer became the dominant mode of inductive demonstration. Invariance has always been, for positivism, the condition for intelligibility. The power of metaphors is also their capacity of persuasion in order to force the widespread acceptance of the methodological errors of their users. So, the answer to the second question is also affirmative, although requiring a definition of criteria in order to discriminate and evaluate metaphors.

The importance and generality of metaphors was recognized by many economists, such as Bentham and Walras, Marshall and Keynes, Schumpeter and Leontief, and by other scientists, like Darwin who systematically used an intelligent combination of metaphors, induction and deduction, implicitly challenging the Baconian preconceptions of his time. None of the great theoretical systems ignores or avoids metaphors for argumentation and, moreover, none ignores the ability and creativity of metaphoric extension from other domains or sciences. This is particularly true in economics, whose marginalist revolution was defined by the metaphor of maximization of energy under constraints in a closed field, the inspiration for the mechanistic explanation of order.

Positivism has been the universal version of the metaphor of the machine, defined by Newton as the locus of the multiplicity of forces, and redefined in the nineteenth century like a propelling motor, with an exogenous source of energy and an internal dynamic principle accounting for order. The example shows that the metaphor may also be a form of metaphysic contamination, just like the apparently more innocent inductive methods: in spite of the danger, it was argued that the progress of science does not need the elimination of metaphors, an impossible and damaging endeavour since the creation of hypotheses is crucial for the development of knowledge, but rather asks for their control and development.

One further and alternative example in the framework of the epistemological debate is that of the evolutionary metaphor discussed in Part One. That metaphor suggests multiple forms of causality (sequential, contemporaneous, exogenous, deterministic, dialectic, intentional) and of hierarchical organization of the 'organic totality' which includes organisms and populations. It is part of the rejection of 'methodological individualism', of the hypothesis of a unified maximizing rationality of all agents and of the idea of a single universal principle organizing all social action.

Variations of the Darwinian metaphor were discussed in the previous chapters: in particular, the idea of the optimization of an objective-function — the egoistic gene — adapted to economics by Alchian, Friedman, Becker and others, was rejected; the version of the genotypic evolution as an analogy for innovation, adopted by Boulding, Nelson and Winter or Faber and Proops, was also rejected; and so was the Lamarckian analogy of the inheritance of acquired characters, so common among evolutionary economists. Evolutionary biology inspires a productive and useful allegory, given that it challenges the onesidedness and causal determinism of positivism and rejects the constitutional metaphor based on nineteenth century physics; but the metaphors based on evolutionary biology must be very general (the 'organic totality' of Marx or Veblen, the 'organic unity' of Moore and Keynes), or else very narrow and concrete. At the same time, the allegory emphasizes the essential difference between biology and economics: from the first, where positive feedbacks have been studied for a long time, relevant indications on the importance, generality and analytic techniques necessary to inspect increasing returns or positive externalities may be derived, but there is still a major difference — in economics the objects are controlled and coordinated processes, and such a coordination is the name for power.

The general conditions for a progressive research program about the historical processes of capitalism are the adoption of a realist epistemology, including a plural and non-deterministic assessment of causality, and the study of the coevolutionary processes accounting for complexity.

II.

Part Two dealt with three more problems in the framework of a discussion about methods for the analysis of cycles and growth: (i) the definition of equilibrium, namely of its properties of existence, uniqueness and stability; (ii) the nature of models and explanatory variables, which is derived from the previous discussion on metaphors; and (iii) the definition of the mode and time of the cyclical fluctuations.

The metaphor of the lake has been since Walras the essential foundation of the static concept of equilibrium: economics is depicted like a lake, always tending to equilibrium and without ever attaining it. J.B. Clark and many others repeated this metaphor, attributed by Schumpeter to 'an old idea of Adam Smith' (HEA: 999 n.), and considered to be the basis for the 'dreamland of equilibrium' (ibid.: 1000 n.). As a matter of fact, the importance of such a metaphor goes well beyond its capacity to affirm the plausibility of a non-observable relation: it supposes the existence of a social equilibrating mechanism and that it is necessarily stable. The rocking horse metaphor, originally formulated by Wicksell and then developed by Frisch,[1] made possible its representation through a technique which is indeed present in every linear regression, explaining the endogenous variables by the composition of systematic (the regressors) and non-systematic factors (the errors or stochastic innovations), fully identifiable and separable and following precise rules so that inference be possible.

These metaphors inspired generations and generations of equivocations. The surface of the lake does not permit any causal proposition on the waves and tides, just as statics does not permit any conclusion about dynamics, since time is not a trivial parameter or variable which may be added or ignored according to the mood or convenience of the modeller. In particular, the statistical methods inspired by a Frischian type of decomposition are unable to discriminate between a stable model and an unstable one with a limit cycle (Blatt, 1983: 147), and confuse complexity with pure extrinsic randomness. On the other hand, the concept of 'error' is deeply ambiguous, given that it originally represented a pure error of measurement in a framework where the causes of fluctuations were considered to be fixed and completely predictable (in astronomy), but since then it represents a mixture of facts and factors including explanatory variables ignored by the model as well as determined and non-determined errors of several origins. Its epistemological status is the Pandora Box of orthodox economics — and, yet, its properties are the decisive condition for parametric estimation.

The development of Bourbakist axiomatization by Koopmans and Debreu, and then by Arrow and others, replacing structural econometrics in the Cowles program from the end of the forties, was in fact a defence strategy following the first crises in the normal science of the day, namely because of that ambiguity, but the solution aggravated the stalemate. The introduction of Keynesian expectations or of uncertainty is not compatible either with the orthodox definition of rationality or with general equilibrium.

From an insight of Kalecki, it was suggested that a new concept, that of semi-autonomous variables, makes possible the recognition of the difficulties and even their partial solution. Kalecki did not generalize his concept, neither did he relate it to other theories, and that is what was done in this book: these semi-autonomous variables are at the core of those theoretical systems and concepts which implicitly or explicitly reject equilibrium — for instance the rate of profit of Marx, innovation in Schumpeter, the 'three psychological laws' in Keynes — and constitute the formal relation between economic determination and the social framework in each one of these models, as well as the articulation between several distinct levels of abstraction. These concepts were used to develop a new classification of the theories of economic cycles following from two constellations of criteria, those related to the mode of conceptualization of the system of hypotheses and those related to the properties of the system of hypotheses. That system of classification was used in order to discuss several contributions to the analysis of cycles and to derive some conclusions identifying the problems inherent to the analysis of movements of different amplitudes and frequencies.

The theoretical conditions for the definition of a program of historical research about real economies are the rejection of the mechanistic definition of models and analogies, and the development of theories based on the social interaction which is represented by semi-autonomous variables.

III.

In Part Three, the theoretical contributions of Schumpeter and Keynes were discussed, and their respective programs compared.

Keynes rejected the physical metaphor, combined induction and analogy in a creative way, assumed an organic and therefore indeterministic vision of economics and discussed the traditional statistical methods, in what was the single most powerful critique of the positivist paradigm in economics in the twentieth century. Throughout his debates with Pearson and Tinbergen, Keynes elaborated major arguments which have not been satisfactorily addressed by the orthodox authors, namely about the specification of the variables, the sampling methods, the technique of multiple correlation and the premises of structural stability and homogeneity through time. At the same time, he stressed the importance of qualitative factors and of the intrinsic uncertainty of economics, namely of the relations not susceptible to description in terms of probability distributions, since they refer to qualitative and non-distributive possibilities (Shackle, 1972: 401).

Schumpeter, in spite of his paradoxical compromise with a vain and inefficient attempt to restore and to generalize Walrasian statics, suggested a revolutionary interpretation of the dynamics of capitalism, based on the starting point of the role of innovation and the entrepreneur. He rejected the notions of a representative agent or of a permanent convergence towards equilibrium: the lake is creative only if it destroys, and it destroys when there are severe storms, these mutations which may be better represented as processes akin to biology.

I argued that the Schumpeterian theory was born from the reunion of two fascinating cultural environments: first, the Methodenstreit between the orthodox Austrians, with whom Schumpeter was educated, and the German historicists, with whom Schumpeter established close links later on and, second, the vivid impression caused by the heterodox work of one of the more conventional of the neoclassicals, J.B. Clark. Like Keynes, Schumpeter argued for the crucial importance of the constructive role of time — but the former did not want to go further than this observation, since he limited his work to the short term, and the latter could not go further, since he revered the Walrasian scheme, a theory he knew to be wrong as such but which he considered to be a part of the future truly general static-dynamic system.

The theoretical condition for the study of historical processes of capitalism is the analysis of disequilibrium and economic and social mutation.

IV.

Part Four is an essay on the theory of emergence, on complexity. It is proved that linear methods and the double decomposition are always dependent on an implicit theory which, in the current versions, is part of the Frischian paradigm: its questions and its methods determine the answers more than empirical data. Such a starting point makes possible three important conclusions.

In the first place, the utilization of statistical methods with no explicit theoretical counterpart is simply rejected; in particular, the punitive methods against empirical data are denied, such as the techniques previously presented for the elimination of heteroscedasticity, which is in the historical series one of the indications of nonlinearity. Defined as a pathology and not as a general symptom of the historical conditions of evolutionary processes in macroeconomic series, the variation in variance has been dealt with, in the examples criticized, by methods of weighting and standardization, interpolation or substitution: after that, it becomes possible to subject the transformed series to the traditional econometric methods, but such a result is really indifferent given that it is not related to reality but to an arbitrary artifact.

Second, there is an unbearable epistemological contradiction in the effort to decompose and discriminate between trend and cycle, or between the random shocks and the damping mechanism which allegedly accounts for the stability of the economy: the shocks, which are the cause of the oscillations, are eliminated in order to become calculable — the determination is relevant if it is ignored. Furthermore, the defence of the plausibility of the theory requires the shocks to be divided into two categories: those which are expelled from the economic sphere since they represent extreme variations (wars, or other facts conveniently considered as disturbingly non random), and those which are accepted since they are small, that is, whose importance depends on their lack of identity. Anyway, it was proved that any decomposition is hopeless: Kalecki easily verified that the distinction between trend and cycle is arbitrary if the project is to explain a complex phenomenon, and the authors of RBC models recognized that it is indeed generally impossible to discriminate between both concepts if the trend is not modelled as purely deterministic. The Frischian paradigm of the rocking horse, with a propelling system which represents causality which should be isolated and ignored and a propagation system to be studied by sophisticated econometric methods, collapses if stability is questioned or if causality is shared. This is exactly what happens in general in nonlinear systems.

Third, there are crucial reasons for believing that economic relations are essentially nonlinear: complexity and chaos may emerge even in very simple models whether there are increasing returns or other externalities, by the simple fact of some agents being price settlers and not price takers, by the dynamics of learning or by complex dynamics in preferences and technologies or by the lag structure (Brock, Dechert, 1991: 2220-1; Brock, Hsieh, LeBaron, 1991: 32-4). The separability between endogenous and exogenous variables, between trend and cycle or between propagation and impulse is really meaningless if the subject of the research is defined as the organic totality of a phenomenon whose relevance is precisely given by the dynamism of complex relations.

In particular, it was proved that the current procedure of statistical aggregation, however necessary and useful it is, may be responsible for hiding the constitutional nonlinearity in economic historical series and that, using new statistical methods, it is possible to detect or to conjecture the presence of non-determined nonlinear structures in the economy. As previously stated, those conclusions are nevertheless nothing more than prudent approximations, since the methods are still embryonic and the quality of data is beyond what would make possible a definitive proof.

This study of nonlinear relations involves severe difficulties. Models defined in that framework are generally impossible to solve, equilibria are multiple and non determined, the exhaustive quantification and parametrization is in general impossible or inadequate. This is why many scientists still prefer the linear models and traditional statistics, whereas others adopt a total agnosticism, but both attitudes are unwise: positivism will not resurrect to give the answers it could not know, nor should science live in a Disneyland where all fantasy is welcome. Analytical rigour depends on the realism of premises and methodologies, just as the progress of knowledge depends on the capacity to explain and understand reality. From this point of view, complexity is not a dead-end but an open door: science will not be unsafer than before, and will eventually be able to progress.

The analytical condition for the study of complex relations in the history of capitalism is the concrete investigation on the switch of regimes, the structural mutations and the evolution of the economies as an indecomposable process which must be explained in its organic totality, with the recourse to nonlinear modelling and statistics.

V.

A large space was until now devoted to the discussion of the characteristics of models of cycles and namely to their ability to explain, to simulate or to represent the historical series. These results can now be summarized, and three criteria are indicated as the building-blocks for a new generation of models rejecting the rocking-horse metaphor:

a. The *principle of simplicity*: the formal model defined in analogy to a theory is bounded to a local dimension.

The formal model explores the plausibility of one specific set of hypotheses, arguing about its internal coherence in relation to reality. But the nature of the relation between the theory, which is a specific set of causal relations claiming to explain reality, and the model, its formal representation, is of the same type as that between the theory itself and reality: it is a metaphorical representation, a narrative which is not generally demonstrative by itself; it is always necessary to check the hypotheses directly against evidence. As a consequence, the formal model and its empirical confirmation cannot prove a theory, given that not all the subsets of the theory are controlled, just like its refutation cannot infirm it, since a supplementary hypothesis may always avoid the negative test, as stated by the Duhem-Quine thesis.

The model can therefore be assessed according to its logical structure, to the realism of its assumptions and to their empirical corroboration: those criteria are indeed indistinguishable, and this is why instrumentalism, dominant in economics, must be rejected. Furthermore, the realism of the premises and the logical construction of the argument impose a severe restriction on the metaphorization: it must be local, since it can only represent a limited set of relations. Finally, the narrative nature of the theoretical argument and realism in science do not collide, since narration is about observable and understandable exterior reality, and the working of the theory is concentrated and depends on its fit to reality.

The principle of simplicity defines the epistemological condition for modelling.

b. The *principle of organization*: the formal model cannot represent more than one in a hierarchy of distinctive levels in the theory, each corresponding to a specific structure of determination and to a precise manifold of variables.

As a consequence of the principle of simplicity, no theory about reality can be exhaustively represented by a formal model. And since the model does not define the theory, the converse being true, the dominant method of construction of a new theory through the logical development of a model, as established within the strict discipline of the neoclassical paradigm, is irrelevant.

The existence of a hierarchy of causal determinations blurs the stability of the distinction between endogenous and exogenous variables and supposes the semi-autonomy of some central variables — as the counter-tendencies in the fall of the rate of profit in Marx, the 'psychological schedules' in Keynes, the concept of innovation in Schumpeter, the 'intermediary levels' in Mandel, the connection between the techno-economic and the socio-institutional sub-systems in Freeman and Pérez, the structure of accumulation in Gordon or the modes of regulation in the Regulationist school — which represent and indeed drive the organicity of the system. This organic nature may only be represented by nonlinear relations, since the observed phenomena of dynamic coupling of events and of positive feedbacks, including increasing returns in economic evolution, are not generally admissible in a linear system. As a consequence, uncertainty is a natural and permanent consequence of nonlinear and multi-level determination. Moreover, the organic nature of the system and of the theory accounting for it impose a realist definition not only of the assumptions — as in the principle of simplicity — but also of the methods to analyse the object of the inquiry: given that the trend and the cycle cannot be meaningfully separated, no model of cycles is relevant that does not consider the inter-relatedness of these dimensions of the evolutionary process.

The principle of organization defines the ontological condition for modelling.

c. The *principle of historical determination*: the realism of the theory to be represented by a formal analogy is inseparable from its historicity.

The principle of organization restricts the models to one of the levels of the theory, and the principle of simplicity restricts the models to local approximations within that level. The principle of historical determination adds now two more general syntactic rules. First, it indicates that time irreversibility is a general feature of evolution, and that the theory must conform to it: time is not a parameter or a coordinate in a space representation, nor is it a simple exogenous factor; it is instead a building factor in complexity. Economic development and morphogenesis occur in time, and this is why indeterminism and uncertainty become crucial in social evolution and why disequilibrium is pervasive and creative. Models of certainty and of equilibrium are irrelevant for evolutionary economics.

Second, it indicates that the hierarchy of the theory is not an impediment to a comprehensive causal explanation, since the historical characteristic of the theory itself represents evolution and metaphorizes complexity. From that point of view, the function of the models becomes clear: they generate hypotheses, inspect their logical plausibility and redefine the conjectures inside the theory. On the other hand, since all empirical confirmation of formal models in social sciences is historical by nature, there is always a source of independent and external confrontation of these new hypotheses with reality, and realism is once again fully compatible with the conjectural metaphorization of models.

The principle of historical determination defines the heuristic condition for modelling.

VI.

Throughout this book, I have tried to define the epistemological, theoretical, analytic and historical conditions for the development of the research program concerning development and fluctuations. I did not adopt as a starting point the need to prove the existence of those short cycles or long waves and tides, or to confirm hypotheses about their behaviour; yet, the theme is sufficiently polemic within the profession that one cannot have the right to consider the matter settled. But two major reasons argue for such a choice.

First, it is sufficient to admit that modern economies have experienced distinct historical phases, and furthermore that there are explainable processes of economic and social mutation that transform the institutional organization, the technological frontiers or the dimension of the capitalist system itself. On the other hand, the traditional forms of legitimization of theories and models are so dependent on the normal science of the last one hundred and twenty years that one must suspect the requirements that impose on a conjecture the need to find statistical evidence of a steady state, of well-behaved random shocks or of battalions of endogenous and exogenous variables disciplined to the functional relations as a priori defined. In short, what is at stake is this acceptance of the Chinese Box strategy, adding successive layers of representation of reality in the formal models on the assumption that the whole is just the sum of these simple parts.

The non-identification of these distinct phases of development and its irregularities is at least as enigmatic as the lack of direct statistical evidence of their existence from the tests, since the economists and historians recognize both these great systemic crises and that the map of the economies changed during and after those events. It is therefore possible to conjecture that, as proven for some specific statistical methods, the mistrust of the profession in relation to this historical vision derives essentially from defective questions and from methods which provide right answers to the wrong questions A statistical argument was provided for this point.

Second, in spite of its advances, the general state of the study of cycles is that of great heterogeneity, of lack of clarity about tools and objectives, of significative influence of orthodox economics, of epistemological paradoxes. A prior work of clarification and synthesis is needed and indeed unavoidable.

As a consequence, it is necessary to draw new avenues for future research, in particular in these two fields which appear as more promising and more demanding: the modelling of the second generation of indeterminism, the concrete study of complex dynamics, and historical inquiry, the complex study of concrete realities. An evolutionary synthesis is still not on the horizon, but the accumulation of signs of the crisis of positivism in economics in the last dozen years and the development of new statistical and historical methods tend to reinforce the project.

The deep roots of the problem were identified many years ago, namely by Maxwell, who noted those 'other classes of phenomena which are more complicated, and in which cases of instability occur, the number of such cases increasing, in an exceedingly rapid manner, as the number of variables increases' (Maxwell, 1876: 442). And Poincaré was certainly right when he stated that:

> observation reveals to us every day new phenomena; they must long await their place and sometimes, to make one for them, a corner of the edifice must be

> demolished. In the known phenomena themselves, where our crude senses showed us uniformity, we perceive details from day to day more varied; what we believed simple becomes complex, and science appears to advance towards variety and complexity. (Poincaré, 1903: 148).

That is precisely what was found by Hurst and then by Mandelbrot: there is weather but there is a climate, there is volatility of prices but there is also a structure of change. And since large scale evolutionary systems tend to self-organize, selecting their selective properties, a study on the interaction of morphogenesis and structure is crucial for the understanding of the large body of factors influencing the economies. The recovery of the status of economics as a social science depends on the viability of that new Sozialokonomie.

VII.

Morphogenesis implies two essential features: change and control, that is, rupture and continuity. Both coexist and are interdependent and inseparable: the onesidedness of the analysis of a single term of that process is indeed responsible for most of the relativist trends in economics, extreme examples of theories of continuity being those defined by the general equilibrium paradigm.[2]

Coordination is the concept that has been previously used in this and other texts in order to interpret and to analyse control systems and cohesive functions in historical development. Coordination, as a social process subjected to complex interactions — and not equilibrium, which is a state — explains the attractors in growth patterns, the weight of social institutions, the relation between the economic system and other parts of society. It establishes, from that point of view, the condition for the viability of morphogenesis, originally defined at the organismic level and then metaphorized to the general evolutionary process of society. In other words, coordination explains why disequilibrium processes exist but are constrained, why different rhythms are mode-locked and why structural instability persist but does not drive the system towards explosion.

Turing, celebrated as a forerunner of computer science but ignored as one of the first discoverers of complexity, modelled morphogenesis with a very simple chemical system of two components diffusing at different rates under random shocks: the system attained the 'onset of instability' given its auto-catalytic properties — the emergence of a pattern of organization or the concentration of a dominant component evidences that in the model it is instability that entails the development of the structure (Turing, 1952: 37). Order out of complexity: like in Waddington's models, as noted by Turing, a biological order emerged not only from natural selection but also from the dynamic properties of the system.

Of course, there are major differences between those simple chemical and organic systems and the social ones. The balance of positive and negative feedbacks in ecological niches, the canalization of development, the selection of the spaces of viability and of stability are mainly driven by naturally coordinated processes, while in economies and in societies there is combination of natural processes and of conscious choice and purposeful action. In this sense, social coordination may be defined as the working of two related sets of variables: (i) the technological, economic, political, institutional and cultural sub-systems; and (ii) the semi-autonomous variables connecting those sub-systems.

VIII.

Real societies and economies, like the other dissipative structures, have the property of self-organization. This recognition has two major implications. First, since the explosion of the myth of reversibility of time, it emphasizes the importance of history and asks for another form of investigation other than that based on artificial models designed for exhibiting the aesthetic qualities of equilibrium. Second, it asks for a political economy, in the classic sense of the words, given that self-organization entails coordination, which is the relation of power.

A long way has been travelled since the first days of the Newtonian universal laws of organized simplicity. With Boltzmann, disorganized complexity was formalized: with the Central Limit Theorem, statistics based its techniques on the assumption that simplicity was created from complexity. What is now suggested is major paradigmatic change: complexity is irreducible to simplicity; complexity and order out of equilibrium account for evolution.

Several processes of coordination were discussed throughout this book. In evolutionist biology, some of the more obvious examples are the social selection of anti-selective processes, the 'reversive Darwinist' effect' of Tort, or the elimination by natural selection of the more unstable species, in order to increase ecological stability. Kauffman, rejecting the idea that natural selection be the only process for the creation of biological order, or that the organism be a purely random and ad hoc ensemble of cells, conceives of forms of group selection and complex dynamics on the edge of chaos, which would explain the creation and the forms of life. This follows Waddington, who suggested a double process of organismic and population selection:

> Natural Selection keeps complex genetic systems within the dynamic range between freedom and fixity in which alone significant evolution can take place, and so explains a deep property of living things that might well be called 'evolvability'. (Depew and Weber, 1996: 21)

These concepts are Darwinian and yet defy the established neo-Darwinist and Weissmannian central dogma: the Darwinian micro evolution is coupled with the dynamic properties of self-organized systems (Bak, 1996: 129), evolving at the edge of chaos (Kauffman, 1995: 15).

In economics, the coordination processes were discussed since the classic authors and, above all, since the marginalist revolution and the very definition of the equilibrium program (the 'invisible hand', the *tâtonnement* process and the lake metaphor): general equilibrium was in economics the first influential formalization of a coordination process. A great part of the impressive intellectual resources at the service of this paradigm was dedicated to the establishment of the conditions of existence and stability of equilibrium; this simultaneously corresponded to a growing abstraction, to the primacy of a relativist epistemology and to the progressive axiomatization of the discipline. And it failed.

The concept of organized simplicity, the invisible hand, was doomed to misrepresent reality wherever complexity and emergent phenomena occurred. New efforts were then addressed to repair the pitfalls of the axioms of equilibrium, certainty and rationality, namely those defined in the framework of game theory and in particular the concept of a Nash equilibrium. Nevertheless, if there are several agents with alternative strategies, there is in that model no guarantee of convergence to a single equilibrium attractor, and if there is some disturbance the new trajectory of the system is not a priori predictable. The results were disproportional to the effort and some of the leaders of the project have taken since then their distance or even openly criticized the failure of the entire program.

Coordination must be explained as a larger process than the one of general equilibrium: Schumpeter defined the dynamics of capitalism as the outcome of the special function of the entrepreneur, which creates disequilibrium. Of course, the historical growth of State regulations and intervention may be interpreted as the expression of the necessity of coordinating societies and economies which grow always more complex and unpredictable. Coordination is meaningless without coevolution.

IX.

For decades, coordination has been identified with general equilibrium. This is understandable, since the neoclassical paradigm assumed from the beginning that such processes dominated the economies and, moreover, that they were representable by the axioms of existence, uniqueness and convergence to equilibrium. Coordination, acting through the auctioneer, the price system and its mode of circulation of information, was defined as extremely simple, automatic and efficient[3] and it became the basis for a comprehensive vision of the world and of human action. Indeed, the impulse-propagation model was so powerful and successful because it incorporated and extended those concepts, it domesticated the stochastic variables emerging from quantum revolution and made possible empirical corroboration and prediction: business cycle analysis could not ask for more. Again, it modelled coordination as the equilibrating mechanism, Frisch dampening the Slutsky's shocks. But soon coordination became just a logical game for orthodox economics, and the abandonment of the econometric program for structural estimation and the consequent adhesion to Bourbakism drove economics to the heights of axiomatization and to the pleasures of instrumentalism.

One of the exceptions and one of the more interesting and contradictory figures of this period to be involved in the debate about coordination was Hayek. Hayek evolved from a traditional general equilibrium position, namely in his model of cycles, to the idea of 'spontaneous order': these two versions of the coordination problem were clearly opposed, to the point that some researchers describe these phases as 'Hayek I' and 'Hayek II'. For the first, the market mechanism was necessary and sufficient for coordination, since it established the perfect price information in a perfectly atomistic society: the markets were consequently supposed to be efficient, and the business cycles were due to errors of monetary overinvestment, exogenous disturbances to be withered away by the price system. But, disillusioned with general equilibrium, Hayek abandoned business cycle analysis by the end of the thirties.

Hayek, as noted in the previous debate on the incorporation of the physical metaphor, was rather critical of the 'scientist' bias in economics, which amounted to the illegitimate transposition of the concepts of optimization of energy and of equilibrium obtained in laboratory experiments to the economic theories, and alternatively defended organic metaphors in order to account for social phenomena. So, his market efficiency was not obtained by the traditional mechanical properties but by an extreme assumption about rationality and methodological individualism. Hayek II, on the other hand, emphasized the rules of conduct mutually imposed as the viable form of coordination, and equilibrium was therefore proxied by the market process, although it was never really attained: those 'spontaneous ordering forces' (Hayek, 1974: 34) were part of the social organism, as the 'spontaneous interplay of the actions of individuals' in an 'organism in which every part performs a necessary function for the continuance of the whole, without any human mind having devised it' (Hayek, 1933: 130-1). In this framework, short-term equilibrium was a fiction abstracting from time (Hayek, 1928: 161-2), had 'no definite meaning' and 'these attempts to give the equilibrium concept a realistic interpretation (the legitimacy of which remains in any case somewhat doubtful) have deprived us of an at least equally important use, which the concept will serve if we frankly recognise its purely fictitious character' (Hayek, 1941: 21). In the long term, equilibrium could eventually be defined as some centre of gravitation, but for practical purposes, this implies the existence of a concrete coordination process without equilibrium since the state of information is in permanent change (Garretsen, 1992: 83-4).

In his last works, Hayek virtually abandoned economics and concentrated in the development of his philosophical views, arguing for the metaphor of natural selection: society evolved through the competition of distinct groups, struggling to impose their own rules — spontaneous order was the emergent property of this structure where complexity dominated societal evolution and it was supposed to be a Lamarckian rather than a Darwinian process. In fact, this organic metaphor was close to sociobiology, the cultural transmission in society being modelled as a selection and self-organizing process (Witt, 1993: xxii). In short, in Hayek I coordination was imposed by the efficiency of the price system of the market, as in the neoclassical story, and in Hayek II coordination was imposed by selection: in both cases, market and natural forces drove the system to the desired state. Kirzner synthesized this approach arguing that entrepreneurship leads to equilibrium: this is just the opposite to the Schumpeterian theory (Kirzner, 1973: 72) — but it is incoherent, since both spontaneous order and disorder are created once out of equilibrium.

As Hayek, Keynes considered the organic metaphors but reached rather different conclusions: coordination was modelled as a specific function of a set of institutions or conventions, namely money and monetary authorities and policies, which were decisive because of the general failure of the market mechanisms. Indeed, his paradox of thrift can be interpreted as the result of an error of coordination: given that saving and investing were decisions taken by different agents at different times with different motivations, they could imply disequilibrium, as they frequently did.

Nevertheless, both the neoclassical approach to the coordination problem and the specific models suggested by Keynes supposed the self-containedness of the system, and consequently explicitly rejected or could not incorporate the historical dimension of open coevolutionary systems. The explanations of coordination developed in the framework of the cycles theories were supposed to deal with the matter, but most of them, given the dominance of Type I models, restricted the answer to the working of one sub-system, either the economic or the technological one. As previously argued, these positions are quite unsatisfactory, since they isolate rather artificially the sub-systems for the sake of some positive causal claim; and, in spite of it, such a claim cannot be meaningfully tested. On the other hand, they tend to ignore the institutional and social features, since they cannot be simply included in the formal model that represents the theory.[4]

X.

In a reference to the Kondratiev cycles, Mandelbrot suggested that the superimposition of different orders of cycles should be assessed by hierarchical descriptions (Mandelbrot, 1987: 126). Such has been the approach of this text: emergence, complexity, and open systems are the central features of the evolutionary theory as applied to social historical processes. In that sense, the basic hypotheses can be stated under the following form: (i) five sub-systems (science, technology, economy, politics, culture) generate a large number of irregular fluctuations, namely cyclical and wave-like movements with different approximate periodicities, caused either by specific sub-system cycles (political business cycles, technological trajectories, cultural movements, life-cycles of products or industries, etc.) or by the lags and feedbacks in the inter-sub-systems connections; and (ii) those streams are combined in some bands of fluctuation by specific coordination processes. The coordination processes are therefore the crucial causal determination for the business cycles and the long wave movements in real historical development.

In some ways, several authors anticipated these hypotheses. Van Gelderen formulated this interpretation when he indicated that periodicity was not simply connected to the technological sub-system but also to the working of the capitalist system as a whole, namely to the construction of social order (van Gelderen, 1913: 45). Fifty years earlier Marx anticipated the importance of the coordination process of 'capitalism as a whole', but explained it as the outcome of profound tendencies and counter-tendencies, and did not consider — at least, not systematically — the undulatory form of their combination. For these authors, coordination was the rationale of the organic metaphor.

Some of the main consequences of this approach are none other than the previously stated epistemological conclusions: given that each sub-system is defined as the heuristics for some social relation, their inter-relations cannot be deterministically discriminated by an exhaustive account of endogeneity or exogeneity of variables. On the other hand, semi-autonomy implies that the variable more relevant to the understanding of historical dynamics is the coordination process itself. In summary, the hierarchical explanation considers the specificity of each of the five sub-systems and of their determination, as well as their organic totality.

This five sub-systems approach (Freeman, 1995), has three central innovative features: (i) it is a description based on the overlapping of sub-systems, since the relationship among the sub-systems is more adequate to explain reality than the artificially isolated description of each of the sub-systems; (ii) it analyses the crises and phase transitions from the viewpoint of the lack of synchronicity and maladjustment between sub-systems, what defines the time band of major fluctuations; and (iii) it suggests that the social conflicts of all types are generated and articulated by the coordination process, that is, by power under all its forms, from the production of legitimacy to strict coercion. This coordination process, it may be added, proceed at several simultaneous levels: the first level is that of the actions embodied in the social working of the economic system, the tension managed in order to integrate the conflicts, in the conventions and institutions; the second level is that of power, strategy and domination, where arbitrary decisions are taken.

In this sense, the fact that there is coordination does not imply that there is harmony or equilibrium, neither in the ideological sense of a general feature of the modern economies, nor in the precise sense of a permanent dynamic stability property prevailing in the markets — which implies some hyper-centralized architecture of decision.[5] In other words, complexity and structural change can only be explained as historical developments, as coevolutionary processes.

XI.

The importance of the metaphoric inspiration by organic models and their superiority in relation to mechanistic alternatives can be assessed in this framework: selection and mutation replace equilibrium, the creation of variety replaces maximization, dissipation and the coordination of development replaces conservation. Self-organization, self-regulation and self-augmenting features are part of this autopoietic system: disequilibrium and coordination are two parts of the same reality — as Hegel wrote, becoming is the truth of being.

Coordination and complexity explain simultaneously the short cycles and the long waves or stages of capitalist development — that is, self-organization — and the characteristics of unpredictability of the same economic processes. Indeed, all societies have material and social limits at every moment of their evolution: complexity does not imply dispersed trajectories in the direction of all azimuths in the horizon. On the contrary, we are dealing with processes which are indeterministic, but which are also subject to constraints, given by the past, by material and factor endowment and by the characteristics of the forces influencing power and coordinating the society. In that case, stability is configurable through well defined institutions for each historical period: the dominant techno-economic paradigm may be represented by an attractor selecting a large band of cyclical processes and coordinating those orbits — that is known in economics as the life cycles of the motive industries, as the constellations of radical innovations, or still may be represented as the evolution of social institutions, including labour organizations, forms of conflict or the cultures which define any society.

Moreover, evolution is the creation of variety and novelty, and *a fortiori* no fixed attractor or strict unchangeable mechanism can represent that process. Irregular waves exist, which cannot be studied under the diktat of the *ceteris paribus* conditions: time is turbulence. The cycle theories tried to encapsulate these phenomena under deterministic equilibrating representations, and then some exogenous noise was added for the sake of the quality of the simulation; they failed to produce either logical explanations or coherent descriptions.

Alternatively, nonlinear complex models address the duality of dynamic stability (around the attractors, in a region bounded by the availability of material resources, labour or technological capacities) of systems which are nevertheless structurally unstable (inducing switches of regimes from changes in the structure). Some central features of real economies are correctly metaphorized by these evolutionary models: capitalism is unstable and contradictory, but it controls its process of accumulation and reproduction. Even more, critical instability generates new developments and new phases of dynamic stability: this morphogenetic feature is the peculiar strength of capitalism that fascinated Schumpeter and was so vividly described in Marx's and Engels's *Communist Manifesto* as the program of modernity, 'all that is solid melts into air'.

Equilibrium represented evolution by unilaterally stressing the role of coordination and metaphysically postulating convergence and stability, ignoring the constructive role of time; it addressed a real problem, but it was unable to solve it. The deficiency of the general equilibrium paradigm becomes obvious when one verifies its radical inconsistency with history and evolution.

The historical and social processes are unstable, and that is the paradoxical and fascinating characteristic which attracted so much attention and investment for deciphering the riddle:[6] why are economies sensitive to the initial conditions, that is, how are they historically determined, and why do these processes not explode, that is, how are they coordinated? Some answers are known by now, in spite of the confusions generated for a long time by the notion of mechanical equilibrium: there are strange attractors and emergent phenomena in the coevolutionary process of the economies. Economics lived for a long period an equivocal affair with the concept of equilibrium, against which the forerunners of the marginalist revolution, in particular Walras, were alerted in vain by Poincaré and afterwards by Volterra: equilibrium describes laboratory situations of one or two dimensional attractors, but these experimental conditions are not reproducible and are indeed inconceivable in real economies. Furthermore, if three or more interdependent oscillators are related, chaos will generally emerge.

Coordination is therefore a more general process in the coevolutionary framework which generates complexity. In particular, coordination is the process of selection of connectivity (of the values, the architecture and the states of the network) so that dynamic stability be maximized: several authors described these processes as being at the edge of chaos, but none of these definitions is very rigorous, given that chaos describes geometrical fixed objects into fixed phase spaces, and live organisms or societies are endowed with other properties of change and intentionality. On the other hand, the concept of deterministic chaos is not satisfactory for social systems, since there are facts and factors outside the control of the system — however defined it may be — which should be considered external and non-determined. Nor is the notion of the edge of chaos well-defined, since the frontiers between the states of pure randomness and pure stasis are not clear (Mirowski, 1996: 15-6) and complexity as a general class of phenomena must include both. But the new paradigm is already a major step forward, since it destroys the traditional orthodox concepts, reinstates history, and restablishes the function of science: to explain and to understand, to formulate and to correct hypotheses and conjectures, to learn. Economics does not achieve the status of a science through its claims on equilibrium, rationality and perfect determinism: instead, what is to be found in open and evolutionary systems is disequilibrium and turbulence, nonlinearities and non-convexities, bounded rationality and purposeful or institutional action. Complexity cannot be reduced to simplicity.

The lack of a deterministic, complete and definitive knowledge is felt by many to be a loss, in the context of the shipwreck of positivism. This book has argued precisely the reverse: such an evolution of economic science, now in more turbulent waters, simply brings it back to the investigation of reality. Nothing is lost and everything is transformed, since that is the condition of the world: stable and mutant, unstable and structural, an indeterministic world of which the scientist is a part. Combining history and analytical techniques, narrative and formal rigour, the task of science is not lessened: ignorant is that certainty about certainty, whereas the wisdom which knows that it only knows something is wiser than that which claims to be able to know everything.

Notes of the Conclusion

1 Frisch, the hero of the story told in Part Two, and one of the founding fathers of the econometric program, the first Nobel winner in economics with Tinbergen, was also one of the first and more lucid scientists to suspect and distance himself from the failures of the structural estimation program and of Bourbakist axiomatization. This topic is a current theme of research for the author.

2 See this new example of the contradictions of the assumption of rationality: 'Why is it that human subjects in the laboratory violate the canons of rational choice when tested as isolated individuals, but in the social context of exchange institutions serve up decisions that are consistent (as though by magic) with predictive models based on individual rationality?' (Smith, 1991: 894). Vernon Smith, who leads the program of experimental economics, designed to repair the averages of the rationality postulate, is right to call for the spell of magic to save the general equilibrium. A somewhat more secular explanation would indicate that institutional or social processes operate in order to coordinate decisions and to avoid extreme tensions and ruptures in the social process.

3 Of course, the coordination problem is addressed by neoclassical economists and indeed it is a central feature of their theories, since it is incorporated in the main assumptions of the paradigm. But it is a hidden assumption, since coordination is considered to be simultaneously the *outcome of competition* — as stated by Mandeville's Fable of the Bees, or by Say's Law — and also the very *nature of the competition* — since the representative agent, whose maximization behaviour is the paradigm of rationality, spontaneously acts in order to coordinate the market. Some neoclassical authors, of course, were fully aware that such simplistic description of reality could not be kept if the assumptions were to be changed: as Marshall indicated, if any historical perspective — a long term vision — is adopted instead of the short term perspective, then positive and negative feedbacks imply nonlinearities (for instance, increasing returns) and coordination would no more correspond to equilibrium and stability.

4 One of the attempts to do so was, of course, that of Ronald Coase, who since 1937 identified the costs of coordination and information: the costs of investigation, of negotiation, the costs of the conclusion of contracts or of the surveillance of contracts. Only imposing implausible restrictions on those processes can general equilibrium be maintained.

5 Contradictorily, the inspiration for the early neoclassicals were extremely centralized markets, where coordination was an obvious feature imposed by regulations and restrictions: for Walras, the model was the Parisian Stock Exchange; at the same time, Edgeworth conceived of a market where re-contracts were immediate, since all the agents were at the same place or at least connected by phone (Witt, 1985: 575-6 n.).

6 Just as in the forties Alfred Cowles financed a research project, driven by the Commission with his name, in order to obtain better predictors for the financial markets, so in the eighties John Reed, of Citicorp, financed the Santa Fé Institute in order to obtain the same type of knowledge about financial evolution: the striking parallel was pointed out by Arrow, who participated in both projects.

References

Adelman, I. and Adelman, F. (1959), 'The Dynamic Properties of the Klein-Goldberger Model', *Econometrica*, 27(4), pp. 596-625
Alchian, Armen (1950), 'Uncertainty, Evolution and Economic Theory', *Journal of Political Economy*, vol. lviii, pp. 211-21
Alchian, Armen (1953), 'Biological Analogies in the Theory of Firm: Comment', *American Economic Review*, 43 (4-1), pp. 600-3
Aldrich, John (1989), 'Autonomy', *Oxford Economic Papers*, 41, pp. 15-34
Allen, R. (1950), 'The Work of Eugen Slutsky', *Econometrica*, 18(3), pp. 209-16
Allen, Robert (1991), *Opening Doors: The Life and Work of Joseph Schumpeter*, New Brunswick: Translation Publishers, 2 vol.
Amariglio, Jack (1990), 'Economics as a Post-modern Discourse', in W. Samuels (ed), *Economics as Discourse*, pp. 15-46
Andersen, Esben (1994), *Evolutionary Economics — Post-Schumpeterian Contributions*, London: Pinter
Anderson, Philip (1988), 'A Physicist Look at Economics: An Overview of the Workshop', in Anderson, P. et al. (eds), *The Economy as an Evolving Complex System*, pp. 265-73
Anderson, P., Arrow, K. and Pines, D. (eds) (1988), *The Economy as an Evolving Complex System*, New York: Addison-Wesley
Antonelli, Cristiano (1992, ed), *The Economics of Information Networks*, Amsterdam: North Holland
Apel, Karl-Otto (1981), *Charles S. Peirce: From Pragmatism to Pragmaticism*, Amherst: University of Massachussets Press
Arbib, Michael and Hesse, Mary (1986), *The Construction of Reality*, Cambridge: Cambridge University Press
Aristotle (CW), *The Works of Aristotle*, ed. by W. Ross, Oxford: Clarendon, ix volume, 1959 edition
Arrow, Kenneth (1984), *The Economics of Information*, Oxford: Basil Blackwell
Arrow, Kenneth (1988), 'Workshop on the Economy as an Evolving Complex System: Summary', in Anderson, P. et al. (eds), *The Economy as an Evolving Complex System*, pp. 275-81
Arrow, Kenneth (1994), 'Problems Mount in Application of Free Market Economics Theory', *Guardian*, 4 January 1994
Arrow, K. and Debreu, G. (1954), 'Existence of Equilibrium for a Competitive Economy', *Econometrica*, 20, pp. 265-90
Arrow, K. and Hahn, F. (1971), *General Competitive Analysis*, San Francisco: Holden-Day
Arrow, K. and Intriligator, M. (1981-1991), 'Introduction', in Arrow and Intriligator

(eds), *Handbook of Mathematical Economics*, 4 vol., Amsterdam: North Holland, pp. 1-14
Arthur, W. Brian (1989), 'Competing Technologies, Increasing Returns, and Lock-In by Historical Events', *Economic Journal*, 99, pp. 116-31
Arthur, W. Brian (1994), *Increasing Returns and Path Dependence in the Economy*, Ann Arbor: University of Michigan Press
Artigiani, Robert (1993), 'From Epistemology to Cosmology: Post-Modern Science and the Search for New Cultural Maps', in Laszlo, E. et al. (eds), pp. 29-60
Ayer, A. J. (1968), *The Origins of Pragmatism: Studies in the Philosophy of Charles Sanders Peirce and William James*, London: Macmillan
Bacon, Francis (1620), *The New Organon and Related Writings*, ed. By F. Andrsen, 1960, New York: Liberal Arts Press
Bak, Per (1996), *How Nature Works — The Science of Self-Organized Criticality*, New York: Springer Verlag
Bak, Per and Chen, Ken (1991), 'Self-Organized Criticality', *Scientific American*, January, pp. 26-33
Barnes, Barry (1982), *T.S. Kuhn and the Social Sciences*, London: Macmillan
Barnett, W. and Chen (1988), 'The Aggregation-Theoretic Monetary Aggregates are Chaotic and Have Strange Attractors: an Econometric Application of Mathematical Chaos', Barnett et al. (eds), *Dynamic Economic Modeling*, pp. 199-245
Barnett, W. and Choi (1989), 'A Comparison between the Conventional Econometric Approach to Structural Inference and the Nonparametric Chaotic Attractor Approach', in Barnett et al. (eds), *Economic Complexity*, pp. 141-212
Barnett, W., Gewecke, J. and Shell, K. (1989), *Economic Complexity: Chaos, Sunspots, Bubbles, and Nonlinearity*, Cambridge: Cambridge University Press
Barnett, W. and Hinich, Melvin (1993), 'Has Chaos been Discovered with Economic Data?', in Day and Chen (eds), pp. 254-65
Barthes, Roland (1953), *Le Degré Zéro de l'Ecriture, Suivi de Eléments de Sémiologie*, Paris: Seuil
Basile, Liliana (1989), 'Uncertainty in Economic Theory: An Epistemological Perspective', *Ricerche Economiche*, 18(1-2), pp. 218-35
Baumol, William (1970), *Dynamic Economics — An Introduction*, London: Macmillan
Baumol, William and Benhabib, Jess (1989), 'Chaos, Significance, Mechanism, and Economic Applications', *Journal of Economic Perspectives*, 3, pp. 77-105
Baumol, William and Quandt, R. (1985), 'Chaos Models and their Implications for Forecasting', *Eastern Economic Journal*, 11, pp. 3-15
Bausor, Randall (1994), 'Qualitative Dynamics in Economics and Fluid Mechanics: A Comparison of Recent Applications', in Mirowski, P. (ed), *Natural Images in Economic Thought*, Cambridge: Cambridge University Press, pp. 109-27
Becker, Gary (1976), 'Altruism, Egoism and Genetic Fitness: Economics and Sociobiology', *Journal of Economic Literature*, 14(2), pp. 817-26
Benassy, J.-P. (1984), *Macroeconomie et Théorie du Désiquilibre*, Paris: Dunod
Benhabib, J. and Nishimura, K. (1984), 'Cyclical Input Demands and the Adjustment Cost Theory of the Firm', in Goodwin et al. (eds), *Nonlinear Models of Fluctuating Growth*, pp. 73-85
Bensaid, Daniel (1995), *Marx l'Intempestif — Grandeurs et Misères d'Une Aventure Critique (XIX-XX Siècles)*, Paris: Fayard
Bentham, Jeremy (1843), 'Essay on Logic', vol. viii of *Works of Jeremy Bentham*, ed. by John Bowring, Edinburgh: W. Tate
Bertalanffy, Ludwig von (1952), *Problems of Life: An Evaluation of Modern Biological Thought*, N. York: Wiley

Betalanffy, Ludwig von (1962), 'General Systems Theory — A Critical View', *General Systems*, 8, pp. 1-20

Beveridge, S. and Nelson, C. (1981), 'A New Approach to Decomposition of Economic Time Series into Permanent and Transitory Components with Particular Attention to Measurement of the Business Cycle', *Journal of Monetary Economics*, 7, pp. 151-74

Black, Max (1962), *Models and Metaphors — Studies in Language and Philosophy*, N. York: Cornell University Press

Black, Max (1981), 'More about Metaphor', in Ortony, Andrew (ed), *Metaphors and Thought*, pp. 19-43

Blanchard, Olivier; Watson, Mark (1986), 'Are Business Cycles all Alike?', in Gordon, R. (ed), *The American Business Cycle*, pp. 123-79

Blatt, John (1978), 'On the Econometric Approach to Business Cycle Models', *Oxford Economic Papers*, 30, pp. 229-300

Blatt, John (1983), *Dynamic Economic Systems: A Post-Keynesian Approach*, Armonk: Sharpe

Blaug, Mark (1982), *La Méthodologie Economique*, Paris: Economica

Blaug, Mark (1986a), *La Pensée Economique — Origine et Développement*, Paris: Economica

Blaug, Mark (1986b), *Great Economists before Keynes — An Introduction to the Lives and Works of One Hundred Great Economists of the Past*, Brighton: Wheatsheaf

Blaug, Mark (1992), 'On the Historiography of Economics', *Actas do Encontro Ibérico sobre História do Pensamento Contemporâneo*, Lisboa: CISEP, pp. xli-liii

Bohm, David (1964), 'On the Problem of Truth and Understanding', in Bunge, Mario (ed), *The Critical Approach to Science and Philosophy — In Honor of Karl R. Popper*, New York: Free Press of Glencoe, pp. 212-23

Boldrin, Michele (1988), 'Persistent Oscillations and Chaos in Dynamic Economic Models: Notes for a Survey', in Anderson, P. et al (eds), *The Economy as an Evolving Complex System*, pp. 49-75

Boldrin, M. and Woodford, M. (1990), 'Equilibrium Models Displaying Endogenous Fluctuations and Chaos: A Survey', *Journal of Monetary Economics*, 25, pp. 189-222

Borges, Jorge Luis (1983), *Novas Inquirições*, Lisboa: Ed. Querco

Bottomore, Tom (1992), *Between Marginalism and Marxism — The Economic Sociology of J.A. Schumpeter*, Hempstead: Harvester Wheatsheaf

Boulding, Kenneth (1981), *Evolutionary Economics*, Beverly Hills: Sage

Boulding, Kenneth (1991), 'What is Evolutionary Economics?', *Journal of Evolutionary Perspectives*, 1, pp. 9-17

Boylan, Thomas and O'Gorman, Paschal (1995), *Beyond Rhetoric and Realism — Towards a Reformulation of Economic Methodology*, London: Routledge

Brock, William (1986), 'Distinguishing Random and Deterministic Systems: Abridged Version', *Journal of Economic Theory*, 40, pp. 168-95

Brock, William (1990), 'Chaos and Complexity in Economic and Financial Series', Furstenberg, G. (ed), *Acting Under Uncertainty: Multidisciplinary Conceptions*, Boston: Kluwer, pp. 423-50

Brock, William (1991), 'Causality, Chaos, Explanation and Prediction in Economics and Finance', in Casti, Kalquist (eds.), *Beyond Belief: Randomness, Prediction and Explanation in Science*, Boca Raton: CRC Press, pp. 230-79

Brock, W. and Dechert, D. (1988), 'Theorems Distinguishing Deterministic from Random Series', in Barnett et al. (eds), *Dynamic Economic Modeling*, pp. 247-65

Brock, William (1991), 'Nonlinear Dynamical Systems: Instability and Chaos in Economics', in Hildenbrand, W. and Sonneschein, H. (eds), *Handbook of Mathematical*

Economics, vol.4, Amsterdam: North Holland, series edited by Arrow, K.; Intriligator, M., pp. 2209-35
Brock, W., Hsieh, D. and LeBaron, B. (1991), *Nonlinear Dynamics, Chaos, and Instability: Statistical Theory and Economic Evidence*, Cambridge/USA: MIT Press
Brock, W. and Malliaris, A. (1988), *Differential Equations, Stability, and Chaos in Dynamic Economics*, New York: North Holland
Brock, W. and Sayers, C. (1988), 'Is the Business Cycle Characterized by Deterministic Chaos?', *Journal of Monetary Economics*, 71-90
Brooks, D. and Wiley, E. (1986), *Evolution as Entropy: Towards a Unified Theory of Biology*, Chicago: Chicago University Press
Buckley, Paul and Peat, F. (1996), *Glimpsing Reality: Ideas in Physics and the Link to Biology*, Toronto: University of Toronto Press
Bunge, Mario (1959), *Causality — The Place of the Causal Principle in Modern Science*, Cambridge, MA: Harvard University Press
Burns, Arthur (1934), *Production Trends in the United States Since 1870*, New York: NBER
Burns, Arthur (ed) (1952), *Welsey Clair Mitchell — The Economic Scientist*, New York: NBER
Burns, Arthur (1969), *The Business Cycle in a Changing World*, New York: NBER
Burns, Arthur and Mitchell, Wesley (1946), *Measuring Business Cycles*, New York: NBER
Caldwell, Bruce (1982), *Beyond Positivism — Economic Methodology in the Twentieth Century*, London: Unwin Hyman
Campbell, John and Mankiw, Gregory (1987), 'Permanent and Transitory Components in Macroeconomic Fluctuations', *American Economic Review*, 77, pp. 111-7
Campbell, Norman (1920), *Foundations of Science — The Philosophy of Theory and Experiment*, New York: Dover, previously named 'Physics, The Elements', ed. 1957
Campbell, Norman (1921), *What is Science?*, New York: Dover, ed. 1952
Capra, Fritjof (1982), *The Turning Point: Science, Society and the Rising Culture*, London: Wildwood House
Carabelli, Anna (1988), *On Keynes' Method*, London: Routledge
Casti, John (1995), 'The Bell Curves and Monkey Languages — When do Empirical Relations become a Law of Nature?', *Complexity*, 1(1), pp. 12-5
Christ, Carl (1994), 'The Cowles Commission's Contribution to Econometrics at Chicago 1939-1955', *Journal of Economic Literature*, 32, pp. 30-59
Clark, John Bates (1899), *The Distribution of Wealth — A Theory of Wages, Interest and Profits*, New York: Kelley
Clark, John Maurice (1927), 'The Relation between Statics and Dynamics', in Hollander, J. (ed), *Economic Essays*, New York: Books for Libraries Press, pp. 46-70
Clark, John Maurice (1952), 'Contribution to the Theory of Business Cycle', in Burns, A. (ed), *W.C. Mitchell...*, pp. 193-206
Clark, Norman and Juma, Calestous (1987), *Long-Run Economics: An Evolutionary Approach to Economic Growth*, London: Frances Pinter
Cohen, Jack and Stewart, Ian (1995), *The Collapse of Chaos*, Harmondsworth: Penguin
Colp, Ralph (1982), 'The Myth of the Marx-Darwin Letter', *History of Political Economy*, 14(4), pp. 461-82
Comte, Auguste (1852), *Catechisme Positiviste*, Paris: Garnier-Flammarion, ed. 1966
Comte, Auguste (1859), *Cours de Philosophie Positive*, Paris: Garnier-Flammarion, ed. 1959
Comte, Auguste (CW), *Oeuvres d'Auguste Comte*, Paris: Anthropos, ed. 1969-1970

Córdon, Faustino (1982), *La Biologia Evolucionista y la Dialéctica*, Madrid: Ed. Ayuso-FIM

Darnell, A.C. and Evans, J.L. (1990), *The Limits of Econometrics*, Aldershot: Edward Elgar

Darwin, Charles (1859), *The Origin of Species by Means of Natural Selection or The Preservation of Favoured Races in the Struggle for Life*, Harmondsworth: Penguin, ed. 1989

Darwin, Charles (1871), *The Descent of Man and Selection in Relation to Sex*, London: Murray, ed. 1901

Darwin, Charles (1876), *Autobiography*, London: Collins, ed. 1956

Davies, Paul (1989), *The Cosmic Blueprint — Order and Complexity at the Edge of Chaos*, Harmondsworth: Penguin

Day, Richard (1983), 'The Emergence of Chaos from Classical Economic Growth', *Quarterly Journal of Economics*, 98, pp. 201-13

Day, R. and Chen, P. (eds) (1993), *Nonlinear Dynamics and Evolutionary Economics*, Oxford: Oxford University Press

Debreu, Gérard (1981), 'Existence of Competitive Equilibrium', in Arrow, K., and Intriligator, M. (eds) (1981-1991), *Handbook of Mathematical Economics*, vol. 2, Amsterdam: North Holland, pp. 697-743

Debreu, Gérard (1991), 'The Mathematization of Economics', *American Economic Review*, 81, pp. 1-7

Depew, David and Bruce Weber (1996), *Darwinism Evolving — Systems Dynamics and the Genealogy of Natural Selection*, Cambridge, MA: MIT Press

Descartes, René (1637), *Discurso do Método e Tratado das Paixões da Alma*, Lisboa: Sá da Costa, ed. 1968

Domar, Evsey (1946), 'Capital Expansion, Rate of Growth, and Employment', *Econometrica*, 14, pp. 137-47

Dore, Mohammed (1993), *The Macrodynamics of Business Cycles — A Comparative Evalution*, Oxford: Blackwell

Dosi, G., Freeman, C. and Fabiana, S. (1994), 'The Process of Economic Development: Introducing some Stylized Facts and Theories on Technologies, Firms and Institutions', *Industrial and Corporate Change*, 9(1), pp. 1-45

Duhem, Pierre (1906), *The Aim and Structure of Physical Theory*, New York: Atheneum, ed. 1977

Durbin, J. (1970), 'Testing for Serial Correlation in Least Squares Regression when Some of the Regressors are Lagged Dependent Variables', *Econometrica*, 38, pp. 410-21

Eberle, Rolf (1970), 'Models, Metaphors, and Formal Intepretations', in C. Turbayne, *The Myth of Metaphor*, Columbia: University of South Caroline Press, Appendix

Edgeworth, F.Y. (1881), *Mathematical Physics — An Essay on the Application of Mathematics to the Moral Sciences*, London: Kegan Paul, ed. 1932

Efron, B. (1979), 'Bootstrap Methods: Another Look at the Jackknife', *Annals of Statistics*, 7(4), pp. 1-26

Eichenbaum, Martin and Singleton, Kenneth (1986), 'Do Equilibrium Real Business Cycle Theories Explain Postwar United States Business Cycles', *NBER Macroeconomic Annual*, pp. 91-135

Engels, Friedrich (1882), *Dialéctica da Natureza*, Lisboa: Estampa, n/d

Epstein, Roy (1987), *A History of Econometric Ideas*, Amsterdam: North Holland

Faber, M. and Proops, J.L.R.(1991), 'Evolution in Biology, Physics and Economics: A Conceptual Analysis', in Saviotti, P. and Metcalfe, J. (eds), *Evolutionary Theories*, pp. 58-87

Feigl, Herbert (1953), 'Notes on Causality', in Feigl, H. and Brodbeck, M. (eds), *Readings in the Philosophy of Science*, New York: Appleton-Century-Crofts, pp. 408-18
Fellner, William (1956), *Trends and Cycles in Economic Activity*, New York: Henry Holt
Fisher, Irving (1893), *Mathematical Investigations in the Theory of Value and Prices*, New Haven: Yale University Press, ed. 1926
Fisher, Irving (1925), 'Our Unstable Dollar and the So-Called Business Cycle', *Journal of Economic Statistics*, 20(2), pp. 181-98
Freedman, David and Peters, Stephen (1984), 'Bootstrapping a Regression Equation: Some Empirical Results', *Journal of the American Statistical Association*, 79(385), pp. 97-106
Freeman, Christopher (1987), 'Technical Innovation, Diffusion and Long Cycles of Economic Development', in Vasko, T. (ed), *The Long Wave Debate*, Berlin: Springer, pp. 295-309
Freeman, Christopher (1992), *The Economics of Hope — Essays on Technical Change, Economic Growth and the Environment*, London: Frances Pinter
Freeman, Christopher (1995), *History, Coevolution and Economic Growth*, mimeo, Falmer: SPRU
Freeman, Christopher and Soete, Luc (1994), *Work for All or Mass Unemployment?*, London: Pinter
Friedman, Milton (1951), 'Comment', in Universities-NBER Conference, NY: NBER
Friedman, Milton (1952), 'The Economic Theorist', in Burns, A. (ed), *W.C. Mitchell...*, pp. 237-82
Friedman, Milton (1953), 'The Methodology of Positive Economics', in Friedman, M., *Essays in Positive Economics*, Chicago: University of Chicago Press
Friedman, Milton (1991), 'Old Wine in New Bottles', *Economic Journal*, 101, 33-43
Frisch, Ragnar (1933), 'Propagation Problems and Impulse Problems in Dynamic Ecnomics', in Koch, Karen (ed), *Economic Essays in Honour of Gustav Cassel*, London: Frank Cass, pp. 171-205
Frisch, Ragnar (1948), 'Repercussion Studies at Oslo', in *American Economic Review*, 38(3), pp. 367-72
Frisch, Ragnar (1951), 'Some Personal Reminiscences on a Great Man', in Harris, S. (ed), *Schumpeter — Social Scientist*, pp. 8-10
Frisch, Ragnar (1970), 'Econometrics and the World of Today', in Eltis, W. et al. (eds), *Induction, Growth and Trade: Essays in Honor of Sir Roy Harrod*, Oxford: Clarendon, pp. 152-66
Gabisch, G. and Lorenz, H.W. (1989), *Business Cycle Theory: A Survey of Methods and Concepts*, New York: Springer Verlag
Gaill, Françoise (1987), 'Organisme', in Stengers, I. (ed), *D'Une Science à l'Autre*, pp. 244-65
Gandolfo, Giancarlo (1980), *Economic Dynamics: Methods and Models*, Amsterdam: North Holland
Garretsen, Harry (1992), *Keynes, Coordination and Beyond — The Development of Macroeconomic and Monetary Theory since 1945*, Aldershot: Edward Elgar
Gelderen, J. van (1913), 'Springvloed: Beschouwingen over Industrieele Ontwikling in Prijsbeweging', *De Nieuwe Tijd*, 18, pp. 177-253, 369-84, 445-64; in Freeman, C. (ed) (1996), *Long Wave Theory*, Aldershot: Edward Elgar
Georgescu-Roegen, Nicholas (1971), *The Entropy Law and the Economic Process*, Cambridge, USA: Harvard University Press
Gervet, Jacques (1985), 'Du Glissement de Sens dans le Discours Scierntifique: A Propos du Modèle Sociobiologique et de son Imprégnation Idéologique', in Tort, P. (ed), pp. 61-91

Giere, Ronald (1988), 'A Natureza da Ciência — Uma Perspectiva Iluminista Pós-Moderna', *Colóquio-Ciências*, 6, pp. 72-84
Goodwin, Richard (1982), *Essays in Economic Dynamics*, London: Macmillan
Goodwin, Richard (1984), 'Disaggregating Models of Fluctuating Growth', in Goodwin et al. (eds), *Nonlinear Models of Fluctuating Growth*, pp. 67-72
Goodwin, Richard (1985a), 'Oral Comments', in Bianchi et al. (eds), *Long Waves, Depression and Innovation*, Laxenburg: IIASA, pp. 271-2
Goodwin, Richard (1990), *Chaotic Economic Dynamics*, Oxford: Clarendon Press
Goodwin, R., Kruger, M. and Vercelli, A. (eds) (1984), *Nonlinear Models of Fluctuating Growth: An International Symposium, Siena, Italy, 1983*, Heidelberg: Springer Verlag
Goodwin, R. and Punzo, L. (1987), *The Dynamics of a Capitalist Economy*, Oxford: Blackwell
Goonatilake, Susantha (1990), *The Evolution of Information: Lineages in Gene, Culture and Artefact*, London: Frances Pinter
Gordon, Robert (1986), *The American Business Cycle*, Chicago: University of Chicago Press
Gordon, Robert and Veitch, John (1986), 'Fixed Investment in the American Business Cycle, 1919-1983', in Gordon, R. (ed), *The American Business Cycle*, pp. 263-357
Gould, Stephen Jay (1987), 'The Panda's Thumb of Technology', *Natural History*, 1, January, pp. 14-23
Grandmont, Jean-Michel and Malgrange, Pierre (1986), 'Nonlinear Economic Dynamics: Introduction', *Journal of Economic Theory*, 40, pp. 3-12
Granger, Clive (1969), 'Investigating Causal Relations by Econometric Models and Cross-Spectral Methods', *Econometrica*, 37(3), pp. 424-38
Granger, Clive (1980), 'Testing for Causality: A Personal Viewpoint', *Journal of Economic Dynamics and Control*, 2, pp. 303-28
Granger, C. and Terasvirta, T. (1993), *Modelling Nonlinear Economic Relationships*, Oxford: Oxford University Press
Grossman, Gene and Helpman, Elhanan (1991), *Innovation and Growth in the Global Economy*, Cambridge, USA: MIT Press
Grossman, Henryk (1943), 'The Evolutionist Revolt against Classical Economics', *Journal of Political Economics*, 51, pp. 381-390, pp. 506-22
Guesnerie, Roger and Woodford, Michael (1992), 'Endogenous Fluctuations', in Laffont, J.J. (ed), *Advances in Economic Theory—Sixth World Congress*, vol. II, Cambridge: Cambridge University Press, pp. 289-412
Gutsatz, Michel (1987), 'Loi et Causalité', in Stengers, I. (ed), *D'Une Science à l'Autre*, pp. 68-87
Haavelmo, T. (1944), 'The Probabilistic Approach in Econometrics', *Econometrica*, 12, supplement
Haberler, Gottfried (1943), *Prospérité et Dépression — Etude Théorique des Cycles Economiques*, Geneve: ONU (3rd edition)
Haberler, Gottfired (1981) 'The Great Depression of the 1930's: Can it Happen Again?', in *The Business Cycle and Public Policy, 1929-80*, A. Compendium of Papers Submitted to the Joint Economic Committee, Congress of the US, 96th Congress, 2nd session, Joint Commitee Print, US Government Printing Office, pp. 1-19
Hahn, F. (1982), 'Stability', in Arrow, K. and Intriligator, M. (eds) (1981-1991), *Handbook of Mathematical Economics*, vol. 2, Amsterdam: North Holland, pp. 745-93
Hahn, F. (1984), *Equilibrium and Macroeconomics*, Oxford: Blackwell
Hahn, F. (1991), 'The Next Hundred Years', *Economic Journal*, 101(1), pp. 47-50
Hansen, Alvin (1941), *Fiscal Policy and Business Cycle*, New York: WW Norton

Hansen, Alvin (1951a), *Business Cycles and National Income*, New York: WW Norton

Hansen, Alvin (1951b), 'Schumpeter's Contribution to Business Cycle Theory', in Harris, S. (ed), *Schumpeter — Social Scientist*, pp. 79-82

Hansen, Alvin (1952), 'Social Scientist and Social Counselor', in Burns, A. (ed), *W.C. Mitchell*, pp. 301-20

Harré, R. (1964), *Matter and Method*, Resenda: Ridgeview

Harrod, Roy (1939), 'An Essay in Dynamic Theory', *Economic Journal*, XLIX, pp. 14-33

Harrod, Roy (1951), *The Life of John Maynard Keynes*, London: Macmillan

Harrod, Roy (1960), 'Second Essay in Dynamic Theory', *Economic Journal*, 70, pp. 277-93

Hausman, Daniel (1984), *The Philosophy of Economics: An Anthology*, Cambridge: Cambridge University Press

Hausman, Daniel (1992), *The Inexact and Separate Science of Economics*, Cambridge: Cambridge University Press

Hayek, F. (1928), 'The System of Intertemporal Price Equilibrium and Movements in the "Value of money"', in Kirzner, I. (ed), *Classics in Austrian Economics*, v. III, London: Pickering, pp. 161-98

Hayek, F. (1933), 'The Trend of Economic Thinking', *Economica*, 13, pp. 121-37

Hayek, F. (1941), *The Pure Theory of Capital*, London: Routledge

Hayek, F. (1974), 'The Pretence of Knowledge', in Hayek, F. (1978), pp. 23-34

Hayek, F. (1978), *New Studies in Philosophy, Politics, Economics and the History of Ideas*, London: Routledge

Hayek, F. (1988), *The Fatal Conceit: The Errors of Socialism*, Collected Works, vol. I, London: Routledge

Hegel, Friedrich (1830), *Logic*, Oxford: Clarendon, ed. 1975

Heilbroner, Robert (1986), 'Economics and Political Economy: Marx, Keynes, and Schumpeter', in Helburn, S. and Bramhall, D. (eds), *Marx, Schumpeter, and Keynes — A Centenary Celebration of Dissent*, Armonk: Sharpe, pp. 13-26

Hendry, David (1980), 'Econometrics — Alchemy or Science?', *Economica*, 47, pp. 387-406

Hendry, D.; Morgan, M. (eds) (1995), *The Foundations of Econometric Analysis*, Cambridge: Cambridge University Press

Hesse, Mary (1966), *Models and Analogies in Science*, Notre Dame, USA: University of Notre Dame Press

Hesse, Mary (1970), *Forces and Fields: The Concept of Action at a Distance in the History of Physics*, Westport: Greenwood

Hesse, Mary (1974), *The Structure of Scientific Inference*, London: Macmillan

Hesse, Mary (1980), *Revolutions and Reconstructions in the Philosophy of Science*, Brighton: Harvester Press

Hicks, John (1937), 'Mr. Keynes and the Classics', in Hicks, J. (1967), *Critical Essays in Monetary Theory*, pp. 126-42, Oxford: Clarendon

Hicks, John (1939), *Value and Capital — An Inquiry into some fundamental Principles of Economic Theory*, Oxford: Oxford University Press

Hicks, John (1950), *A Contribution to the Theory of the Trade Cycle*, Oxford: Oxford University Press

Hicks, John (1976), 'Some Questions of Time in Economics', in Tang, E., et al. (eds), *Evolution, Welfare, Time in Economics — Essays in Honor of Nicholas Georgescu-Roegen*, Lexington: Lexington Books, pp. 135-57

Hicks, John (1979), *Causality in Economics*, Oxford: Blackwell

Hodgson, Geoffrey (1993), *Economics and Evolution — Bringing Life Back into*

Economics, Cambridge: Polity Press
Hollander, Jacob (1927), "John Bates Clark as an Economist", in Hollander, J. (ed), *Economic Essays contributed in Honor of John Bates Clark*, New York: Books for Library Press, pp. 1-5
Hollander, Samuel (1985), *The Economics of John Stuart Mill*, Oxford: Blackwell
Homan, Paul (1952), 'Place in the Contemporary Economic Thought', in Burns, A. (ed), *W.C. Mitchell*, pp. 155-92
Hoover, K. D. (1991), 'Review of Mirowski's More Heat than Light', *Methodus*, III (1), pp. 139-45
Horsburgh, H. (1958), 'Philosophers against Metaphor', *Philosophical Quarterly*, 8, July, pp. 231-45
Hull, David (1988), *Science as a Process: An Evolutionary Account of the Social and Conceptual Development of Science*, Chicago: University of Chicago Press
Hume, David (1740), 'A Treatise of Human Nature, Being an Attempt to Introduce the Experimental Method of Reasoning into Moral Subjects', in *The Philosophical Works of David Hume*, vol. I, Boston: Little, Brown & Co.
Hume, David (1748), 'An Enquiry Concerning Human Understanding', in *Enquiries Concerning Human Understanding and Concerning the Principles of Morals*, Oxford: Clarendon, 1992 reprint
Hurst, H. (1952), *The Nile — A General Account of the River and the Utilization of its Waters*, London: Constable
Hurst, H., Black, R. and Simaika, Y. (1966), *The Nile Basin*, vol. X of *The Major Nile Projects*, Cairo: Government Printing Offices
Hutchison, Terence (1938), *The Significance and Basic Postulates of Economic Theory*, New York: Kelley, ed. 1960
Hutchison, Terence (1992), *Changing Aims in Economics*, Oxford: Blackwell
Ingrao, Bruna (1989), 'The Hidden Epistemology of Rational Expectations and the Paradoxes of Rationality', *Ricerche Economiche*, 43 (1-2), pp. 100-28
Ingrao, Bruna and Israel, Giorgio (1985), 'General Equilibrium Theory — A History of Ineffectual Paradigm Shifts', *Fundamenta Scientiae*, 6(1), pp. 1-45; 6(2), pp. 89-125
Jantsch, Erich (1980), *The Self-Organising Universe: Scientific and Human Implications of the Emerging Paradigm of Evolution*, Oxford: Pergamon
Jarsulic, Marc (1993), 'Recent Developments in Business Cycle Theory', *Review of Political Economy*, 5(3), July, pp. 344-64
Jenner, Richard (1994), 'Schumpeterian Growth, Chaos, and the Formation of Dissipative Structures', *Journal of Evolutionary Economics*, 4, pp. 125-39
Jevons, William (1870), *Elementary Lessons in Logic: Deductive and Inductive*, London: Macmillan
Jevons, William (1877), *The Principles of Science*, New York: Dover Publications, ed. 1905
Jones, Roger (1982), *Physics as Metaphor*, New York: New American Library
Jones, Lamar (1989), 'Schumpeter versus Darwin: in re Malthus', *Southern Economic Journal*, 56(2), pp. 410-22
Juglar, Clément (1862), *Des Crises Commerciales et de Leur Retour Périodique en France, en Angleterre et aux Etats-Unis*, Roma: Edizione Bizarri (ed. 1968)
Juglar, Clément (1893), *A Brief History of Panics — And Their Periodical Occurrence in the United States*, New York: A.M. Kelley Publishers (adapted in 1916, by De Courcy Thom, authorized by the author; ed. 1966)
Kaldor, Nicholas (1985), *Economics without Equilibrium*, Cardiff: University of Cardiff Press
Kalecki, Michal (1936), 'Some Remarks on Keynes' Theory', in Targetti, F. and Kinda-

Hass, B., 'Kalecki's Review of Keynes' General Theory', *Australian Economic Papers*, 1982, pp. 244-60
Kalecki, Michal (1943), 'Political Aspects of Full Employment', *Political Quarterly*, 14, pp. 322-30
Kalecki, Michal (1966), *Théorie de la Dynamique Economique — Essai sur les Variations Cycliques et à Long Terme de l'Economie Capitaliste*, Paris: Gauthier-Villars
Kalecki, Michal (1968), 'Trends and Business Cycles Reconsidered', *Economic Journal*, 78, pp. 263-76
Kalecki, Michal (1970), *Estudios sobre la Teoria de los Ciclos Economicos*, Barcelona: Ariel
Kalecki, Michal (1971), *Selected Essais on the Dynamics of the Capitalist Economy, 1933-1970*, Cambridge: Cambridge University Press
Kant, Immanuel (1790), *Critique of Judgement*, London: Macmillan, ed J. Bernard, 1914
Katouzian, Homa (1980), *Ideology and Method in Economics*, London: Macmillan
Kauffman, Stuart (1988), 'The Evolution of Economic Webs', in Anderson, P. et al. (eds), *The Economy as an Evolving Complex System*, pp. 125-46
Kauffman, Stuart (1993), *The Origins of Order — Self-Organization and Selection in Evolution*, Oxford: Oxford University Press
Kauffman, Stuart (1995), *At Home in the Universe — The Search for Laws of Self-Organization and Complexity*, Oxford: Oxford Univ. Press
Kenkel, James (1974), *Dynamic Linear Economic Models*, New York: Gordon and Breach
Keynes, John Neville (1890), *The Scope and Method of Political Economy*, New York: Kelley (ed) (1963)
Keynes, John Maynard (TP, 1921), *A Treatise on Probability*, London: Macmillan, ed. 1978
Keynes, John Maynard (TM, 1930), *Treatise on Money*, London: Macmillan
Keynes, John Maynard (GT, 1936), *The General Theory of Employment, Interest and Money*, London: Macmillan
Keynes, John Maynard (1937), 'The General Theory of Employment', *Quarterly Journal of Economics*, 51, pp. 209-23
Keynes, John Maynard (1939), 'Professor Tinbergen's Method', *Economic Journal*, pp. 558-68 (also compiled in *The Collected Writings of John Maynard Keynes*, vol. xiv, Cambridge: Macmillan, pp. 285-321)
Keynes, John Maynard (CW, 1971-1989), *The Collected Writings of John Maynard Keynes*, ed. D. Moggridge, London: Macmillan for the Royal Economic Society, v. i to xx
King, R., Plosser, C., Stock, J. and Watson, M. (1987), *Stochastic Trends and Economic Fluctuations*, NBER Discussion Paper 2229
Kirman, Alan (1989), 'The Intrinsic Limits of Modern Economic Theory: The Emperor Has No Clothes', *Economic Journal*, 99, pp. 126-39
Kirzner, I. (1973), *Competition and Entreprise*, Chicago: Chicago University Press
Knight, Frank (1921), *Risk, Uncertainty and Profit*, Boston: Houghton Mifflin, ed. 1940
Knight, Frank (1935), 'Statics and Dynamics — Some Queries Regarding the Mechanical Analogy in Economics', in Knight, F., *The Ethics of Competition — And Other Essays*, London: Unwin, pp. 161-85
Knight, Frank (1956), *On the History and Method of Economics — Selected Essays*, Chicago: University of Chicago Press
Kondratiev, Nikolai (1992), *Les Grands Cycles de la Conjoncture*, Paris: Economica
Koopmans, T.C. (1937), *Linear Regression of Economic Time Series*, Haarlem: Erven

Bohn

Koopmans, T.C. (1947), 'Measurement Without Theory', in *Readings in Business Cycle Theory*, 1966, ed. by American Association of Economists, London: Unwin, pp. 186-203

Koopmans, T.C. (1957), 'The Interaction of Tools and Problems', in *Three Essays on the State of Economic Science*, New York: McGraw

Koopmans, T.C. (1979), 'Economics among the Sciences', *American Economic Review*, vol. 63, pp. 1-13

Krugman, Paul (1996), *The Self-Organizing Economy*, Oxford: Blackwell

Kuhn, Thomas (1975), *La Estructura de las Revoluciones Cientificas*, Ciudad México: Ed. Fondo de Cultura

Kuhn, Thomas (1979), 'Reflexões sobre os Meus Críticos', in Lakatos and Musgrave (eds), *A Crítica e o Desenvolvimento*, pp. 285-343

Kuhn, Thomas (1981), 'Metaphor in Science', in Ortony, A. (ed), *Metaphor and Thought*, pp. 409-19

Kuhn, Thomas (1989), *A Tensão Essencial*, Lisboa: Edições 70

Kuznets, Simon (1955), 'Economic Growth and Income Inequality', *American Economic Review*, 45, pp. 1-28

Lackoff, G. and Johnson, M. (1980), *Metaphors we Live By*, Chicago: Chicago University Press

Lakatos, Imre (1979), 'O Falseamento e a Metodologia dos Programas de Investigação Científica', *in* Lakatos, I. and Musgrave, A. (eds), *A Crítica e o Desenvolvimento do Conhecimento*, pp. 109-243

Lamarck, J. B. (1809), *Zoological Philosophy—An Exposition with Regard to the Natural History of Animals*, New York: Hafner, ed. 1963

Laplace, Marquis de (1812), *Théorie Analytique des Probabilités*, Paris: Gauthiers-Villiers, vol. vii, Oeuvres Complètes, ed. 1886

Lawson, Anthony (1989), 'Realism and Instrumentalism in the Development of Econometrics', in *Orford Economic Papers*, 41, pp. 236-258

Leamer, Edward (1987), *Specification Searches: Ad Hoc Inferences with Nonexperimental Data*, New York: Wiley

Leamer, Edward (1983), 'Let's Take the Con out of Econometrics', *American Economic Review*, 73, March, pp. 31-43

Leontief, Wassily (1966), *Ensayos sobre Economia*, Barcelona: Planeta Agostini, ed. 1986

Leontief, Wassily (1971), 'Theoretical Assumptions and Nonobserved Facts', *American Economic Review*, 61, pp. 1-7

Li, Tien-Yen and Yorke, James (1975), 'Period Three Implies Chaos', *American Mathematical Monthly*, 82, pp. 985-92

Lorenz, W.-H. (1989), *Nonlinear Dynamical Economics and Chaotic Motions*, Berlin: Springer-Verlag

Losee, John (1980), *A Historical Introduction to the Philosophy of Science*, Oxford: Oxford University Press

Lotka, A. (1956), *Elements of Mathematical Biology*, New York: Constable

Louçã, Francisco (1987), 'A Economia como Ciência (Notas sobre Definições)', *Revista de Economia*, 8(1),pp 109-13

Louçã, Francisco (1993), *A Transição de 'Paradigma Tecno-Económico' como Factor das Ondas Longas do Desenvolvimento Capitalista — O Caso da Revolução Micro-Electrónica*, mimeo, ISEG/UTL, July 1993

Louçã, Francisco (1994), *A Maldição de Midas — A Cultura do Capitalismo Tardio*, Lisboa: Cotovia

Louçã, Francisco (1997a), *Frisch, Schumpeter and the Intriguing Pendulum*, mimeo, Lisbon: ISEG
Louçã, Francisco (1997b), 'Ernest Mandel and the Pulsation of History', in Achcar, G. (ed.), *The Contribution of Ernest Mandel to Marxist Theory*, London: Verso
Lucas, Robert (1975), 'An Equilibrium Model of the Business Cycle', *Journal of Political Economy*, 83, pp. 1113-44
Lucas, Robert (1977), 'Understanding Business Cycles', *Journal of Monetary Economics*, 5, pp. 7-29
Lucas, Robert (1980), 'Methods and Problems in Business Cycle Theory', *Journal of Money, Credit and Banking*, 12, pp. 696-715
Lucas, Robert (1981), *Studies in the Business Cycle Theory*, Oxford: Blackwell
Lukacs, John (1971), 'Quantum Mechanics and the End of Scientism', in Olson, R. (ed), *Science as Metaphor*, pp. 293-300
Machlup, Fritz (1951), 'Schumpeter's Economic Methodology', in Harris, S. (ed), *Schumpeter—Social Scientist*, pp. 95-101
Machlup, Fritz (1956), 'On Indirect Verification', in Hausman, D (ed) (1984), *The Philosophy of Economics*, pp. 198-209
Maddison, Angus (1977), 'Phases of Capitalist Development', *Banca Nazionale del Lavoro Quarterly Review*, pp. 103-37
Maddison, Angus (1991), *Dynamic Forces in Capitalist Development*, Oxford: Oxford University Press
Maddison, Angus (1993), 'Standardised Estimates of Fixed Capital Stock: A Sixth County Comparison', *Inovazione e Materie Prime*, April, pp. 1-29
Maki, Ulskali (1989), 'On the Problem of Realism in Economics', *Ricerche Economiche*, 43, pp. 176-98
Maki, Uskali (1993), 'Two Philosophies of the Rhetoric of Economics', in Henderson et al. (eds), *Economics as Language*, pp. 23-50
Malthus, Thomas Robert (1798), *An Essay on the Principle of Population and A Summary View of the Principle of Population*, Harmondsworth: Penguin, ed. by Antony Flew, ed. 1970
Mandel, Ernest (1985), 'Partially Independent Variables and Internal Logic in Classical Marxist Economic Analysis', *Social Science Information*, xxiv(3), pp. 485-505
Mandelbrot, Benoit (1972), 'Statistical Methodology for Non-Periodic Cycles', *Annals of Economic and Social Measurement*, 1, pp. 259-90
Mandelbrot, Benoit (1983), *The Fractal Geometry*, New York: Freeman
Mandelbrot, Benoit (1987), 'Towards a Second Stage of Indeterminism in Science', *Interdisciplinary Science Reviews*, 12, pp. 117-27
Mandelbrot, Benoit (1989), 'Chaos, Bourbaki, and Poincaré', *Mathematical Intelligencer*, 11(3), pp. 10-12
Mani, G.S. (1991), 'Is There a General Theory of Biological Evolution', in Saviotti, P. and Metcalfe, J. (eds), *Evolutionary Theories*, pp. 31-57
De Marchi, N. and Gilbert, P. (1989), 'Introduction', *Oxford Economic Papers*, 41, pp. 1-11
Marshall, Alfred (1885), 'The Present Position of Economics', in Pigou, A. (ed), *Memorial of Alfred Marshall*, ed. by A. Pigou, London: Macmillan, ed. 1925
Marshall, Alfred (1890), *Principles of Economics—An Introductory Volume*, New York: Macmillan (including the 1898 and 1907 prefaces)
Marquez, Gabriel Garcia (n.d.), *Cem Anos de Solidão*, Lisboa: Europa América
Marx, Karl (1857-8), *Fondements de la Critique de l'Economie Politique*, Paris: Anthropos, ed. 1967
Marx, Karl (1867, 1885, 1894), *Le Capital—Critique de l'Economie Politique*, 3 vol.,

Paris: Editions Sociales, ed. 1977
Marx, Karl (1873), *Postface à la Deuxième Edition Allemande du Capital*, in Marx, K. (1867), pp. 15-22
Marx, Karl and Engels, Friedrich (1988), 'Letters', in *Collected Works of Marx and Engels*, v. 41, v. 43, London: Lawrence and Wishart
Masterman, Margaret (1979), 'A Natureza de um Paradigma', *in* Lakatos and Musgrave (eds), *A Crítica e o Desenvolvimento do Conhecimento*, pp. 72-108
Maxwell, J.C. (1876), 'Does the Progress of Physical Science Tend to Give any Advantage to the Opinion of Necessity (or Determinism) over that of Contingency of Events and the Freedom of Will?', in Campbell, L. and Garnett, W., *The Life of James Clark Maxwell, With a Selection from his Correspondence and Occasional Writings and a Sketch of His Contributions to Science*, London: Macmillan, ed. 1882, pp. 434-44
Maxwell, J.C. (1890), *The Scientific Papers of James Clerk Maxwell*, New York: Dover, ed. by W. Niven, 1965
May, Robert (1973), *Stability and Complexity in Model Ecosystems*, Princeton: Princeton University Press
Maynard Smith, John (1972), *On Evolution*, Edinburgh: Edinburgh University Press
Maynard Smith, John (1993), *The Theory of Evolution*, Cambridge: Cambridge Univ.ersity Press
Mayr, Ernst (1982), *The Growth of Biological Thought — Diversity, Evolution, and Inheritance*, Cambridge /USA: Harvard University Press
Mayr, Ernst (1988), *Toward a New Philosophy of Biology — Observations of an Evolutionist*, Cambridge/USA: Harvard University Press
McCloskey, D. (1983), 'The Rhetorics of Economics', *Journal of Economic Literature*, 21, pp. 481-517
McCarthy, Georges (1988), *Marx's Critique of Science and Positivism — The Methodological Foundations of Political Economy*, Dordrecht: Kluwer
McCloskey, D.(1985), *The Rhetorics of Economics*, Madison: University of Wisconsin Press
McKenzie, Lionel (1989), 'General Equilibrium', in Eatwell, J. et al. (eds), pp. 1-35
Medio, Alfredo (1992), *Chaotic Dynamics*, New York: Cambridge University Press, with Giancarlo Gallo
Meyerson, Emile (1908), *Identity and Reality*, London: Allen & Unwin, ed. 1964
Mill, John Stuart (1843), *A System of Logic, Ratiocinative and Inductive — Being a Connected View of the Principles of Evidence and the Methods of Scientific Investigation*, Toronto: University of Toronto Press, ed. 1973
Mill, John Stuart (1844), 'On the Definition and Method of Political Economy', in Hausman, D. (ed), pp. 52-69, ed. 1984
Mill, John Stuart (1848), *Principles of Political Economy with Some of Their Applications to Social Philosophy*, London: Longmans, ed. 1909
Mill, John Stuart (1865), *Auguste Comte and Positivism*, London: Trubner
Mill, John Stuart (CW) *Collected Works of John Stuart Mill*, Toronto: University of Toronto Press-Routledge
Mill, John Stuart (1873), *Memórias*, Lisboa: Gleba, n/d
Mills, Frederick (1952), 'A Professional Sketch', in Burns, A. (ed), *W.C. Mitchell*, pp. 107-24
Mini, Piero (1974), *Economics and Philosophy*, Gainsville: University of Florida Press
Mini, Piero (1991), *Keynes, Bloomsbury and the General Theory*, London: Macmillan
Mini, Piero (1994), *John Maynard Keynes: A Study in the Psychology of Original Work*, New York: St. Martin's Press
Mirowski, Philip (1985), *The Birth of the Business Cycle*, New York: Garland Press

Mirowski, Philip (1988), *Against Mechanism: Protecting Economics from Science*, Totowa: Rowman and Littlefield

Mirowski, Philip (1989a), 'The Rise and Fall of the Concept of Equilibrium in Economic Analysis', *Recherches Economiques de Louvain*, 55(4), pp. 447-68

Mirowski, Philip (1989b), *More Heat than Light: Economics as Social Physics, Physics as Nature's Economics*, Cambridge: Cambridge University Press

Mirowski, Philip (1989c), ''Tis a Pity Econometrics Isn't an Empirical Endeavor: Mandelbrot, Chaos, and the Noah and Joseph Effects', *Ricerche Economiche*, 18(1-2), pp. 76-99

Mirowski, Philip (1989d), 'The Probabilistic Counter-Revolution, or How Stochastic Concepts Came to Neoclassical Economic Theory', *Oxford Economic Papers*, 41, pp. 217-35

Mirowski, Philip (1992), 'Nicholas Georgescu-Roegen', in Samuels, W. (ed), *New Horizons in Economic Thought—Appraisals of Leading Economists*, Aldershot: Elgar, pp. 58-85

Mirowski, Philip (ed) (1994), *Natural Images in Economic Thought — 'Markets Read in Tooth and Clay'*, Cambridge: Cambridge University Press

Mirowski, Philip (1995), 'Mandelbrot's Economics After a Quarter Century', *Fractals*, 3(3), pp. 581-600

Mitchell, B.R. (1992), *International Historical Statistics — Europe 1750-1988*, New York: Stockton

Mitchell, Wesley (1927), *Business Cycle — The Problem and its Settings*, New York: NBER (ed 1956)

Moggridge, Donald (1992), *John Maynard Keynes—An Economist's Biography*, London: Routledge

Moore, G.E. (1903), *Principia Ethica*, Cambridge: Cambridge University Press

Moore, G.E. (1922), *Philosophical Studies*, London: Routledge

Moore, G. and Zarnowitz, V. (1986), 'The Development and Role of the National Bureau of Economic Research's Business Cycle Chronologies', in Gordon, R. (ed), *The American Business Cycle*, pp. 735-79

Morgan, Mary (1990), *The History of Econometric Ideas*, Cambridge: Cambridge University Press

Morgan, Mary (1994), 'Marketplace Morals and the American Economists: The Case of John Bates Clark', in De Marchi, N. and Morgan, M. (eds), *Higgling*, Durham: Duke University Press, pp. 229-52

Morishima, Michio and Catephores, George (1988), 'Anti Say's Law versus Say's Law: A Change in Paradigm', in Hanusch, H. (ed), *Evolutionary Economics—Applications of Schumpeter's Ideas*, Cambridge: Cambridge University Press, pp. 23-52

Moura, Francisco (1981), 'Ciclos Políticos e Modelos Político-Econométricos', *Estudos de Economia*, 1(3), pp. 263-88

Mullineux, Andy and Peng, WenSheng (1993), 'Nonlinear Business Cycles Modelling', *Journal of Economic Surveys*, 7(1), pp. 41-83

Nasmyth, George (1971), 'A Critique of Social-Darwinism', in R. Olson (ed), *Science as Metaphor*, pp. 133-44

Nelson, C. and Plosser, C. (1982), 'Trends and Random Walks in Macroeconomic Time Series', *Journal of Monetary Economics*, 10, pp. 139-62

Nelson, Richard (1995), 'Recent Evolutionary Theorizing about Economic Change', *Journal of Economic Literature*, 33(1), pp. 48-90

Nelson, Richard and Winter, Sidney (1982), *An Evolutionary Theory of Economic Change*, Cambridge, USA: Harvard University Press

Newton, Isaac (1704), *Opticks or A Treatise of the Reflections, Refractions, Inflections and Colours of Light*, New York: Dover Publications, ed. 1952

Newton, Isaac (1687), *Mathematical Principles of Natural History*, Berkeley: University of California Press, 2 vol., ed 1962

Newton, Isaac (1713), Letters, in Thayer, H. (ed.), *Newton's Philosophy of Nature — Selection from his Writings*, New York: Hafner, 1953

Nordhaus, William (1975), 'The Political Business Cycle', *Review of Economic Studies*, 42, pp. 169-90

Oackley, Allen (1990), *Schumpeter's Theory of Capitalist Motion—A Critical Exposition and Reassessment*, Aldershot: Edward Elgar

O'Donnell, R. (1989), *Keynes' Philosophy, Economics and Politics—The Philosophical Foundations of Keynes' Thought and their Influence on his Economics and Politics*, London: Macmillan

Olson, Richard (1971a), *Science as Metaphor—The Historical Role of Scientific Theories in Forming Western Culture*, Blemont: Wadsworth

Palmer, Richard (1988), 'Final Plenary Discussion', in Anderson, P. et al (eds), *The Economy as an Evolving Complex System*, pp. 257-61

Pasinetti, Luigi (1974), *Growth and Income Distribution—Essays in Economic Theory*, Cambridge: Cambridge University Press

Patinkin, Don (1976), 'Keynes and Econometrics: On the Interaction between the Macroeconomic Revolutions of the Interwar Period', *Econometrica*, 44(6), pp. 1091-123

Pearson, Karl (1892), *The Grammar of Science*, London: Dent

Peirce, Charles (1934), *Collected Papers of Charles Sanders Peirce: From Pragmatism to Pragmaticism*, vol. 5, Cambridge, USA: Harvard University Press, ed. by C. Hartshorne and P. Weiss

Penrose, Edith (1952), 'Biological Analogies in the Theory of the Firm', *American Economic Review*, vol. 42 (4), pp. 804-19

Pessis-Pasternak, Guitta (ed) (1993), *Será Preciso Queimar Descartes? Do Caos à Inteligência Artificial: Quando os Cientistas se Interrogam*, Lisboa: Relógio d'Água

Peters, Edgar (1991), *Chaos and Order in the Capital Markets —A New View of Cycles, Prices, and Market Volatility*, New York: Wiley

Peters, Edgar (1994), *Fractal Markets Analysis —Applying Chaos Theory to Investment and Economics*, New York: Wiley

Phelps, Edmund (1962), 'A New View of Investment: A Neoclassical Analysys', *Quarterly Journal of Economics*, 78(4), pp. 548-67

Poincaré, Henri (1903), *Science and Hypothesis*, New York: Walter Scott Pub. Co.

Poincaré, Henri (1905), *The Value of Science*, in Poincaré, *Foundations of Science*, Washington: University Press of America, ed. 1982

Poincaré, Henri (1908), *Science et Méthode*, Paris: Flammarion

Popper, Karl (1962a), *The Logic of Scientific Discovery*, London: Hutchison

Popper, Karl (1962b), *The Open Society and its Ennemies*, London: Routledge

Popper, Karl (1972), *Objective Knowledge — An Evolutionary Approach*, Oxford: Clarendon

Popper, Karl (1974), *The Unended Quest — An Intellectual Autobiography*, London: Fontana

Popper, Karl (1982), *Conjecturas e Refutações — O Progresso do Conhecimento Científico*, Brasília: Editora Universidade de Brasília

Popper, Karl (1983), 'The Rationality Principle', in Miller, D. (ed), *A Pocket Popper*, Oxford: Fontana, pp. 357-65

Possas, Mário (1987), *Dinâmica da Economia Capitalista — Uma Abordagem Teórica*, São Paulo: Brasiliense

Prigogine, Ilya (1981), 'Time, Irreversibility, and Ramdomness', in Jantsch, E. (ed), *The*

Evolutionary Vision, pp. 73-82
Prigogine, Ilya (1986), 'Science, Civilization and Democracy — Values, Systems, Structures and Affinities', *Futures*, August, pp. 493-507
Prigogine, Ilya (1989), 'The Philosophy of Instability', *Futures*, August, pp. 396-400
Prigogine, I. and Stengers, I. (1985), *Order Out of Chaos — Man's New Dialogue with Nature*, London: Fontana
Quine, W. (1979), 'A Postscript on Metaphor', in Sacks, S. (ed), *On Metaphor*, Chicago: University of Chicago Press
Radner, Roy (1968), 'Competitive Equilibrium under Uncertainty', *Econometrica*, 36(1), pp. 31-58
Radner, Roy (1982), 'Equilibrium under Uncertainty', in Arrow, K. and Intriligator, M. (eds) (1981-1991), *Handbook of Mathematical Economics*, vol. 2, Amsterdam: North Holland, pp. 923-1006
Radner, Roy (1989), 'Uncertainty and General Equilibrium', in Eatwell et al. (eds), *General Equilibrium*, London: Macmillan, pp. 304-23
Ricoeur, Paul (1977), *The Rule of the Metaphor — Multi-Disciplinary Studies of the Creation of Meaning in Language*, Toronto: University of Toronto Press
Ricour, Paul (1979), 'The Metaphorical Process as Cognition, Imagination and Feeling', in Sacks, S. (ed), *On Metaphor*, Chicago: University of Chicago Press, pp. 141-57
Robbins, Lionel (1932), *An Essay on the Nature and Significance of Economic Science*, London: Macmillan, ed. 1984
Robinson, Joan (1962), *Economic Philosophy*, London: Watts
Robinson, Joan (1973), 'A Lecture Delivered at Oxford by a Cambridge Economist', in *Collected Economic Papers*, vol. IV, Oxford: Blackwell, pp. 254-63
Robinson, Joan (1979a), 'History versus Equilibrium', in *Collected Economic Papers*, vol. V, Oxford: Blackwell, pp. 48-58
Robinson, Joan (1979b) 'What are the Questions?', in *Collected Economic Papers*, vol. V, Oxford: Blackwell, pp. 1-31
Robinson, Joan (1980), *What are the Questions? And Other Essays — Further Contributions to Modern Economics*, New York: Sharpe
Rosenberg, Alexander (1992), *Economics — Mathematical Politics or Science of Diminishing Returns?*, Chicago: University of Chicago Press
Rosenberg, Nathan (1994), 'Joseph Schumpeter: Radical Economist', in Shionaya, Y. and Perlman, M. (eds), *Schumpeter in the History of Ideas*, Ann Arbor: University of Michigan Press, pp. 41-57
Rosenberg, N. and C. Frischtak (1986), 'Technological Innovation and Long Waves', in Freeman, C. (ed), *Design, Innovation and Long Cycles in Economic Development*, pp. 5-26
Rosser, J.B. (1992), 'The Dialogue Between the Economic and the Ecological Theories of Evolution', *Journal of Economic Behaviour and Organization*, 17, pp. 195-215
Rousseau, Jean Jacques (1755), 'Discours sur l'Economie Politique', in *Ecrits Politiques*, Paris: UGE, ed. 1972
Ruelle, David (1988), 'Can Nonlinear Dynamics Help Economists?', in Anderson, P. et al. (eds), *The Economy as an Evolving Complex System*, pp. 195-204
Ruelle, David (1990), 'Deterministic Chaos: The Science and the Fiction', *Proceedings of the Royal Society of London*, A427, pp. 241-48
Ruelle, David (1991), *Hasard et Chaos*, Paris: Odile Jacob
Ruse, M. (1982), *Darwinism Defended: A Guide to the Evolution Controversies*, Reading: Addison-Wesley
Russell, Bertrand (1913), 'On the Notion of Cause', *Proceedings of the Aristotelian Society*, 13, pp. 1-26

Sacarrão, Germano (1989), *Biologia e Sociedade — Crítica da Razão Dogmática*, Mem Martins: EuropaAmérica
Samuelson, Paul (1943), 'Dynamics, Statistics, and the Stationary State', *Review of Economics and Statistics*, 25, pp. 58-68
Samuelson, Paul (1947), *Fundamentos daAnálise Económica*, São Paulo:Abril, ed. 1983
Samuelson, Paul (1951), 'Schumpeter as a Teacher and Economic Theorist', in Harris, S. (ed), *Schumpeter — Social Scientist*, pp. 48-53
Samuelson, Paul (1974), 'Rememberances of Frisch', *European Economic Review*, 5, pp. 7-23
Santos, Boaventura Sousa (1987), *Um Discurso sobre as Ciências*, Porto: Afrontamento
Santos, Boaventura Sousa (1989), *Introdução a uma Ciência Pós-Moderna*, Porto: Afrontamento
Sargent, Thomas (1993), *Bounded Rationality in Macroeconomics*, Oxford: Clarendon
Saviotti, P. and Metcalfe, J. (eds) (1991a), *Evolutionary Theories of Economic and Technological Change: Present Studies and Future Prospects*, Reading: Harwood Academic Publishers
Saviotti, P. and Metcalfe, J (1991b), 'Present Development and Trends in Evolutionary Economics', in Saviotti and Metcalfe (eds), *Evolutionary Theories*, pp. 1-30
Say, Jean-Baptiste (1803), *A Treatise of Political Economy — Or the Production, Distribution and Consumption of Wealth*, London: Longman, ed. 1821
Scheinkman, José (1990), 'Nonlinearities in Economic Dynamics', *Economic Journal*, 100, pp. 33-48
Scheinkman, J. and LeBaron, B. (1989),' Nonlinear Dynamics and Stock Returns', *Journal of Business*, 62, pp. 311-37
Scheinkman, José and Woodford, Michael (1994), 'Self-Organized Criticality and Economic Fluctuations', *American Economic Review*, 84(2), May, pp. 417-21
Schlanger, Judith (1971), *Les Métaphores de l'Organisme*, Paris: Harmattan, ed. 1995
Schrodinger, Erwin (1944), *What is Life? The Physical Aspect of the Living Cell*, Cambridge: Cambridge University Press
Schumpeter, Joseph (1906), 'Professor Clark Verteilungstheorie', *Zeitschrift fur Volkswirtschaft, Sozialpolitik und Verwaltung*, 15, pp. 325-33
Schumpeter, Joseph (TED, 1911), *Teoria do Desenvolvimento Económico — Uma Investigação sobre Lucros, Capital, Crédito, Juro e o Ciclo Económico*, S. Paulo: Abril, ed. 1982
Schumpeter, Joseph (EDM, 1914), *Economic Doctrines and Method — An Historical Sketch*, New York: Oxford University Press, ed. 1954
Schumpeter, Joseph (1927), 'The Explanation of the Business Cycle', *Economica*, pp. 286-311
Schumpeter, Joseph (1930), 'Mitchell's Business Cycles', *Quarterly Journal of Economics*, pp. 150-72
Schumpeter, Joseph (1933), 'The Common Sense of Econometrics', *Econometrica*, 1(1), pp. 5-12
Schumpeter, Joseph (1935), 'The Analysis of Economic Change', *Review of Economic Statistics*, 17(4), May, pp. 2-10
Schumpeter, Joseph (1936), 'Review of Keynes' General Theory', *Journal of the American Statistical Association*, December, pp. 791-5
Schumpeter, Joseph (1937), 'Preface to the Japanese Edition of Theorie der Wirtschaftlichen Entwicklung', in Schumpeter, J. (1951), pp. 158-63
Schumpeter, Joseph (BC, 1939), *Business Cycles—A Theoretical, Historical and Statistical Analysis of the Capitalis Process*, 2 vol., New York: McGraw Hill
Schumpeter, Joseph (CSD, 1942), *Capitalism, Socialism and Democracy*, London: Unwin

University Books, ed. 1974
Schumpeter, Joseph (1949), 'The Historical Approach to the Analysis of Business Cycles', in Schumpeter, J. (1951), pp. 308-15
Schumpeter, Joseph (1951), *Essays on Economic Topics of Joseph Allois Schumpeter*, New York: Kennikat Press (edited by R. Clemence)
Schumpeter, Joseph (1952), 'The General Economist', in Burns, A. (ed), *W.C. Mitchell*, pp. 321-40
Schumpeter, Joseph (HEA, 1954), *History of Economic Analysis*, London: Routledge, ed. 1994
Schumpeter, Joseph (TGE, 1990), *Diez Grandes Economistas: De Marx a Keynes*, Madrid: Alianza Editorial
Scoon, Robert (1968), 'The Rise and Impact of Evolutionary Ideas', in Persons, S. (ed), *Evolutionary Thought in America*, pp. 4-42
Shackle, George (1967), *The Years of High Theory—Invention and Tradition in Economic Thought, 1926-1939*, Cambridge: Cambridge University Press
Shackle, George (1968), *A Scheme of Economic Theory*, Cambridge: Cambridge University Press
Shackle, George (1972), *Epistemics and Economics: A Critique of Economic Doctrines,* Cambridge: Cambridge University Press
Shackle, George (1982), 'Sir John Hicks' IS-LM: An Explanation, a Comment', *Journal of Post-Keynesian Economics*, 4(3), pp. 435-8
Shackle, G. (1990), *Time, Expectations and Uncertainty in Economics—Selected Essays of G.L.S. Shackle*, ed J. Ford, Aldershot: Edward Elgar
Sheehan, R. and Grieves, R. (1982), 'Sunspots and Cycles: A Test of Causation', *Southern Economic Journal*, January, pp. 775-7
Sherman, Howard (1991), *The Business Cycle—Growth and Crisis Under Capitalism*, Princeton: Princeton University Press
Silvestre, Joaquim (1993), 'The Market-Power Foundations of Economic Policy', *Journal of Economic Literature*, 31(1), pp. 105-41
Simon, Herbert (1952), 'On the Definition of the Causal Relation', *Journal of Philosophy*, 49(16), pp. 517-28
Simon, Herbert (1953), 'Causal Ordering and Identifiability', in Hood and Koopmans (eds), *Studies in Econometric Methods*, New York: Wiley, pp. 49-74
Sims, Christopher (1980), 'Macroeconomics and Reality', *Econometrica*, 48(1), pp. 1-48
Sismondi, Sismonde de (1819), *Nouveaux Principes d' Economie Politique—Ou de la Richesse das ses Rapports avec la Population*, v. I, Paris: Calmann-Lévy, ed. 1971
Skidelsy, Robert (1983), *John Maynard Keynes—Hopes Betrayed 1883-1920*, London: Macmillan
Skidelsky, Robert (1992), *John Maynard Keynes—The Economist as a Saviour, 1920-1937*, London: Macmillan
Slutsky, Eugen (1927), 'The Summation of Random Causes as the Source of Cyclical Processes', *Econometrica*, 5(1937), pp. 105-46
Smale, Stephen (1980), *The Mathematics of Time: Essays on Dynamical Systems, Economic Processes and Related Topics*, Berlin: Springer Verlag
Smith, Adam (1758),' The Principles which Lead and Direct Philosophical Enquiries: Illustrated by the History of Astronomy', in *Works*, vol. V, ed. D. Stewart, London: Cadell
Smith, Adam (1776), *The Wealth of Nations*, New York: Modern Library, ed. Edwin Canan, 1937
Smith, Vernon (1991), 'Rational Choice: The Contrast Between Economics and

Psychology', *Journal of Political Economy*, 99(4), pp. 877-97
Smithies, Arthur (1951a), 'Schumpeter and Keynes', in Harris, S. (ed), *Schumpeter — Social Scientist*, pp. 136-42
Solow, Robert (1959), 'Investment and Technical Progress', in Arrow, K. et al. (eds), *Mathematical Methods in the Social Sciences*, Stanford: Stanford University Press, pp. 89-104
Solow, Robert (1963), *Capital Theory and the Rate of Return*, Amsterdam: North Holland
Solow, Robert (1988), 'Comments from Inside Economics', in Klamer, A, et al. (eds), *The Consequences of Economic Rhetoric*, pp. 31-7
Solow, Robert (1990), 'Goodwin's Growth Cycle: Reminiscense and Rumination', in Velupillai, K. (ed), *Nonlinear and Multisectoral Macrodynamics*, pp. 31-41
Sontag, Susan (1983), *Illness as Metaphor* and *AIDS and its Metaphor*, Harmondsworth: Penguin
Stadler, George (1990),'Business Cycle with Endogenous Technology', *American Economic Review*, 80(4), pp. 763-78
Stengers, Isabelle (1987), 'Complexité — Effet de Mode ou Problème?', in Stengers (ed) (1987), *D' Une Science a l'Autre — Des Concepts Nomades*, Paris: Seuil, pp. 331-51
Stengers, I. and Schlanger, J. (1991), *Les Concepts Scientifiques — Invention et Pouvoir*, Paris: Gallimard
Stenseth, N. (1985), 'Darwinian Evolutions in Ecosystems: The Red Queen View', Greenwood et al. (eds), *Evolution*, pp. 55-72
Stiglitz, Joseph and Uzawa, Hirofumi (eds) (1969), *Readings in the Modern Theory of Economic Growth*, Cambridge, USA: MIT Press
Stock, J. and Watson, M. (1988a), 'Testing for Common Trends', *Journal of the American Statistical Association*, 83(404), pp. 1097-107
Stock, J. and Watson, M. (1988b), 'Variable Trends in Economic Time Series', *Journal of Economic Perspectives*, 2(3), pp. 147-74
Stolper, Wolfgang (1994), *Joseph Allois Schumpeter — The Public Life of a Private Man*, Princeton: Princeton University Press
Streissler, Erich (1994), 'The Influence of German and Austrian Economics on Joseph A. Schumpeter', in Shionoya, Y. and Perlman, M. (eds), *Schumpeter in the History of Ideas*, Ann Arbor: University of Michigan Press, pp. 13-38
Sugihara, George, Grenfell, Bryan and May, Robert (1990), 'Distinguishing Error from Chaos in Ecological Time Series', *Philosophical Transactions of the Royal Society of London*, B330, pp. 235-25
Sugihara, G. and May, R. (1990), 'Nonlinear Forecasting as a Way of Distinguishing Chaos from Measurement Error in Time Series', *Nature*, 344, pp. 734-41
Swedberg, Richard (1991), *Joseph A. Schumpeter — His Life and Work*, Cambridge: Polity Press
Tarski, Alfred (1953), *Undecidable Theories*, Amsterdam: North Holland, in collaboration with Andrzej Mostowski and Raphael Robinson, ed. 1968
Tarski, Alfred (1965), *Introduction to Logic and to the Methodology of Deductive Sciences*, New York: Oxford University Press
Tinbergen, Jan (1939), *Statistical Testing of Business Cycles Theories*, Geneva: League of Nations, Economic Intelligence Service
Tinbergen, Jan (1940), 'Econometric Business Cycle Research', *Review of Economic Studies*, 7, pp. 73-80
Tinbergen, Jan (1951), 'Schumpeter and Quantitative Research in Economics', in Harris, S. (ed), *Schumpeter — Social Scientist*, pp. 59-61
Tinbergen, Jan (1990), 'The Specification of Error Terms', in Velupillai, M. (ed), *Nonlinear and Multisectoral Macrodynamics*, pp. 201-6

Tort, Patrick (1985), 'Modèles Biologiques et Idéologies Sociales dans l'Histoire des Sociétés Libérales', in Tort, P. (ed), *La Misère de la Sociobiologie*, pp. 171-91
Toulmin, Stephen (1953), *The Philosophy of Science: An Introduction*, London: Hutchison
Toulmin, Stephen (1972), *Human Knowledge*, Princeton: Princeton University Press
Toulmin, Stephen (1982), *The Return to Cosmology — Postmodern Science and the Theology of Nature*, Berkeley: University of California Press
Turing, Alan (1952), 'The Chemical Basis of Morphogenesis', *Philosophical Transactions of the Royal Society of London*, 237(B641), pp. 37-72
Varian, Hal (1981), 'Dynamical Systems with Applications to Economics', in Arrow, K. and Intriligator, M. (eds) (1981-1991), *Handbook of Mathematical Economics*, vol. 1, Amsterdam: North Holland, pp. 93-110
Veblen, Thorstein (1898), 'Why is Economics not an Evolutionary Science?', *Quarterly Journal of Economics*, 12, July, pp. 379-97
Veblen, Thorstein (1899), *The Theory of the Leisure Class*, New York: Mentor
Veblen, Thorstein (1908), 'Professor Clark's Economics', *Quarterly Journal of Economics*, 23, pp. 180-230
Vercelli, Alessandro (1982), 'Is Instability Enough to Discredit a Model?', *Economic Notes*, 11, 73-90
Vercelli, Alessandro (1990), 'Probability Causality and Economic Models: Suppes, Keynes and Granger', in Velupillai, K. (ed), *Nonlinear and Multisectoral Macrodynamics*, pp. 224-46
Vercelli, Alessandro (1991), *Methodological Foundations of Macroeconomics: Keynes and Lucas*, Cambridge: Cambridge University Press
Veuille, Michel (1987), 'Corrélation — Le Concept Pirate', in Stengers, I. (1987), pp. 36-67
Voltaire (1733), *Lettres Philosophiques*, Paris: Garnier-Flammarion, ed. 1964
Voltaire (1752), *Micromégas, une Histoire Philosophique*, Paris: Hachette, ed. 1972
Voltaire (1759), *Candide*, New York: Viking Press, ed. 1949
Waddington, Conrad (1975), *The Evolution of an Evolutionist*, Edinburgh: Edinburgh
Waldrop, Mitchell (1992), *Complexity: The Emerging Science at the Edge of Order and Chaos*, New York: Simon and Schuster
Walker, D.A. (1991), 'Economics as Social Physics', *Economic Journal*, 101, pp. 615-31
Wallace, Alfred (1858), 'On the Tendency of Varieties to Depart Indefinitely from the Original Type', *Journal of the Proceedings of the Linnaen Society* (reproduced in Olson, R. (ed, 1971), pp. 112-20)
Walras, Léon (1874), *Elements of Pure Economics or The Theory of Social Wealth*, London: Allen & Unwin, ed. 1954
Walras, Léon (1883), *Compêndio dos Elementos de Economia Política Pura*, São Paulo: Abril Cultural, ed. 1983
Ward, B. (1972), *What's Wrong with Economics?*, New York: Basic Books
Weintraub, R. and Mirowski, P. (1994), 'The Pure and the Applied: Bourbakism Comes to Mathematical Economics', *Science in Context*, 7(2), pp. 245-72
White, Alan (1958), *G.E. Moore — A Critical Exposition*, Oxford: Blackwell
Wilson, E.O. (1975), *Sociobiology: The New Synthesis*, Cambridge, USA: Harvard Univ. Press.
Winter, Sidney (1971), 'Satisficing, Selection and the Innovating Remnant', *Quarterly Journal of Economics*, 85(2), pp. 237-61
Witt, Ulrich (1985), 'Coordination in Individual Economic Activities as a Evolving Process of Self-Organization', *Economie Appliquée*, 37, pp. 569-95
Witt, Ulrich (ed) (1993), *Evolutionary Economics*, Aldershot: Edward Elgar

Witt, Ulrich (1995), 'Schumpeter versus Hayek: Two Approaches to Evolutionary Economics', in Meijer, Gerrit (ed), *New Perspectives on Austrian Economics*, London: Routledge, pp. 81-101

Wullwick, Nancy (1990), 'Comment', in W. Samuels (ed), *Economics as Discourse*, pp. 216-24

Xia, Zhihong (1994), 'Arnold Diffusion and Oscillatory Solutions in the Planar Three-Body Problem', *Journal of Differential Equations*, 110, 289-321

Young, Robert (1985), *Darwin's Metaphor — Nature's Place in Victorian Culture*, Cambridge: Cambridge University Press

Yule, George (1897), 'On the Theory of Correlation', *Journal of the Royal Statistical Society*, pp. 812-54

Yule, Georges (1899), 'An Investigation in the Causes of Changes in Pauperism in England', *Journal of the Royal Statistical Society*, March, pp. 243-86

Yule, Georges (1921), 'On the Time-Correlation Problem, with Special Reference to the Variate Difference Correlation Method', *Journal of the Royal Statistical Society*, 84, pp. 497-526

Yule, Georges (1926), 'Why do we Sometimes Get Nonsense Correlations between Time Series? A Study on Sampling and the Nature of Time Series', *Journal of the Royal Statistical Society*, 89, pp. 1-64

Zarnowitz, Victor (1985), 'Recent Work on Business Cycles in Historical Perspective: A Review of Theories and Evidence', *Journal of Economic Literature*, 23, pp. 523-80

Zarnowitz, Victor (1992), *Business Cycles: Theory, History, Indicates and Forecasting*, Chicago: Chicago University Press

Zellner, Arnold (1979), 'Causality and Econometrics', in Brunner, K. and Meltzer, A. (eds), *Three Aspects of Policy and Policy-Making*, Amsterdam: North Holland, pp. 9-54

Index